Gay and Gaia

Gay and Gaia

Ethics, Ecology, and the Erotic

Daniel T. Spencer

The Pilgrim Press
Cleveland, Ohio

The Pilgrim Press, Cleveland, Ohio 44115

Printed in the United States of America on acid-free paper

01 00 99 98 97 96 5 4 3 2 1

Library of Congress Cataloging-in-Publication Data
Spencer, Daniel T. (Daniel Teberg), 1957–

Gay and Gaia : ethics, ecology, and the erotic / Daniel T. Spencer.

p. cm.

Includes bibliographical references and index.

ISBN 0-8298-1149-4 (alk. paper)

1. Ecology—Religious aspects—Christianity. 2. Environmental ethics. 3. Homosexuality—Religious aspects—Christianity. 4. Sex—Religious aspects—Christianity. 5. Sexual ethics. 6. Christian ethics. 7. Spencer, Daniel T. (Daniel Teberg), 1957– . I. Title.

BT695.5.S7 1996
241'.691—dc20 96-38899
CIP

For my parents,

Ann and Lorin Spencer,

who first taught me to love

both Gay and Gaia

Contents

Acknowledgments

This book has emerged over a period of years, and has been inspired, nurtured, encouraged, and critiqued by many valued friends, colleagues, and communities in many different settings and times. The initial ideas and outlines sprang from an evening of food, wine, and spirited conversation with four dear friends from the "Gang of Five" doctoral students with whom I began my doctoral studies at Union Theological Seminary in 1989: M. B. Walsh, Diane Moore, Mary Solberg, and Ann Wetherilt. I thank these four friends in particular for showing that doctoral work and academia can be collegial and mutually supportive instead of competitive and isolating.

Conversations with co-students at Union informed early parts of this work to a degree impossible to assess. Kathy Talvacchia, Robin Gorsline, Ann Gilson, and Union alumnus and friend Carter Heyward discussed or read early drafts of chapters and provided me with wisdom and encouragement on how to move through the process. A special word of thanks goes to my dear friends Jim and Karen Martin-Schramm and their sons, Joel and Joshua, for a decade-long journey of friendship. From our original days working together at the Center for Global Education in Minneapolis, to our weekly dinners and conversations during three years together at Union, each of these four individuals added in immeasurable ways to the reflection that has gone into this project, as well as making the whole task more human and (at times) fun.

Richard Brown, my editor at Pilgrim Press, believed in this project from its early days and encouraged me through the long editing process to the publishing stage. I appreciate his persistence and patience. Support from the Drake University Center for the Humanities helped me to prepare the final manuscript.

I have benefited and grown from time immersed in several distinctive faith communities. Following my undergraduate years as a geologist, journeying in 1980 to Holden Village, a Lutheran ecumenical retreat community in the North Cascades of Washington State, provided the first opportunity to connect faith and justice issues. The site of much of my original coming out as both gay and Christian, the beauty of Holden's spectacular setting, and the rhythms of intentional community life also opened the door to early reflections on the connections between sexuality, spirituality, and the earth. I am particularly grateful to John and Mary Schramm and Dan and Karen Erlander as mentors and friends who first modeled for me active commitment to justice and peace issues with intellectual vitality and a strongly rooted faith. Five years with the members of the Community of St. Martin in Minneapolis helped me to deepen and develop these commitments in an urban setting, while at the same time my colleagues at the Center for Global Education enabled me to draw connections with global justice issues. Finally, I thank Dan Geslin and other early members of Spirit of the Lakes Ecumenical Community Church in Minneapolis for providing a faith and theological context rooted in gay, lesbian, and bisexual experience to pursue these ideas. The support and encouragement of so many there helped me to believe that I was pursuing something vitally important to our communities.

Four theological friends and mentors have been particularly important. I am grateful to James Cone for completely reorienting my theological perspective from the moment I first sat in his "Introduction to Theology" class as a somewhat bewildered geologist starting seminary in the fall of 1980, through working as his graduate teaching assistant a decade later during 1991 and 1992. He, more than anyone else, has sharpened my antennae for the way dynamics of visibility and invisibility function in social mechanisms of domination and exploitation, including in my own work. J. Michael Clark has broken so much ground as an openly gay theologian when such an identity guaranteed career unemployment in academia. His work in gay ecotheology and our conversations around these issues have greatly aided the development of my own reflections. In addition to forging a way for a generation of feminist and profeminist scholars in ethics and religion, Beverly W. Harrison has been a consistent source of critical encouragement to me as I have developed my own work. Bev has had an uncanny ability to see critical threads and implications in my academic work and to suggest ways that I could develop them further.

Larry Rasmussen has been an extraordinary friend and advisor in every conceivable way. Since we first met at Holden Village in 1983, Larry has gently, persistently, and insightfully nudged my development in ethics, and has provided warm and delightful friendship and mentoring along the way. In addition to his companionship in a mutual exploration of many of the issues developed here, I am grateful for Larry's ability to provide guidance and ask the right question at the right time without imposing his own agenda. He has been a model of being an advisor, mentor, teacher, and friend.

Finally, three others deserve special mention. M. B. Walsh has been an extraordinary friend and companion throughout this project. She read drafts of every chapter and offered astute critical feedback and well-timed encouragement throughout the seemingly endless process of turning ideas into text. For her willingness to listen to both my excitement and discouragement in long discussions over coffee in her apartment, and her ability to help me turn both of them into workable ideas, as well as months of companionship in teaching and dissertating together, I am forever grateful to M.B.

Lastly, my parents, Ann and Lorin Spencer, have given me their unwavering and enthusiastic support from start to finish of this project. It is they who first and most thoroughly taught me to love the earth and its creatures, human and otherkind, and to respect and revere our sexualities as among the most profound of gifts. They provided the early contact with diverse ecological communities and instilled in me a respect for all aspects of the natural world that has grown into my deep love for the earth. It is to them that I gratefully dedicate this book.

Part 1

Toward an Ethic of Gay and Gaia: Guides for the Journey

1 Starting the Journey: Initial Reflections

She say, My first step from the old white man was trees. Then air. Then birds. Then other people. But one day when I was sitting quiet and feeling like a motherless child, which I was, it come to me: that feeling of being a part of everything, not separate at all. I knew that if I cut a tree, my arm would bleed. And I laughed and I cried and I run all round the house. I knew just what it was. In fact, when it happen, you can't miss it. It sort of like you know what, she say, grinning and rubbing high up on my thigh.

—Alice Walker, *The Color Purple*

Springtime, 1973. A teenage boy wanders out onto the Colorado prairie, seeking the solace and comfort in its wildflowers and gentle breezes that seem denied him in the boarding school environment he has left behind. Struggling with sexual and emotional feelings too confused for words, taunts of "faggot" and "homo" ringing in his ears, he heads toward the familiar cottonwoods and streams, seeking out a place where he feels embraced, a place where he feels connected to others and they to him, where he is able to love and be loved in return.

Summer, 1990. Now older, the boy relaxes in the arms of his lover's strong body, surrounded by the high North Cascades peaks, still mantled in snow, serenaded by distant waterfalls and wind whispering through tamarack and pine. Embraced by erotic love and all of nature, he marvels at the gift of connectedness to others and otherkind.

Gay and *Gaia*, two parts of who I am, two ways through which I know and am known. Gay—the erotic life force welling up within me, seeking connection and intimacy with others. Gaia—the ecological life force surrounding me and flowing through me, showing me the connection of all things with each other. Gay and Gaia tell me much about who I am and how I relate to others, human and otherkind, with whom I share this planet. Gay and Gaia tell me about where and how I am located within the human and earth ecology, when that location feels right and when it is fundamentally at odds with whom I am convinced we are created to be.

Gay and Gaia also tell me about *dis*location: how as a gay man I have been dislocated from both human and natural ecologies by societal beliefs that discount homosexuality as "unnatural," outside the bounds of natural, acceptable human behavior; how as a member of human communities growing voraciously out of control with respect to our nonhuman neighbors I have become dislocated from ecologically sustainable patterns and relationships in the broader earth community; how nonhuman members of the earth's biotic communities increasingly are dislocated and extinguished as human activities threaten the continuity of all life on the planet.

Paired together, Gay and Gaia may reveal connections that as a teenager I could only intuit: that the societal forces and dynamics involved in the relentless destruction of the prairie ecosystem where I lived, in turning it into a rapidly spreading suburban sprawl, are related to those that tried to undermine and destroy my emerging gay identity and those of countless other lesbians and gay men. Similarly, Gay and Gaia can tell us what I later experienced with my lover in a mountain meadow free from homophobia: that the erotic energy that most deeply connects us with others also can point to a deeper ecological connection with all of creation.

In recent years, thinking about these ecological and erotic locations, dislocations, and connections has shaped much of my ethical reflection. Near the end of my first year of teaching, I was asked by some of my students to give a talk under the rubric "Last Lecture"—to speak about whatever I would like, as if it were the last time I could address my students and colleagues. While finding it a bit daunting to sum up my thoughts after so brief an academic career, I took it as an invitation to speak about what is most important to me, which I knew ultimately would come down to three of my main passions: ethics, theology, and the earth. I used the well-known passage from Alice Walker's book *The Color Purple* to help make the connections between these three.

Besides the obvious point of being a trained ethicist, I chose to talk about my passion for ethics because who we are and where we are going—both as a human community and as a much broader and more diverse earth community—are important to me. We humans are but one member of the earth's diverse communities; we are an extraordinary member with many remarkable skills and powers, but nevertheless a member that is thoroughly dependent in our ecological interdependence with other members of the earth's ecosystems.

Also I wanted to talk about theology, because theology and religion give me ways to ground my ethics, to ground questions of human and ecological well-being in questions of ultimate concern—what different religious traditions have referred to as God or the sacred or the divine or the Great Spirit. Where we find the sacred within, among, and between us tells us much about the shape of our ethics—who we are and how we act in light of who we claim to be and want to act.

Finally, I knew I would talk about the earth and our place on it. I had loved the earth long before my days as a geologist gave me tools to understand the earth's story more deeply, and I continue to be fascinated by it. I love everything about this marvelous planet—discovering how it works, learning what its history has been, watching how life evolves on it, wondering where it is headed and for how long we will accompany it.

Gay and Gaia is an effort to reflect deeply and critically on these themes: how we choose to live and be, where we find the sacred in our midst and how we discern it and name it, and what all this has to do with the earth community and our place within it. As an ethicist, I am by vocation about the business of looking at the world around us, and at who we are in this world, and helping to discern and clarify the choices, values, and actions that can make our world a better place for its human and nonhuman members. As a way of moving into this, rather than starting with the traditional language and method of ethics, I begin by sketching some of the ways I see ethics, theology, and the earth grounded in story—sacred story, if you will—and what for me unites them and gives them coherence.

I start here with a section of the story of Celie and Shug from *The Color Purple*. The story is not my own story; in fact, it is a story rooted in realities that in some ways are very different from those that have shaped my story. Yet part of what I believe is necessary to ensure the survival and thriving of all our stories and of our planet is to learn to listen to the stories of others so that we can learn to see the world and our place in the world differently, more fully, in richer shades of color and involving more intricate webs of relationship.

I go back to the story of Celie and Shug often to become grounded again in my own story. Though the experiences that shape it are very different from my own, the things that Celie and her lover, Shug, come to love are very similar to the things that I love, that shape my values and commitments as I move within a very different social and ecological fabric, yet one also shaped largely by race, gender, sexuality, class, and geography.

Second, this story helps me to better understand my vocation as an ethicist, one who looks at who we are and how we act in the world in light of our deepest values. My ethical and academic commitments are centered most strongly in three areas: ecological ethics, global justice issues, and gender and sexuality issues in ethics. Some of my colleagues often seem bemused by this seemingly disparate collection of ethical issues and questions, but I see them as fundamentally interrelated—grounded as interwoven parts of what I have come to know theologically as God. In one remarkable section toward the end of *The Color Purple*, Celie and Shug talk about these issues by talking about God and what God looks like. How we image God, the sacred, that place/person/spirit where our deepest commitments lie—that is, how we do theology—will tell us a lot about who we are and how we live—that is, our ethics.

At one point in their conversation about God, Shug asks Celie, "Tell me what your God look like, Celie." Celie answers, "He big and old and tall and greybearded and white," to which Shug responds that Celie's God is "the one that's in the white folks' white bible." Celie is shocked by this assertion, and Shug goes on to describe her very different understanding of God.

> Here's the thing, say Shug. The thing I believe. God is inside you and inside everybody else. You come into the world with God. But only them that search for it inside find it. And sometimes it just manifest itself even if you not looking, or don't know what you looking for. Trouble do it for most folks, I think. Sorrow, lord. Feeling like shit. . . .
>
> She say, My first step from the old white man was trees. Then air. Then birds. Then other people. But one day when I was sitting quiet and feeling like a motherless child, which I was, it come to me: that feeling of being a part of everything, not separate at all. I knew that if I cut a tree, my arm would bleed. And I laughed and I cried and I run all round the house. I knew just what it was. In fact, when it happen, you can't miss it. It sort of like you know what, she say, grinning and rubbing high up on my thigh.
>
> *Shug!* I say.
>
> Oh, she say. God love all them feelings. That's some of the best stuff God did. And when you know God loves 'em you enjoys 'em a lot more. You can just relax, go with everything that's going, and praise God by liking what you like.
>
> God don't think it dirty? I ast.
>
> Naw, she say. God made it. Listen, God love everything you love—and a mess of stuff you don't. But more than anything else, God love admiration.
>
> You saying God vain? I ast.

> Naw, she say. Not vain, just wanting to share a good thing. I think it pisses God off if you walk by the color purple in a field somewhere and don't notice it....
>
> Well, us talk and talk about God, but I'm still adrift. Trying to chase that old white man out of my head. I been so busy thinking bout him I never truly notice nothing God make. Not a blade of corn (how it do that?) not the color purple (where it come from?). Not the little wildflowers. Nothing.
>
> Now that my eyes opening, I feels like a fool. Next to any little scrub of bush in my yard, Mr. ——————— 's evil sort of shrink. But not altogether. Still, it is like Shug say, You have to git man off your eyeball, before you can see anything a'tall.
>
> Man corrupt everything, say Shug. He on your box of grits, in your head, and all over the radio. He try to make you think he everywhere. Soon as you think he everywhere, you think he God. But he ain't. Whenever you trying to pray, and man plop himself on the other end of it, tell him to git lost, say Shug. Conjure up flowers, wind, water, a big rock.
>
> But this hard work, let me tell you. He been there so long, he don't want to budge. He threaten lightening, floods, and earthquakes. Us fight. I hardly pray at all. Every time I conjure up a rock, I throw it.[1]

What can we discern from this story to begin to point us toward an ecojustice ethic of Gay and Gaia? What does looking for the color purple, rubbing high on my lover's thigh, paying attention to trees have to do with the complex and important issues of ecological ethics, global justice, and gender and sexuality issues in ethics? Quite simply, quite a lot. For Shug knows in a deeply intuitive way that good theology and good ecology ground good ethics. That is, Shug knows that how we *image* what is most sacred to us—the theo-logos, if you will—and the words we use to understand life on this earth—that is, the eco-logos—shape who we are and how we live.

Shug's understanding of the sacred, that which she holds in awe, reverence, and respect, is very similar to my own. Like Shug, the ways I have been touched by the divine lead me to believe in a world where everything is deeply interconnected. In other words, God/the sacred is *in* everything that ever was or ever will be. Native Americans know this, too; they know that if they cut a tree their own arm bleeds, and so their ancestors prayed for forgiveness and in gratitude whenever they used other parts of nature's web for their own sustenance.

Theology and ecology both have shaped my views strongly on this point. Theologically, my view of the divine has been influenced markedly by the insights of process and feminist theologies that God is found in the interwoven processes that shape our lives. And ecologically speaking, far from being independent and autonomous Enlightenment creatures, human beings and all living creatures are interdependent with each other and the larger ecosystems for our well-being and survival. We live foolishly—and unethically—when we live in ignorance and isolation from other parts of the earth community: after all, the earth and most of its members can get along very well without us, but we cannot live without them. We ignore the earth or damage it at our peril. As Chief Seattle is recorded to have said so prophetically over one hundred years ago, "Contaminate your bed and you will suffocate in your own waste."[2]

It is this sense of the sacred found in the interconnectedness of all that leads to the central conviction of this book, that ecological ethics must become the grounding for all ethics, whether business ethics, biomedical ethics, sexual ethics, or international ethics. In other words, in terms of ethics, we need a fundamental shift from an anthropocentric, human-centered worldview to an ecocentric, all-of-life-centered worldview. Who we are and how we act must always be seen within the larger ecological context in which we live, not just our human communities. Concern for the well-being of future generations must include nonhuman as well as human beings and communities.

To help us begin to see and think from an ecocentric perspective, I think theologian Sallie McFague is right on target when she calls us to image the earth as God's body—that what we do to the earth, whether acts of tender care or of ruthless exploitation, we do to the divine, to what is sacred. For too long we in the West have seen the earth as separable from the divine, and in some traditions, not just separate, but antithetical. God has been a distant sky monarch who reigns over the earth, commanding our allegiance, demanding that our eyes turn skyward, away from the earth. Not to do so, not to deny the earth for the sake of divine demands—or in its secular, capitalistic versions, for the sake of economic "progress" and "development"—was the primary sin. Yet as McFague says, when we image the earth as God's body, sin—that which is unethical, which alienates us from the good and the sacred—is seen differently. Sin is now the refusal to be *part of the body*, to refuse to accept with gratitude our interconnectedness with all that is. McFague writes, "To sin is not to refuse loyalty to the king, but to refuse to take responsibility for nurturing, loving, and befriending the body and all its parts."[3] Yet as long as our view of the sacred and the world remains anthropocentric, we will not see this. Like Shug says, "You have to git man off your eyeball, before you can see anything a'tall."

Also like Shug, my understanding of the sacred is one where pleasure, especially erotic pleasure, is central to our moral calling. As Shug says, "And when you know God loves [them feelings], you enjoys 'em a lot more. You can just relax, and go with everything that's going and praise God by liking what you like." The joy and pleasure of the erotic found in the mutuality of relationship is one of the deepest sources of knowing our interconnectedness with others. As Shug says about experiencing God in connectedness, "when it happen, you can't miss it. It sort of like, you know what, she say, grinning and rubbing high on my thigh."

Celie's response to this is what our society's and religious communities' responses too often have been. "Shug! . . . God don't think it dirty?" Too often we've been taught to fear the erotic as a source of sin and temptation, rather than to celebrate and respect it as a powerful source of connecting with others. So we try to box it up and control it, keep it in the bedroom and out of sight, and certainly not on library shelves where "pure, uncontaminated" children might encounter it. It's that wonderful mixed message so many of us received: "Sex is dirty and nasty and filthy, so save it for the one you love." Do we wonder, then, that this repressed erotic energy too often comes out distorted as abuse rather than mutuality in connection?

Only when we are able to reintegrate our sexuality with our spirituality—to reintegrate the energy that flows from within with the energy that draws us from without—will we have the moral grounding and maturity needed to live our ethical values. This is the central claim of an ethic of Gay and Gaia, and one of the lessons that Celie learns at the end of her life journey. A story that begins with the violent rupture of her sexuality and spirituality through the brutal rape by her stepfather culminates with her saying, "Lately I feel like me and God make love just fine anyhow."[4] How ironic that a Western religious tradition centered on love too often has made "making love" to appear to be the opposite of God!

This is what I believe is at stake, finally, in many of the issues of gender and sexuality in ethics. It is the violation of right relationship along lines of gender and sexuality through fearing and distorting the erotic, channeling it only into safely controlled norms—that is, compulsory heterosexual marriage—that leads to the abuses of sex and power that have engulfed our society. Because religious and societal attempts to control the erotic have been the source of much of the oppression of those of us who are lesbian and gay, our voices are critical to revisioning what right relationship might look like at all levels of our human and natural communities.

Hence, central to an ecojustice ethic of right relationship is the realization that the erotic and the ecological are deeply intertwined. They both have to do with our fundamental interconnectedness, the energies and activities that bind us into ecological and human communities. Those who oppose this view of the sacred understand this point all too well. It is not coincidental that the same people who oppose civil rights for gay men and lesbians and procreative choice for women also oppose legislation to protect the environment and label their opponents "feminazis" and "ecofascists."[5] Yet the truth is that the more we are in touch with the erotic within ourselves—connecting to the deepest parts of ourselves as we connect with other, what Audre Lorde calls "self-connection shared"[6]—the more we are able to live ecologically, living from within outward, interdependent, interconnected, and interwoven with each other and all the earth's fabric.

This is the central thesis of *Gay and Gaia*. I am convinced that only when we reintegrate these two powerful forms of connecting, the erotic and the ecological, will we have the moral grounding and vision needed for living in and through these perilous times. I use the signifier *Gay* to draw attention to the presence of the erotic in our lives as a pointer toward relationship, for me most deeply experienced in gay relationship. As Carter Heyward has pointed out, because of our historical exclusion at the point of erotic love, it may be the special privilege of lesbians and gay men to help lead the way in revising deeply rooted right relation in our communities and relationships.[7] *Gaia*, from the name of the Greek earth goddess, draws attention to the metaphor of the earth as a living organism, now threatened by ongoing human assault.

Finally, Gay and Gaia claims to be not only an ecological ethic, but also a liberationist ethic. My understanding of the divine is of a God who is not neutral when confronted with suffering, injustice, and oppression. From my Judeo-Christian heritage, I draw on the God of the Exodus who leads the Hebrew people from slavery to liberation, the God the peasant girl Mary magnified for putting down the mighty from their thrones and exalting those of low degree, the God whom Jesus testifies has sent him to preach good news to the poor and to set at liberty the oppressed. What I hold to be most sacred in the universe, that which has my awe, reverence, loyalty, and friendship, is not indifferent when faced with injustice. Rather it is an understanding of ultimate commitment that responds to injustice with "the power of anger in the work of love."[8] It is this view of God, of the sacred, that led an anonymous Latin American songwriter to pen the following:

> Yo no puedo callar, no puedo pasar indiferente
> Ante el dolor de tanta gente, yo no puedo callar.

No, no puedo callar. Me van a perdonar amigos mios.
Pero yo tengo un compromiso, y tengo que cantar . . .
la realidad.

I cannot remain silent, I can't go on indifferently,
Faced with the pain of so many people, I cannot remain silent.
No, I can't remain silent. My friends will have to forgive me,
But I have a commitment, and I must sing out reality.

In Latin America, Christians call this God's preferential love for the poor and the oppressed. God loves everybody, everything, but God has a special love and concern for the most needy, the most marginalized. Latin American Christians believe God calls us to love like this, actively to seek out as our neighbor those who have fallen among thieves. Here McFague's image of the body can be helpful. We value all parts of our bodies for the many different things each contributes to the whole, but when one part is sick or injured or threatened, we focus our healing, protective efforts there, giving it preferential attention, in order to restore health to the whole again. So is this liberationist understanding of the sacred—it prioritizes the well-being of the most needy in order to bring health and wholeness to the whole. Without this preferential perspective, this "epistemological privilege of the oppressed,"[9] we cannot see well, let alone act justly. And changing our way of seeing is never easy and meets with lots of resistance, both external and internal. As Celie says about gittin' the man off her eyeball so she can see the world in its fullness, "But this hard work, let me tell you. He been there so long he don't want to budge. He threaten lightening, floods, and earthquakes." Yet this deconstructive task—calling into question familiar ways of seeing—is central to opening up new and more liberating images, and it forms a key starting point for an ethic of Gay and Gaia.

What follows in the pages of *Gay and Gaia* is an effort to spell this out, to move from the intuitive level expressed in the language of narrative to a critical/reflective level expressed in the language of critical theo-ethical reflection and analysis. Grounded in a liberationist approach that draws on lesbian and gay male constructionist theory and ethics but is attendant to broader liberation concerns, I begin by asking: What are the critical contributions liberationist insights have to make to ecological ethics? More specifically, how may understandings of reality and methodological insights that emerge from lesbian and gay male experiences and theorizing contribute to the development of Christian ecological ethics so as to avoid reproducing and reinforcing the social dynamics of oppression? And what do the insights of lesbians and gay men engaged in a praxis of liberation contribute to the revolution in terms of ecological consciousness and praxis so needed to confront contemporary social and ecological crises?

I argue that emerging ecological theologies and ethics not informed adequately by liberationist insights—including but not confined to those of lesbian and gay male experiences and theorizing—risk reproducing and reinforcing several dynamics of social oppression. Because lesbian and gay male experiences of being considered unnatural and outside both human and natural ecologies reveal important insights to the ways societies construct nature, human nature, and the nature/culture boundary, I emphasize the voices of lesbians and gay men in this project, while remaining open to the insights of others.

As the central ethical norms of this project, Gay and Gaia serve as paired criteria of ecojustice to draw attention to location and power in the social/ecological web of relations by specifically grounding ecological awareness in lesbian and gay male realities and insights. *Gay* calls attention to the centrality of lesbian and gay male (and bisexual and transgendered) perspectives in developing a liberationist perspective on ecological ethics in this project. It specifically incorporates attention to homophobia and heterosexism in its "ecological hermeneutics of suspicion"; that is, it examines how oppressive historical and contemporary attitudes and practices toward lesbians and gay men relate to exploitative attitudes and practices toward nature and the land. Gay also draws attention to the positive role of the erotic as a powerful force for connection in right relation, which is so urgently needed as part of an ecological ethic.

Gaia was first employed by biologists James Lovelock and Lynn Margulis to underscore their belief that the entire planet operates as a self-regulating living system or single organism.[10] Here the term functions specifically to remind us that all ethics must develop an ecocentric framework that incorporates the ecosystem as part of its community of accountability.

In light of this, several questions guide this project: What are the connections between the historical oppression of homoerotic behaviors and persons and the oppression and exploitation of the earth? How does awareness of these influence an ecological ethic's assumptions about nature, human nature, and their relation? What does an ecological ethic need to learn and include from liberationist perspectives and praxis to avoid reproducing oppressive social dynamics? What are the specific contributions of a liberationist approach rooted in lesbian and gay male realities to ecological ethics, particularly in the areas of *hermeneutics* (how we interpret our traditions and current realities), *epistemology* (the ways we know and understand the world around us), and *praxis* (how we engage one another and the world in light of our ethical reflection)?

The location of Gay and Gaia specifically within the Western Christian tradition coincides with my own location and is part of the broader community to which I hold myself accountable. For this reason, discussion and assessment of models of ecological ethics are limited to four approaches rooted in a Western Christian context, though I draw on theories and insights of others outside this context for my critique.

I seek to be attentive to both the promises and liabilities of my social and ecological locations as a white, highly educated, gay man raised in the upper-middle stratum of late twentieth-century capitalist United States society and in a variety of ecosystems that have shaped, enabled, and constrained the ways I view both society and the broader natural world. These locations, combined with the voluntary and involuntary dynamics of being in or out of the closet as a gay man in academia and the church,[11] result in a complex web of social relations structured by social privilege and exclusion that infuse my relations with others. Working from a profeminist stance, I hold myself accountable to engaging in power-sensitive conversation with others, particularly with lesbian and other feminists, while observing differences of race and class within these communities. I seek to learn from and incorporate the insights of others without reproducing patterns of invisibility and exclusion through coopting their voices or conflating our differences. Hence, in this project I speak from my location as a gay man, but *not* as a "representative" of lesbian and gay male communities. I provide the grounding for *a* liberationist ecological ethic rooted in and sensitive to lesbian and gay male realities and concerns; I make no claim that the ecojustice ethic of Gay and Gaia I propose here is *the* definitive approach, though I believe it is substantially informed by theories rooted in both communities.[12]

Gay and Gaia is organized both as a reflection on ecological ethics in general and as the specific formulation of an ecojustice ethic rooted in lesbian and gay experience. Any ethic is shaped largely by the sources of material it chooses to draw on for guidance and articulation of central values, as well as in its decisions of how these sources should be read, interpreted, used, and appropriated. Part 1 grounds the ethic by exploring in depth these methodological issues. Chapter 1 introduces the ethic and frames the overall project. Chapter 2 develops a liberationist hermeneutics rooted in lesbian and gay experience and theory. Chapters 3, 4, and 5 draw on social constructionist theory to explore three central questions, respectively, that underlie and inform ecological theo-ethics: how we understand "human nature" for what is termed natural versus unnatural; how we view what we term "nature" and what should be our understanding of the humanity/nature relationship; and how our understandings of God or the sacred relate to the humanity/nature relationship and what divine attributes need to be highlighted in a liberationist ecological ethic.

The rest of the book reflects the dual function of Gay and Gaia: to provide both a lens for critical assessment of ecological ethics from a liberationist perspective (Part 2), as well as a constructive grounding for working out a liberationist ecological ethic (Part 3). In chapters 6 through 9, I use the critical lens of Gay and Gaia to examine and assess four contemporary approaches to Christian ecological ethics, drawing particularly on lesbian and gay male liberationist theorizing to assess its adequacy for an ecological ethic broadly inclusive of and rooted in liberationist insights and praxis. Each of these models—biblical stewardship, Christian realism, process theology, and feminist theology—has been influential in the contemporary discourse on Christian ecological ethics, and each presents an array of potential contributions and obstacles toward developing a liberationist ecological ethic.[13] All four models take seriously the need to integrate issues of social justice and ecological integrity, a premise also at the heart of *Gay and Gaia*. At the personal level, all four authors are personally committed to ecojustice *and* to social justice for lesbian and gay persons. A guiding question for this work, then, is how well these dual commitments are reflected in the ecological ethics they formulate.

Chapters 10 through 12 turn to the second function of *Gay and Gaia:* providing a constructive basis for developing a liberationist ecological ethic grounded in and sensitive to lesbian and gay male concerns, with a focus on three primary areas. Chapter 10 looks at the implications of the liberationist methodology of an ethic of Gay and Gaia for ecological ethics. In particular, I explore the implications of expanding the notion of social location to ecological location and examine some of the ramifications of this within ecological and lesbian and gay ethics. Chapter 11 gives attention to the themes and insights an ethic of Gay and Gaia draws from lesbian and gay experience and theorizing, focusing particularly on connecting bodily integrity to ecological integrity, valuing diversity versus lesbian and gay experiences of social disposability and dispensability, and problematic aspects of appropriation of experiences and resources without reciprocity and accountability. Chapter 12 explores applications of an ethic of Gay and Gaia, particularly within (but not limited to) the lesbian and gay communities with respect to issues such as consumerism, relational dynamics, and the use of animals and animal products. While in chapters 2 through 5 primary attention is given to how the norm Gay reshapes the way we view Gaia, the ecological whole, here I reverse this to ask how Gaia in turn reshapes our thinking and acting within lesbian and gay contexts, and what the implications of this are for society as a whole.

With this initial narrative and methodological grounding, let us turn now to exploring the particular claims of Gay and Gaia, an ecojustice ethic rooted in the experience and insights of lesbian and gay people. The first step is to rethink how we see the world, and from where. What would it look like to take Shug's advice "to git man off your eyeball, before you can see anything a'tall"? The following chapters seek to address that question.

2 Crossing Bridges: Hermeneutics and Connection from the Margins

Only connect!

—E. M. Forster, *Howards End*

No people liberation without gay liberation,
No gay liberation without people liberation.
The struggle against
Defoliation of the earth—and gay liberation—
are the same struggle.
Our blood was spilled at Stonewall, Berkeley
Attica, Vietnam—all over the world for all time.
NO MORE.

—From an anonymous gay liberation poster,
New York Public Library, Stonewall '94 exhibit

If ethics is to be liberationist as well as ecological, as concerned for social justice as for ecological sustainability, then the shift from an anthropocentric to ecocentric worldview must be accompanied by close attention to the insights and commitments of liberationist approaches to avoid inadvertently perpetuating human injustices. As early gay activists intuited, "The struggle against defoliation of the earth—and gay liberation—are the same struggle." One important contribution that liberationist thinking, in general, and lesbian and gay approaches, in particular, have to make to this shift is in the area of hermeneutics—how we approach and use the sources and resources that inform and give shape to ethics, what commitments, presuppositions, and assumptions guide this process, and why.

E. M. Forster's aphorism, "Only connect," provides a helpful entry for linking insights from lesbian and gay hermeneutics with ecological ethics. Both ecological and lesbian and gay ethics are about connection, about interrelation. Both in a sense focus on right relation, and what happens when these relations are distorted or disrupted. For lesbians and gay men, "only connect" goes to the heart of coming out: connecting first with ourselves and then with others, we come out of isolation and into interconnected community, forming our community and individual identities as we go. Lesbian and gay hermeneutics are grounded in these newly constructed identities. As J. Michael Clark has noted, being gay "is not just a matter of what I do in bed with a lover, but my way of perceiving, interpreting, interacting, and being in the world."[1] Yet all too often our attempts to connect with others out of our lesbian and gay experiences and identities are either rejected or not valued in the dominant heterosexual ethos of mainstream society. Reflection on the experience of being lesbian or gay in a homophobic culture has provided us with a critical lens for examining hermeneutical issues in ethics.

Lesbian and gay identities are shaped in part by the ways we connect and are prevented from connecting with others.[2] Coming out as lesbian or gay is a process of connecting by overcoming the barriers society and particularly the churches have maintained to keep us excluded and render us invisible. In formulating a heterosexist ethics to justify this, the churches have appealed to science, to the Bible, and to church teaching and tradition as authoritative sources, weapons in the battle to keep lesbians and gay men in the closet. Yet within each of these areas—scientific efforts to describe and understand the natural world, the biblical testimony to the ongoing struggle for justice and liberation in relation to God, the witness of countless diverse Christian communities ever since to be faithful to the promise of the liberating good news—lesbians and gay men have found not only sources of our oppression, but also resources that help us to connect with each other and with others in the world in liberating ways. In addition, we have discovered and learned from voices whose social location differs from our own and whose cultures construct sexual identities differently than the bipolar compulsory heterosexual roles of the industrialized West.

These same sources—science, the Bible, church teaching and tradition—are central to formulating an ecological ethic, yet they have the same ambiguous heritage of functioning both as resource and obstacle in doing so. While any ecological ethic must be grounded in the insights of the natural sciences for how the natural world functions, such historians of science as Carolyn Merchant and Donald Worster have demonstrated clearly that one source of the current ecological crisis has been the emergence during the Enlightenment of a mechanistic scientific worldview that continues to be the dominant paradigm for understanding nature.[3] Similarly, ever since Lynn White in 1967 published his now classic essay accusing the Judeo-Christian biblical tradition of harboring the roots of the contemporary ecological crisis, there has been an ongoing assessment of the degree to which that tradition is responsible for, or can be a constructive part in addressing, the current ecological situation.[4]

How then can we draw on these examples to develop an ecologically sustainable ethic that does not inadvertently perpetuate injustice in other social arenas? Here the experience of lesbian and gay theory and praxis can contribute. In the process of recovering and reclaiming liberating sources from the midst of those used to exclude us, lesbians and gay men have developed *critical hermeneutics of suspicion and appropriation* that ground and guide our efforts. A *hermeneutics of suspicion* reminds us always to begin by asking what is hidden or distorted in pictures of reality we receive from dominant sectors, and why.[5] Whose interests are advanced by such views (typically portrayed as "objective fact") and at whose expense? A *hermeneutics of appropriation* reminds us of the need to be aware of and accountable to the sources of the insights and ideas we borrow from others. Such resources emerge from communities and contexts often very different from our own, and often reflect great struggle and cost; we cannot simply expropriate them like "natural resources" free for the taking, oblivious to their communities of origin and the role they play in them. These hermeneutics are *critical* in that they constantly pose the questions from lesbian and gay standpoints of what is *not* true, *not* authoritative, and therefore *not* to be believed about the claims of science and the Christian tradition. Because they emerge from the crucible of liberationist praxis and theory in the struggle for connection and holistic right relation, these hermeneutical insights may have important contributions to make to a liberationist ecological ethics.

Hence this chapter focuses on the hermeneutical and methodological insights of lesbian, gay, and feminist writings that reflect on the ways we use sources drawn from the natural and human sciences, the Bible and Christian tradition, and across lines of cultural, racial, ethnic, economic, and sexual difference. While science and the Bible function in very different ways in ethics, the common element that links them with respect to lesbians and gay men and ethics is that both have been used with an uncritical sense of their authority, resulting in a scientific positivism or biblicist theology that has had disastrous results for lesbian and gay people. These same dynamics can be problematic in ecological ethics if we do not from the outset tend to the critical issues of how we draw on the sources we use for ethics and in what ways they operate as authoritative.

Lesbian and gay theorizing together with feminist insights have begun to point to ways to move beyond uncritical use of science and the biblical tradition without ignoring the positive resources they offer. Learning to see science as an authoritative narrative that weaves together fact and fiction to understand the world rather than as a positivistic claim for "objective truth" about the world allows lesbians and gay men to engage and contest scientific claims to understand what is trustworthy and authoritative while also exposing heterosexist and other problematic assumptions that too often uncritically inform these claims. Reshaping our views of authority by grounding them in our liberating praxis also provides a standpoint from which to assess, critique, and appropriate constructive strands from the Bible and Christian tradition.[6]

This critical approach to science and the Bible is a necessary starting point, but is inadequate on its own. Once we have deconstructed the mystified authority of science and the Bible, lesbians and gay men, in order to be able to fully connect with each other and supportive friends, need constructive resources to break through the compulsory invisibility that society demands of homosexual persons. In order to give guidance and energy to our own efforts, we need to recover the variety of ways homoerotically oriented persons have existed historically and cross-culturally. Drawing on the insights and experiences of others different from ourselves offers us new possibilities in our own locations to revision a liberationist ethics and praxis. It also raises important hermeneutical issues about how we learn from the experience of others to whom we relate across multiple lines of socially constructed power differences such as race, class, and gender. As in lesbian and gay ethics, in formulating an ecological ethic we also have much to learn from voices traditionally excluded from the conversation, such as indigenous cultures and the urban poor. We also must pay attention to the hermeneutical issues that arise out of such conversation and appropriation of sources from cultures and locations that may be different from our own.

TOWARD A LIBERATIONIST HERMENEUTICS OF SCIENCE

If ethics is in part thoughtful reflection on who we are and how we act in light of our deepest values and commitments, science can play an indispensable role in giving us deeper insight into who we *really* are and how we *actually* act. Yet as environmental historian Donald Worster reminds us, "We are facing a global crisis today, not because of how ecosystems function but rather because of how our ethical systems function. Getting through the crisis requires understanding our impact on nature as precisely as possible, but even more, it requires understanding those ethical systems and using that understanding to reform them."[7] Science as a discipline aims for objective accounts to describe and explain the world, yet the accounts it gives must never be confused with the reality they seek to describe. Science always remains a human discipline and discourse that takes place within and draws upon the conceptual and value frameworks of its historical and social context. As I explore further in the next chapter, lesbian, gay, and feminist scholars have devoted considerable attention to examining the character of scientific discourse to show the ways it incorporates and reproduces societal dynamics of power, often to the detriment of lesbians, gay men, and others considered marginal (including parts of nature, and at times, nature itself). Yet the natural and human sciences also have provided important constructive resources for lesbians and gay men engaged in struggles against oppression. Also, in revealing the interwoven and interdependent patterns of ecology, science provides an indispensable resource for ecological ethics. An important challenge, therefore, is to develop a liberationist hermeneutics that enables us to take seriously the claims of science without being tyrannized by them.

The work of lesbian, gay, and feminist theorists suggests at least three components that should be part of this hermeneutics for it to address these issues. First, it must take a critical, rather than positivistic or reductionist, perspective on knowledge and truth claims. All descriptive or narrative accounts of the world make choices in what they point to as important; science always involves an ongoing process of discernment about what merits our attention. Constructionist studies of science show that this process of discernment inevitably, though usually unconsciously, draws on the values and assumptions of the scientist in making choices about what is important and how best to interpret results. Hence, in examining scientific claims, attention must be paid as much to what is hidden or not revealed as to what is claimed to be discovered or uncovered—a point I will develop further in the next chapter when we turn to the question of how to understand humanity and human nature.

Second, while taking a critical stance with respect to scientific claims, a liberationist hermeneutics must avoid pursuing a pluralistic relativism, reducing science and scientific authority to simply one voice among many, no more, no less valid than other claims. Science is too powerful, too authoritative, and too important for that. At one very basic level, "science works." It produces truth claims that wide varieties of people across cultures find persuasive in understanding how the world works. The question is not really whether science as a discipline has authority in our postmodern world; despite some loss of confidence and bravado in recent years, our world today and in the future is inconceivable without an ongoing central role for science. The more important questions are *how* and *in what ways* can science play a constructive role as part of a liberationist ethic? For this, science must be understood critically, but not dismissively.

Finally, in a liberationist ethic, scientific knowledge must be engaged with and held accountable to other ways of knowing that shape our praxis. Separating scientific knowledge from other ways of knowing results in a reductionistic epistemology characteristic of the Enlightenment mechanistic worldview—an epistemology that threatens the diversity of experiences and meanings in our social and ecological worlds. For lesbians and gay men this means insisting that the use of science take our experiences and ways of knowing seriously rather than reducing the complexities of our experiences and identities to an "objective" biological factor in formulating its truth claims.[8]

In this area, the work of such feminist cultural critics and historians of science as Donna Haraway and Sandra Harding goes furthest in helping us to formulate an adequate methodological grounding for a liberationist hermeneutics of science. Haraway views science as narratives on the history of nature that interweave fact and fiction. In this framework, fact and fiction are interrelated, but distinguishable. As "original, irreducible nodes from which a reliable understanding of the world can be constructed," the truth of facts should be discovered rather than simply constructed, yet facts also emerge from human actions, and are always mediated by cultural meanings.[9] Hence the words, metaphors, and concepts that are chosen, as well as the context in which they are read and heard, mean that narratives of scientific facts are never simply "objective truth," but simultaneously are culturally bound and reflect cultural constructs and values.

Fiction represents reality in ways very different than fact, but it too can be true, and like fact, fiction is known to be true by appeals to nature. Ursula Le Guin captures this sense of the truth of fiction in her discussion of the subversive potential of fantasy: "Fantasy is true, of course. It isn't factual, but it is true. Children know that. Adults know it too and that is precisely why many of them are afraid of fantasy. They know that its truth challenges, even threatens, all that is false, all that is phony, unnecessary, and trivial in the life they have let themselves be forced into living."[10] Fiction that resonates with our experience and reveals new things to us about the world, like fact, is involved in the scientific production of knowledge. Thomas Kuhn has shown how science advances through paradigm shifts that combine fact and fiction to provide new and persuasive accounts of the world until the paradigm no longer fits well enough and is replaced by a new construct with a more persuasive combination of fact and fiction.[11]

Hence both fact and fiction are grounded in epistemologies that appeal to human experience. But they function differently in our interactions with them; as Haraway explains, "The word *fiction* is an active form, referring to a present act of fashioning, while *fact* is a descendant of a past participle, a word form which masks the generative deed or performance. A fact seems done, unchangeable, fit only to be recorded; fiction seems always inventive, open to other possibilities, other fashionings of life."[12] Building knowledge on both fact and fiction entails risk for ethics, but for different reasons. Facts appear solid and uncontestable, but they hide the histories and political struggles that make their generation possible while reifying worldviews that make alternative perspectives seem impossible. This is why it is so important to take a critical approach to science, to reveal the contested social histories that yield scientific knowledge, while at the same time being clear to distinguish the "truth" of scientific narratives and account from the ever-more complex realities they attempt to describe. Fictions are powerful because they destabilize current "realities" and open up new possibilities, but they also always contain the threat of deceiving us, of merely feigning "reality," not telling the true form of things.

As a human activity with culturally mediated meanings, science contains both fact and fiction and hence both enables and constrains our possibilities of interacting with it to construct reliable knowledge. Looking at science through the lenses of four theoretical approaches—liberationist theories and commitments, social constructionism, Marxist-feminist standpoint theory, and scientific realism—reveals important insights about the possibilities and limits of scientific study of the world. It is helpful to keep each approach in dialogue with the others to avoid one position silencing the others, resulting in a reductionist view with loss of key insights.

Commitment to liberationist politics and theories, such as feminism, antiracism, and lesbian and gay liberation, grounds the first approach. Particularly important here is the hermeneutics of suspicion that grounds liberationist approaches, calling into question dominant views of reality that have served to justify oppressive social structures and exploitative practices. Haraway's work on primatology demonstrates both the particularity and the power of this approach, and shows how each of the perspectives shaped by liberationist commitments are necessary for understanding, reading, and writing scientific texts. Although race, gender, and sexuality operate differently than science, they also are historically constructed categories and narratives that interweave fact and fiction and fundamentally shape the way we see and understand the world. "Race and gender are not prior universal social categories—much less natural or biological givens. Race and gender are the world-changing products of specific, but very large and durable, histories. The same is true of science."[13] Beginning with a hermeneutics of suspicion that calls into question reified "objective" views of reality is a critical first step for a liberationist hermeneutics of science.

The hermeneutics builds on these liberationist commitments through social constructionist perspectives on science. Constructionist approaches open new possibilities through their critique of positivist or realist epistemologies that reduce what can be known to empirically testable phenomena and conflate the truth of the scientific account with the reality it tries to describe. Positivist scientific approaches have been important to maintaining the social, political, and economic hegemony of dominant societal sectors because they make it appear that current social arrangements are simply "the way the world is," and therefore attempts to change such arrangements are futile. Constructionist approaches are important to a liberationist hermeneutics because they pay attention to hidden issues of power involved in the naming and interpreting of the world that are then presented as objective scientific "discovery"—facts that hide the history of how they are conceptualized and presented. Constructionist approaches also highlight important issues of power present in the practice of science: who can enter the contest, play the game, open or close debate, and how these social dynamics in turn shape the resultant scientific theories and knowledge that emerge. By paying close attention to power, science is revealed "as effective belief and world-changing power to enforce and embody it."[14]

Because science is so effective in this role of shaping and enforcing beliefs about the world, and because one consequence of this is to reinforce a problematic subject/object split between the scientist as objective subject who names the world and the studied "other" (whether other as homosexual or other as nature) as passive object without agency in the relationship, it is particularly important that we take a critical stance toward science.[15] The problem with a sole emphasis on constructionism, however, is that it can lead to a kind of postmodern relativism where every stance is seen merely to reflect the interests and power of the individual or community holding it. Grounds for making evaluative judgments—central to a liberationist ethic—end up destabilized along with the dominant discourse. Here Marxist-feminist approaches, which argue for "the historical superiority of particular structured standpoints for knowing the social world, and possibly the 'natural' world as well,"[16] provide an important complement to constructionism.

Standpoint theory is based on the insight that the social relations of production and reproduction of daily material life allow people in some social locations to see these relations more clearly, hence giving them an "epistemologically privileged" stance. For reasons of survival, African slaves had to understand how race and gender structured the culture and world of their white slave masters who could afford to remain ignorant—and usually were—of the slave world. Lesbians and gay men who must operate, and often pass as straight, in a heterosexually defined world quickly gain insight into how sexual orientation structures social expectations and relations to which most heterosexual persons, as the beneficiaries of heterosexism, are oblivious. Epistemological privilege results from critical reflection on these dynamics in order to gain a deeper understanding of the social world.[17] It is tied to the ability to generate critical knowledge; dominant members of society typically are precluded by their social location from authentic knowing because of "their inability to generate the most critical questions about received belief."[18] This position has been particularly important for lesbian and gay liberationist hermeneutics that take seriously knowledge generated out of lesbian and gay experience and praxis.

Finally, with these groundings and cautions, the claims of science itself must be taken seriously. Scientific knowledge is not simply about power and control, as the most radical constructionist would argue, but it also is about producing believable accounts of the natural world. Its power lies in the believability of scientific narratives: "Scientists are adept at providing good grounds for belief in their accounts for action on their basis. Just how science 'gets at' the world remains far from resolved. What does seem resolved, however, is that science grows from and enables concrete ways of life, including particular constructions of love, knowledge, and power."[19] Science succeeds in its power to convince and reorder societies worldwide. As stated earlier, science works. It is therefore that much more important that science be understood critically so that it can function in socially liberating and ecologically sustainable ways. Hence this critical evaluation must engage and take seriously the methodologies that produce scientifically credible accounts:

> One grating consequence of my argument is that the natural sciences are legitimately subject to criticism on the level of "values," not just "facts." They are subject to cultural and political evaluation "internally," not just "externally." But the evaluation is also implicated, bound, full of interests and stakes, part of the field of practices that make meanings for real people accounting for situated lives, including highly structured things called scientific observations. *The evaluations and critiques cannot leap over the crafted standards for producing credible accounts in the natural sciences because neither the critiques nor the objects of their discourse have any place to stand 'outside' to legitimate such an arrogant view.*[20]

What this means is that a liberationist hermeneutics must take science seriously enough to engage it on its own grounds while bringing to the table constructionist and epistemological insights from our own distinctive locations, whatever those are. To engage science critically, we much understand both the methodological playing field on which it operates, as well as our own interested locations and commitments that inform our critiques.

The combination of these four approaches provides an "aesthetic and ethics" of scientific practice as storytelling that highlights both the "pleasure and responsibility in the weaving of tales."[21] Critical to a liberationist ecological ethic, it can move us beyond seeing science as inevitable capitulation to progress or scientific knowledge as the passive reflection of "the way things are," while also avoiding the reductionist tendency to view science only as a discourse of power. The task for ecological ethics is to build better accounts of the world that emerge from and make possible better living for the earth's human and nonhuman members. In this task, science has an indispensable—and dangerous—role to play.

To summarize, this methodological approach to scientific discourse provides a powerful grounding for a liberationist hermeneutics informing an ecological ethic. Central to such an ethic is finding ways to take science seriously, especially the biological sciences, without subsuming liberationist insights and claims as secondary or less grounded. This four-fold approach takes scientific claims seriously while engaging them with constructionist insights. Critical to a lesbian and gay ethic, it provides for privileging those perspectives that emerge from socially marginalized locations in constructing knowledge, while linking these to the praxis of struggle against oppression. Scientific knowledge becomes one resource—albeit, a very important resource—in shaping and reshaping ethics, without dominating or tyrannizing other knowledge claims and the practices that produce them. These constructionist insights and new understandings of what makes narrative accounts authoritative may in turn inform our stance toward other sources such as the Bible and tradition, to which we now turn.

TOWARD A LIBERATIONIST HERMENEUTICS OF THE BIBLE AND TRADITION

Perhaps even more so than with science, the relationship of homoerotically oriented persons to the Christian Bible and tradition is complex and ambiguous. The biblical tradition has been both a source of power and a source of oppression for lesbians and gay men in the Christian tradition. Historically, a few select verses in the Bible have been used in Western societies as primary justification for the exclusion and repression of men and women who expressed homoerotic affection, but these same women and men have heard in the biblical message a call from a God who sides with the marginalized, who calls us to be freed from the oppression of the world, a God from whose love nothing can separate us.

This ambiguous relationship has caused lesbians and gay men to wrestle with hermeneutical issues such as the role and authority of the Bible in our faith and theology, the relation between the biblical canon and Scripture, and the place and authority of Christian tradition. We have sought a critical hermeneutics to help us root out the sources of oppression and reclaim as authoritative biblical sources of our power and liberation. Increasingly, lesbian and gay scholars have articulated criteria that can shape such a liberating hermeneutics to guide our engagement with the Bible and tradition. As scholars such as Lynn White Jr. and Paul Santmire have demonstrated, the Bible and Christian tradition have had a similarly ambiguous relationship to developing an ethics of caring for the earth.[22] To this point there has been relatively little effort among those working in Christian ecological ethics to develop a critical biblical hermeneutics; hence the criteria being articulated in lesbian and gay scholarship may have helpful contributions to make in formulating a liberationist hermeneutics for ecological ethics.

Because of the historically damaging role uncritical biblical scholarship and theology have played in the lives of lesbians and gay men, as in the case with science discussed above, the first step in approaching the biblical tradition must be critical and deconstructive. Biblical passages and interpretations that have been used to buttress heterosexist and other oppressive conclusions must be examined critically to reveal the social factors and power interests at play, both in their original formulation as texts as well as in their subsequent interpretation in the life of the church. Such an approach must also be engaged and accountable to the well-being of lesbians, gay men, women, and others who have been socially marginalized. Here the work of feminist biblical scholars provides important methodological considerations for approaching and interpreting scripture.

Feminists have argued that biblical historiography and exegesis are always selective tasks, and therefore objectivity in scholarship can only be approached by naming and reflecting critically on one's theoretical presuppositions and political allegiances. All methods thus employ—explicitly or implicitly—theological and hermeneutical presuppositions that shape the work. For example, New Testament scholar Elisabeth Schüssler-Fiorenza names critical/historical analysis and a clearly specified commitment to women as the common ground between academic biblical scholarship and a feminist theology of liberation.[23] A lesbian and gay liberationist hermeneutics is similarly grounded in this dual commitment to critical/historical analysis and a clearly specified commitment to gay and lesbian people.

While a number of gay and pro-gay scholars have dealt with issues of biblical exegesis and hermeneutics,[24] lesbian feminist biblical scholar Bernadette Brooten is perhaps the most clear and explicit in working from a liberationist stance. Brooten has focused her attention on the attitudes of early Christian communities and the surrounding cultural milieu toward same-sex relations between women. To do so she has had to develop approaches that enable her to distinguish between several levels in the text, such as the recorded canonical voices that survive today, voices and views opposed by the canonical witness, and other voices excluded from the canon. Her research, along with that of other contemporary biblical scholars, has revealed an early Christian world much more diverse and complex in its communities and theologies than previously imagined. Such reconstructions can help explain the early debates and struggles among these communities, which views eventually gained ascendancy and why, and at what cost to the voices that were excluded. This has proved critical to lesbian and gay liberationist efforts that have sought to move beyond fundamentalist and biblicist interpretations of the few biblical texts that mention same-sex behavior to examine the cultural and theological assumptions of gender and sexuality that inform the worldview of the biblical writers. Such an approach will also be important for an ecological ethics in that it allows us to better understand the assumptions and attitudes toward nature and humanity's relationship to the rest of nature that inform biblical texts, and that lie at the heart of an ecological ethics.

Several features of Brooten's methodology are helpful to developing the deconstructive phase of a liberationist biblical hermeneutics.[25] First, the development of a critical view of the historical context of the text reveals how issues of power in social relations may have shaped the text and its interpretations. Constructionist insights can help here to examine societal assumptions of gender and sexuality that influence the text. An example of this is how purity attitudes and property assumptions formed within the patriarchal Jewish and Hellenistic communities of the Mediterranean shaped biblical attitudes toward sexuality in general and homosexuality specifically.[26] These attitudes and assumptions need to be examined critically in light of our ethical commitments and contemporary scientific insights to evaluate the extent to which they remain valid for us today. A liberationist ecological ethics needs to take the same approach to biblical texts that deal with nature and humanity's relationship with the rest of nature.

Second, Brooten ilustrates how approaching the Bible and non-biblical materials through a hermeneutics of suspicion grounded in gender and sexuality insights enables us not only to understand the biblical writers within their cultural context, but also to distinguish their (predominantly male) thought from those voices excluded from Scripture—such as many early Christian women— and to discern what they may have thought or how they may have lived. Recovering these voices may mean reading between the lines of the text, turning to nonbiblical materials, or applying imaginative acts of reconstruction to the few sources available. It also means that we set aside the presumed authority of the received canonical text and its message in order to recover the struggles and debates that informed these texts and may in turn inform our own debates. As will be discussed below, biblical authority lies more in the ongoing engagement of the community with God and with each other to wrestle with meaning and insight, than in taking the received message of the canonical text and trying to apply it to our own context.

Finally, a liberationist approach explicitly acknowledges the role one's own social location and community of accountability play in scholarship. Just as the social location of the biblical authors is reflected in the way their texts were shaped and formed, so too does the social location of its subsequent interpreters—including readers today—influence the life of the text and its meanings. Brooten, for example, reflects critically on her own location and commitments as a lesbian feminist and maintains that it is not enough for biblical scholars simply to identify the cultural constructions and influences on biblical texts, but they must also assess and evaluate the impact of these texts on women, especially lesbians, and gay men. The active exclusion and repression of these voices in much of the Bible leads her to reject assumptions about the canonical authority of the Bible as it currently stands and to formulate other ways of relating to the biblical texts.

The problem of the assumed authority of a biblical canon that contains numerous texts that have been interpreted to be anti-women and antihomosexual has led lesbian and gay scholars to reflect on and define new understandings of authority in theology and ethics. As feminists have argued, the problem with simply assuming canonical authority as the canon currently stands is that the process of canonization of biblical texts was not neutral with respect to women (or homoerotically oriented people), but in many cases was used actively to suppress women's leadership and voices. Similar issues about the role and authority of the Bible and tradition are at stake in the models of ecological ethics. The Bible emerged in different times and historical contexts with different issues and questions than many of the ecological issues and contexts that confront us today—in what sense is the biblical witness authoritative in a Christian ecological ethics for addressing these questions? Because this has been a central and ongoing issue in lesbian and gay theology and ethics, the reflections of lesbian and gay scholars on these issues provide numerous insights that can contribute to a liberationist biblical hermeneutics for an ecological ethic.

Lesbian theologian Ann Wetherilt, for example, begins her examination of the notion of *authority* by noting meanings captured in its etymology. The root *augere* means "to make grow or increase." To be truly authoritative, a source or text should foster an increase in moral agency and responsibility among adults in community: "no loss of freedom or agency is involved in placing oneself in relationship with . . . authority." This understanding of authority does not deny the reality of limits on behavior, so important to an ecological perspective, but it calls into question any conceptions of authority as sovereignty, as unidirectional "power over." Hence, "a quest for political and ecclesial systems which are just and participative mandates the development and implementation of alternative understandings of authority which will truly 'augment and enable to grow.' When authority is understood in growth-producing ways, it is neither coercive nor dictatorial."[27]

Similarly, Carter Heyward argues that authority lies in that which proves trustworthy in relation. We find stories, texts, or other persons authoritative not because they impose an extraneous set of expectations upon us, but because they engage us and evoke something already present within ourselves or among us. As opposed to force or coercion that function by diminishing, "authority is that which (or those whom) we can trust to help us to become more, not less, ourselves." Because authentic authority is experienced in the context of mutual relation, "no person, religion, tradition, profession, rule, or resources should be inherently authoritative for us."[28] Rather, authority is discerned by asking first if it helps to realize our interconnectedness and the shape of our identities as persons-in-relation.

From these understandings of authority rooted in communitarian values of right relation, liberation theologians locate the authority for theological reflection in the community's efforts toward wholeness and the struggle against oppression: "Our authority to speak is actually borne out of our experience of oppression. Our exclusion as gay men and lesbians in a heterosexist and homophobic ethos stands in judgment upon both canon and tradition and explodes their boundaried exclusivity."[29] It is the *content* and *process* of a *liberationist praxis* that provides the criteria for discerning what is authoritative. Hence "ecclesiastical, civil, or professional laws and rules hold genuine authority in our lives only insofar as we experience them as forged in our actual struggles toward right relation. To be creative rather than coercive, real and not rhetorical, authority must be shaped in the context of our movement into mutually empowering relationship."[30]

Therefore, authority that is liberating is experienced communally, in the lived relationships of our communities. Mary Hunt suggests further that it is found in the "activity of justice-seeking friends who do theology in order to right injustice."[31] This understanding of authority stands in continuity with those parts of the Christian tradition that "began with the disciples and members of the Jesus movement who were, if nothing else, friends seeking ways to express their faith in a hostile environment."[32] Authority experienced communally also implies accountability. Hence lesbian, gay, and feminist theologians are explicit in naming their accountability to the well-being of lesbians and gay men, as well as other marginalized women and men.[33]

Finally, lesbian and gay theologians are explicit in taking an unapologetic stance toward the Bible and tradition, as reflected in the titles of books such as *A Place to Start: Toward an "Unapologetic" Gay Liberation Theology* by J. Michael Clark, and *Gay Theology "Without Apology"* by Gary Comstock (quotation marks added for emphasis). Apologetic efforts to rescue the Bible and to fit into homophobic churches are self-defeating and leave gay men and lesbians drained from having to respond constantly to the dominant heterosexual (and heterosexist) agenda. Liberationists argue, instead, "that we cannot wait for institutions to sanction our efforts to theologize in liberationally sound ways. We must instead find our own voice and assert our *assumed* authority to speak, unabashedly and unapologetically. Ultimately, neither gay men and lesbians, nor women, nor people of color, nor Native Americans, nor any other oppressed people can afford to wait for a white heteromale conferral of authority to speak. . . . We must assume and assert our own *a priori* prophetic authoritativeness to speak."[34]

The same follows for approaching the Christian tradition: "The 'gay theology without apology' that I develop here examines the Bible and Christianity not with the purpose of fitting in or finding a place for them, but of fitting them into and changing them according to the particular experiences of lesbian/bisexual/gay people. Christian Scripture and tradition are not authorities from which I seek approval; rather they are resources from which I seek guidance and learn lessons as well as institutions that I seek to interpret, shape, and change."[35] This approach implies serious engagement with Scripture and tradition both to criticize those parts that have condemned lesbian, gay, and bisexual persons, as well to find liberating and affirming resources that have been obscured by traditional interpretations.

The work of scholars such as Comstock, Heyward, Clark, Wetherilt, and Hunt suggest that the stories and texts in the Bible and tradition may be authoritative to the extent they shed light and meaning on *our* stories and struggles. When we experience them as empowering resources in our struggles for justice and life, we feel their authority in our lives. That these texts have been experienced this way by communities in the past (with which we experience both continuity and discontinuity), which have granted them canonical authority, causes us to take their claims seriously, without, however, subsuming our own experiences and insights to a presumed inherent and nonrelational understanding of their authority.

One ramification of this is the need to move away from regarding the Bible as a parental authority from which we seek approval and permission, toward seeing our relationship with the Bible and tradition modeled on friendship where each engages the other in a process of mutual growth and change. Out of the destructive realities of rejection and isolation in so many gay and lesbian lives, Comstock, for example, names *acceptance* and *connection* as two criteria that shape what parts of the Bible are meaningful and therefore authoritative for him. For Comstock, "to be accepted, supported, and producing/contributing are what define me as most fully human and lend ultimate meaning to my life."[36] When parts of the Bible as friend support this process, we experience them as having authentic authority, for, as noted above, they foster freedom and agency.

Another criterion suggested by these authors is paying attention to voices that speak from the margins because they often reveal critical insights that are overlooked or not available in more dominant perspectives. This is the central task of Wetherilt's work on reclaiming the voices of women's struggles, and it also can guide our approach to the Bible. Voices that speak to us from the margins in the biblical text will have particular meaning and authority for lesbian, gay, and bisexual persons who find ourselves at the margins in many areas of our lives. Listening to the marginalized voices in the Bible and tradition can help us to hear those voices around us. We cannot control when and where these prophetic voices emerge in our lives; "we only know that the voices usually come from the margins, from those who are not usually taken seriously, from those desperate enough to see clearly. But we can place ourselves in their midst; we do not have to place our faith in the normal course of events or in the word from on high; we do have the option of listening to the voices from below and outside."[37]

These authors' understandings of authority, Scripture, and tradition suggest several features of a liberationist hermeneutics that emerge from lesbian and gay experience and reflection on these issues. Central to a liberationist hermeneutics is a *relational* understanding of biblical authority. Metaphors such as friendship and stress on mutuality in relation emphasize human agency and interactiveness in encountering biblical messages. Wetherilt, for example, draws on the mutuality of hearing and speaking in the metaphor of giving and finding voice to shift away from "the rigid 'Word'" toward "a mutual dialogue that includes both scripture and tradition *and* diverse human voices. Authority is freely granted to those sources that have come to be trusted in their capacity to enhance the growth and life in the world of the community."[38]

A key feature of this relational hermeneutics grounded in the integrity of lesbian and gay lives is that it looks to the Bible and tradition critically as providing *resources for moral agency* rather than rules for behavior. This means shifting the authority of the Bible and tradition from a vertical, hierarchical relation of "authority over" to an engaged, horizontal relationship of "authority with." It opens up theological reflection to draw on other resources, including the multiplicity of lesbian and gay experiences: "Religious tradition mediated by scripture and church doctrine or dogma can now take a 'back seat,' informing theological work as a valuable resource, but as one no longer holding any oppressively binding authority over us."[39]

A relational stance implies an *engaged* hermeneutics that is neither apologetic nor dismissive in its stance toward the Bible. It respects the integrity of both sides of the relation while insisting that both be open to critique, growth, and change. Thus it recognizes the Bible as a document that through time has been proven to have both resources and obstacles, not as an external authority from which we seek approval. Comstock gives a helpful example of this in discussing the danger of an apologetic approach in lesbian and gay readings of the Leviticus texts condemning male homosexual acts: "To offer excuses for the Leviticus passage is to fail to grasp the seriousness with which our sexuality threatens patriarchy and the measures that those who benefit from patriarchy will take to secure and protect it."[40] Beverly Harrison has noted the dangers of the other temptation, of withdrawing from encounter with the Bible, thereby abdicating its interpretation—and power—to our enemies. What is required instead is engagement to transform the way the Bible is used in a liberation praxis.[41]

An engaged, relational hermeneutics is not relativistic, but recognizes the *epistemological privilege of the oppressed* in engaging and interpreting biblical messages. A specifically lesbian and gay hermeneutics built on this epistemological privilege names the falsehood of such anti-gay texts as Leviticus 18:22 and 20:13 and Romans 1:26–27, while finding new promise in old stories such as the relationships of Naomi and Ruth, and Jonathan and David.[42]

Finally, a liberationist hermeneutics insists on engaging the canon and tradition with other voices that have been silenced or excluded. While many canonical scholars do not understand the biblical canon to be the exclusive Word of God, in many Christian circles it continues to function this way, particularly when the issue at stake is homosexuality. In these cases lesbian and gay voices commonly are dismissed out of hand as "nonbiblical" and "biased."

Lesbian and gay scholars have approached this issue from many angles. Clark follows the example of feminist theologians such as Carol Christ to argue that the exclusion of our experience from the original canon means it must be reopened: "Our very exclusion, whether as women or as gay people, becomes a criticism of scripture and tradition. Revaluing minority experience, therefore, means penetrating/resolving the conflict of experience and tradition by forcibly reopening the canon."[43] Wetherilt approaches the issue of a closed canon from a relational stance. In light of a relational understanding of authority, she suggests, "As in any mutual dialogue . . . *both* parties would be changed in the process—and canonical texts would be no exception."[44]

Once authority is no longer seen as inherent in any one thing, including the Bible, the canon ceases to be seen as *the* Word of

God. Rather, "when interpreted by our lives-in-relation as a word of love, the bible is *a* word of God."[45] For Heyward, the question is no longer whether or not the Bible is Scripture, but on what basis is the Bible a *part* of sacred Scripture? Comstock agrees with this expanded understanding of Scripture, and finds encouragement within the Bible itself to keep the canon of Scripture open: "Actually, the Bible encourages me to enlarge my recognition and appreciation of special stories outside of it; as a document compiled of writings from a variety of times, places, and experiences, it seems to be not so much a closed book as one that pushes its own seams." He therefore has begun to assemble and name as Scripture a body of literature "in which I find myself accepted for who I am."[46]

A liberationist hermeneutics as outlined above has several implications for ecological ethics. Perhaps most importantly, emphasizing a relational understanding of authority is fundamentally an *ecological* approach. It recognizes and assumes as normative interrelatedness and interdependency for that which is authoritative. To be authoritative, both sides of the relation—or better, each strand in the web of relationships—must be enabled to grow and mature. Ecologically as well as socially, then, that which is in or fosters "right relation" is authoritative.

Emphasis on the communal nature of hermeneutics is critical to shifting out of the destructive individualism that characterizes much of Western society and has been so damaging to the earth. Ecologically it opens the way to extending a communal hermeneutics to including, or being grounded in, the biotic communities that shape and nurture our most fundamental web of relationships. We can extend Wetherilt's understanding of authority as a "mutual dialogue with both scripture and tradition in the service of diverse human voices"[47] to include the voices of other creatures as well, engaging with and granting authority to the voices and experience of otherkind as all creatures, including the human creature, interact to preserve life and thrive.

The shift in the authority of the Bible and tradition from a vertical "power over" to a horizontal, engaged "power with" stance means seeing them as resources for ecological liberation and well-being rather than rule books mandating the nature of our relation to the rest of the planet. The earth and the health of the earth become the context and norm for ethics; we experience the Bible and Christian tradition as authoritative to the extent they engage us in caring for and living justly with one another and the wider earth community.

This shift to an understanding of authority grounded in communitarian values of right relation opens up the possibility of developing an ecological hermeneutics of suspicion that actively engages the Bible and tradition without falling into either temptation—apologetics or dismissal and withdrawal. An ecological hermeneutics of suspicion questions anything that may lead to a harmful relation with the earth, examining anthropocentric and androcentric texts and claims for the hidden interests they reflect and their ramifications for the interweaving of social and ecological justice and well-being.

Part of this shift is accomplished by paying explicit attention to the socially constructed nature of all knowledge and truth claims, including biblical categories such as dominion, stewardship, justice, and love. It does not deny that revelatory experiences of the sacred may have contributed to formulating these insights and claims, but acknowledges that all descriptions of experiences of the divine are also culturally mediated and will reflect the interests and biases of those groups within that culture that have access to and control the discourse—especially discourse about the divine. Attention to constructionist insights also allows us to examine how meanings and interpretations change over time, and how texts that in one context may have had a liberating dimension may be oppressive in another time and place, and vice versa.[48]

Constructionist insights allow ecological ethics to pay attention to power differences at several points within the human/nonhuman web of relations. They recognize that not all human/nonhuman relations are alike, that power differences *within* the human community also affect power arrangements in the multiplicity of human/nonhuman relations.[49] A liberationist hermeneutics privileges diversity and inclusiveness—fundamentally ecological values—while rejecting a liberal relativism of different viewpoints and locations. It pays attention to the impact of power differences by granting epistemological privilege to the most marginalized and oppressed, a stance with rich possibilities for ecological ethics in extending this "privilege" to nonhuman nature. "Think like a mountain," advises Joanna Macy, "and like a chicken," adds vegetarian ecofeminist Carol Adams.[50]

Finally, a liberationist hermeneutics moves beyond suspicion and deconstruction to the reconstructive task of remembrance and recovery. An ecological ethics needs continually to examine the biblical text and tradition for the earth-friendly voices that have been silenced or excluded, and to ask why. What are the hidden resources in the text that a liberationist ecological hermeneutics can help recover? This hermeneutics calls for a stance of openness, opening up the canon and looking at nonbiblical sources to help shape our ethics. Indeed, ecofeminists such as Anne Primavesi have argued that a central feature of ecological paradigms is keeping open pathways to other ways of thinking, imaging, and acting, what she terms an interactive "openness within constraints."[51]

The reconstructive task and openness to learning from others different than ourselves raises important methodological questions for a liberationist hermeneutics. It is to that question I turn in the next section.

TOWARD A LIBERATIONIST HERMENEUTICS OF APPROPRIATION AND RECIPROCITY ACROSS LINES OF SOCIAL DIFFERENCE

We live in the era of the global village. We have only to turn on CNN to have images of peoples, contexts, and environments vastly different from our own beamed into our homes. Learning from others different from ourselves not only can be important in shaping ethics today, increasingly it is unavoidable. Learning from the experiences of persons whose sexual identities and lived experience fall outside dominant Western constructs of normative and compulsory heterosexuality has been a critical part of lesbian and gay liberationist thinking. In addition to expanding the resources available for rethinking ethics in creative and new ways, it has helped us understand the socially constructed dimension of all sexual identities.[52] Yet drawing on the insights of persons from other traditions, cultures, and social locations has critical methodological implications for persons in Western societies with our histories of imperialistic appropriation of "resources"—human, "natural," and otherwise—from other lands and cultures.

Hence we must ask, what methodological insights inform a liberationist perspective of learning from and respecting the experience and insights of others? Here the emerging dialogue between womanist, mujerista, and feminist theologians and scholars has much to contribute to a liberationist hermeneutics. The participation of lesbians and, to a lesser degree, profeminist gay men, has played an important role in this dialogue. A fascinating and insightful discussion of "Appropriation and Reciprocity in Womanist/Mujerista/Feminist Work" took place in the Women and Religion Section at the 1991 annual meeting of the American Academy of Religion in Kansas City and can serve us well in developing this portion of a liberationist ecological hermeneutics.[53]

The 1991 discussion of how women read and use each other's work was the third such meeting between womanist, feminist, and mujerista theologians on the theme of how to understand women's experience and theologizing from multiple locations.[54] The central theme of the third meeting on appropriation and reciprocity was an "exploration of the ethics of listening and responding among women of all colors."[55] The diverse panel of two presiders, four presenters, and three respondents included the voices of African American, Asian, white, Hispanic, and heterosexual and lesbian-identified women from a variety of Christian and Jewish communities and traditions. Numerous issues critical to developing a liberationist hermeneutics of respectful listening and appropriation were raised and discussed.

The grounding assumption of the discussion emerged from years of feminist dialogue and also reflects a fundamentally ecological value: learning from and using each other's work, experience, and insights must be *relational and mutual.* It involves searching for ways that respect and honor the integrity of each other's work, which entails paying attention to the differences in location and experience that inform each person's reflection and analysis. A central element of these discussions is *power*: "how we read and make use of each other's work in a society based on relationships of domination and subordination—relationships that are reproduced within the community of womanists/feminists/mujeristas working in religion."[56] Paying attention to the power dynamics involved in using each other's work involves recognizing the multiple locations and relations women occupy with respect to each other.

Judith Plaskow cites three different locations in her own situation as an example. As a Jew learning from Christian feminists, she is a member of a dominated group borrowing from the dominant culture. As a Jew whose tradition and work are used by Christians, she is a member of a dominated group whose experience is appropriated by the dominant culture. And as a white middle-class woman who draws on the work of women of color, she is a member of the dominant culture who uses the work of women from dominated cultures.

Stemming from sensitivity to power is the need to pay attention to *boundaries* as a way to preserve the distinctness of different perspectives. Awareness of boundaries must happen on both sides of the power relation, but for different reasons. As a Jew working with Christian feminists, Plaskow notes, "When I use Christian materials, I must continually be conscious of boundaries in order to preserve my identity—whether I find difference or similarity on the other side of the boundary. . . . Because of the power differential between Jews and Christians, I want Christian women to draw boundaries from their side just as I need to draw them from my side. I want Christian women not to assume a commonalty with my experience or my theory, but to acknowledge this commonalty when it exists."[57] Attention to boundaries by Plaskow is a means of maintaining, recovering, and creating a minority Jewish identity within a dominant Christian context. Attention to this boundary by Christians implies recognizing the partial and particular perspective of

Christian locations and avoiding perpetuating the invisibility of Jewish perspectives and their exclusion from the dominant discourse by assimilating them without recognition of their distinctive location.

Attention to boundaries and not erasing or silencing the distinctive voices that produce theological reflection means *naming the roots and sources* of the work we use. Our theological and ethical reflection is deepened by learning from and naming the particularities of the work on which we draw, noting the differences and similarities with our own contexts and experiences. Mary Hunt argues that it is particularly important for women to name the roots of their work when they take it to other locations, "to avoid the exportation of theology from the United States as if it were all done by white women."[58]

To be liberating, appropriation must be genuinely *interactive.* It involves accountability to and a willingness to be informed, engaged, and changed by the work one appropriates. In examining the history of the efforts of African American women to get white feminists to pay attention to difference in moving beyond the universalized term "women's experience," womanist theologian Toinette Eugene notes that many white feminists now recognize the problem of difference, but not necessarily difference itself. She argues that their acknowledgment is noninteractive. Acknowledgment of difference gets used not as a recognition of one's own partiality in order to genuinely learn from and engage with others, but as a disclaimer, as a reason for not engaging someone else's experience. Disclaimers such as "I can only speak from my own experience" are heard by those outside that context as "I've been acknowledged and then dismissed: 'I'm not talking about you. I've acknowledged your difference and then we'll go on.' The interactive step is what is missing."[59]

Mujerista theologian Ada María Isasi-Díaz agrees with Eugene that it is the element of interactiveness that distinguishes between recognizing the problem of difference and recognizing differences per se. Drawing from her experience of often feeling invisible as a Hispanic woman in feminist dialogue, she cites three criteria for testing whether there is genuine interactiveness in appropriating the work of others. First, it must involve hearing the other on the other person's terms. "For difference to be positive we must allow the person who is different to be herself, and not require or demand that she be or act or present herself in a way that is intelligible to us. We need to enter into each other's world view as much as we can and help others open up to new perspectives. Unless we are willing to do this, the self we present to people who are different from us is a 'pretend self.' We will hide our real selves in order to protect ourselves from other's projections of us."[60] Second, evaluation of each other's work is not done primarily based on how it fits into one's own work. "If we recognize differences, we allow each other's work to be important in itself and do not evaluate its importance by how much I can use it in my own work. It requires a commitment to know each other's work even if it does not fit into our present work." Finally, appropriation that is genuinely interactive means "allowing differences to impact and change you and your work."[61] By paying attention to these criteria, women can better protect against coopting, conquering, and assimilating the work of other women, thereby reproducing oppressive dynamics of exclusion and invisibility.

Epistemological issues around *self-knowledge, knowledge of others*, and a more accurate *knowledge of social realities* are at stake in issues of appropriation and reciprocity. Forms of oppression both result in and require lack of knowledge in these three areas, particularly lack of real self-knowledge in the oppressor. True knowledge comes only from engaging those different from ourselves, learning from them in order to learn about ourselves. This is self-knowledge that can be learned in no other way.

Hispanic feminist philosopher María Lugones argues that those who are different "are mirrors in which you can see yourselves as no other mirror shows you."[62] Engagement across difference can help persons from dominant cultures to recognize the partial, particular character of their experience and location, common knowledge to those who must straddle both cultures in order to survive: "Racial/ethnic women have always known that what we say about ourselves can't be universalized."[63]

Self-knowledge through interaction and learning from the experience of others enables us to see things about ourselves we otherwise might miss; Judith Plaskow explains, "Just as Valerie Saiving argued (I think rightly) that women's experience can open up aspects of the human situation that have been neglected or ignored, so black or mujerista or other women's experiences can open to me as a white woman neglected dimensions of myself."[64] Yet womanist theologian Emily Townes cautions that learning from others must be paired with doing the critical hard work on one's own history if it is to avoid being a utilitarian expropriation of the suffering, struggle, and work of others:

> I *do* want you to hear my story
> take it in
> consider how you have been a part of it
> or not
> but do this in contrast to your own traditions and cultures
> to understand and consider
> how your lives and history
> are a part of the fabric of creation
> with mine
> guard against setting
> my story in your script
> having me illuminate points
> you must/should make on your own
> through the integrity of your witness
> your analysis
> your ability to critique and analyze
> from your perspective
> each of us must begin with our own cognitive dissonance
> we cannot appropriate each other's
> and have a truly articulate and pithy analysis
> we have much to learn from one another
> as we appropriate and reciprocate
> but we must not use each other up or down.[65]

Townes's caution is related to another theme common to the panelists: interactive appropriation means *active solidarity of praxis with and among the oppressed,* particularly with those from whom one draws work and insights. Sharing a common cause is what fosters dialogue. This requires "pivoting the norm" when we come together to speak, so that traditionally subjugated knowledges and groups are given voice to speak from the center.

Kwok Pui-lan identifies *attentive listening and solidarity with the pain and struggle of women in other traditions* as criteria for respectful appropriation of their religious symbols and stories. After listening to a lecture by Judith Plaskow, she found herself reflecting, "As a Christian theologian I asked myself what right I have to use the Jewish myths, legends, and scriptures without sharing the pain Jewish women have experienced through the centuries. On what grounds can we Christians claim their stories as our own?" Authentic sharing and reciprocity between women of different cultures and traditions requires that "collectively we develop our capacity of listening, so that we can have shared pain, shared anger, and shared strength."[66] With respect to Christian use of Jewish materials, Kwok suggests:

> As Christians we can learn from them or appropriate their work only if we are prepared to be in solidarity with their pain and struggle. Otherwise we will again steal the sacred symbols from the Jews, appropriate them for our consumption, and use their symbols against them, as we have done in the past. Religious symbols, sacred stories, great novels and literature emerge from peoples' yearnings, struggles, and visions. There is a holy dimension in them, because they illuminate the meaning of existence and the destiny of a people. If we take them out of context, we not only do injustice to the stories and symbols, we also violate the integrity of the people.[67]

A liberationist commitment means being attentive to the problems and dangers involved in appropriating the work of others across lines of difference in ways that do not violate the integrity of the work, author, or people. One danger is taking work out of context to make a point different from the message of the text. A key criterion here is a *nonutilitarian approach*: acknowledging and respecting the central message and integrity of a work when using it to support or illuminate one's own.

Another danger is a too-easy adoption of another perspective that leads to romanticization and trivialization, forms of dismissal of the serious character of a work and the struggle that has produced it. Emily Townes argues that this often results from a too-easy identification with another that ignores or minimizes important boundaries:

> the notion that we are aware of another person's feelings and
>   experiences
> on the basis of empathic inferences from our own
> veers into solipsism
> self-consciousness and awareness of others
>   are not natural dance partners
> understanding the other
>   is not predicated on how the individual (or the group)
>   makes the
>   shift from the certainty of her inner experiences
>   to the unknowable person
> when we make this kind of tenuous shift
>   the outcome generally falls into two categories
>   romanticization
>   or trivialization
> what we must be about as we approach one another's work
> is *care-filled listening and observing and engagement*
> this takes time
>   energy
>   resources
>   fortitude
>   and a stout will-to-comprehend[68]

As a convert to Christianity, Kwok Pui-lan recognizes "the religious and spiritual need for people to appropriate myths and stories that are not their own. . . . We sometimes recognize ourselves in other people's myths more vividly than we have ever recognized ourselves in the myths of our own culture."[69] Yet she warns that the adoption of someone else's tradition must be done with extreme caution and with a critical awareness of the limitations and shortcomings that each tradition has. She recalls her feelings of anger on one occasion when she sat and listened to a North American scholar talk about the wonderful male and female images in Buddhism. Kwok raised her hand and asked, "'Where on earth can you find a truly egalitarian Buddhist community?' I wondered if she had heard the cry and anguish of Asian women living in a predominantly Buddhist society. The suffering and pain of our Asian sisters becomes even more invisible if we simply romanticize the Asian religious traditions without critical discernment."[70]

Three dangers have to do with the way one emphasizes difference in appropriating the work of others. One is the danger of reading the work of others *simply* in terms of difference, thereby reducing the person and work to the category of essential other, obscuring or missing points of connection and commonalty. From the perspective of a Jew whose work is read by Christians, Judith Plaskow states, "I want my difference acknowledged, and I believe we have a great deal to learn from differences, both critically and constructively. But I also believe that to see me simply in terms of difference—especially to see me primarily in terms of my oppression—continues to turn me into an Other. . . . I think knowledge of [our] commonalties is every bit as important as knowledge of our differences and that these commonalties need to be named."[71] As is discussed in more detail in the following chapters, the social construction of the other, whether homosexual as other, African as other, or nature itself as other, has been instrumental in objectifying and exploiting groups of people and otherkind by dominant sectors of society. It is particularly critical, thus, that recognizing our difference does not prevent also recognizing our points of connection so that we inadvertently contribute to the exploitative and objectifying practice of "othering."

A second danger is that of *reifying* difference, ignoring the ways social differences and identities are historically generated and evolve and change. This can freeze other identities into a moment of time, turning them into essentialized others, which denies them their own process of change, development, and growth. Along these lines, Emily Townes cautions, "In our conversation about appropriation and reciprocity / take care when you name my reality / for i am still discovering it myself."[72] In this sense appropriation and reciprocity are ecological: they allow for and anticipate change and growth along all strands of the relational web.

Finally, there is the danger of ignoring differences *within* one's communities and categories. Within any community or group, important differences exist in terms of power, culture, class, and sexuality, to name a few. Paying attention to commonalties and differences within our communities enables us to acknowledge the parts of our identities that are most central. It can also make us conscious of our responsibility to take advantage of situations of *relative privilege* to lift up and make visible and audible the experiences and voices of less-privileged members.[73]

What are some of the implications of the above discussion for a liberationist ecological ethics? If, as Charlene Spretnak maintains, we need to explore the richness and depth of the diversity of human spiritualities and religious traditions to confront the ecocide that is killing the planet,[74] how might this discussion shape our approach to learning from others outside our own communities, cultures, and locations?

First, with Primavesi, Spretnak, and others, we can affirm that openness to learning from others outside our traditions is an essential ecological value, but it must be grounded in the liberationist emphasis on attention to power and difference. Our style of drawing on the insights and experiences of others to formulate ecological ethics must be genuinely ecological, that is, relational, and authentically interactive rather than exploitative. In learning from others we need to pay attention to differences in power in both our current and historical relationships, and how that power difference has influenced our current location in the total ecology of relationships. It means paying attention to boundaries.

An example of how this has been violated is in the selective appropriation of much Native American spirituality by members of the dominant Euro-American culture with no regard to the ongoing oppression of Native Americans within the United States. The answer is not to ignore the wisdom of native traditions, for as Spretnak has observed, "Although Native Americans understandably find an intrusive raiding of their spiritual culture (especially by members of the dominant society that has treated them so abominably for centuries) to be offensive, many of them also feel that sharing part of their spiritual heritage of ecological Wisdom could benefit Western civilization."[75] In examining clashes between environmentalists and Native Americans over some native religious practices, Robert Warrior notes the irony that "a group of New Agers is demanding freedom to imitate American Indian religious practices those they are imitating do not enjoy the freedom to practice."[76] Warrior argues cogently that appropriation of native religious tradition without active solidarity with Native Americans in their current political struggles to overcome a five-hundred-year legacy of oppression and genocide is simply continued exploitation by the dominant sectors of society. Haraway points to a similar dynamic when First World environmentalists paint indigenous persons in the Amazon rainforest or the African savannah as environmental threats while ignoring the history of North Atlantic colonialism and imperialism that in large part created the conditions of the current ecological crisis threatening these regions.[77]

A related criterion is hearing and learning from the ecological wisdom of others on their own terms rather than selecting the parts that fit our predetermined agenda. It means avoiding romanticization, trivialization, selective use, and a too-easy adoption of another tradition through commitment to critical learning: engaging seriously, questioning, being open to being changed through power-sensitive interaction that recognizes historical and current power differentials in the relation. This means acknowledging the central insight or message of a tradition, even if it is different from our emphasis or current need.

A liberationist ecological learning from others pays attention to learning through interaction at many levels, including increased self-knowledge, knowledge of others, and a more accurate reading of the social (and ecological) reality and web of relationships. For those in positions of relative social privilege it means attention to gaining genuine self-knowledge through the eyes and experiences of others, rather than seeing one's own self-image reflected back in the work of others, filtered through one's preset agenda.

Genuine self-knowledge and learning depends on critical awareness of what one brings to the conversation, including acknowledging up front what is at stake personally and professionally, what evokes a passionate response and commitment in one, and why. This means taking responsibility for critical learning about one's own cultures and histories, reclaiming and owning them in all their complexity for the resources they provide and the obstacles they contain.

As I will develop further in chapter 10, drawing on the experience and insights of others means paying close attention to our social (and ecological) locations, to where we are variously situated with respect to other human beings (and nonhuman creatures), and to the histories and power relations that have shaped these locations. This includes attention to differences both between and within communities, understanding our identities as fluid historical constructs that may serve constructive strategic ends, rather than fixed biological or essentialized boundaries of "who's in, who's out." Wendell Berry's thoughtful writings on race relations in his native Kentucky give examples of how these multiple levels of historically constructed power shape the different relationships of whites, blacks, and Native Americans to the land (and of the different land relationships *within* those communities along lines such as class and gender)—and the destructive ecological consequences of relations of human inequality.[78]

Finally, one implication of paying attention to difference and particularity is that, just as Rosemary Ruether calls for "many ecofeminisms,"[79] we will need a plurality of ecological ethics appropriate to different social and ecological locations, communities and religious traditions, rather than one universalized or monolithic approach. This is an ecological insight—recognizing that each "ecological niche" or location must develop an ecologically sound ethics appropriate to that web of relationships. What makes it liberationist is the insistence that this pluralism of different ecological ethics is not simply relativistic, but engaged, interactive, and with continued attention to how shifting power relations between communities and different social/ecological locations affect and reshape the resulting ethics.

With this methodological grounding, let us turn in the next chapter to one of the central questions in ecological ethics: how should we understand humanity and "human nature"? What can lesbian and gay constructionist approaches tell us about how we have understood what is natural or unnatural about humanity in its relation to the rest of nature, and what difference does this make in an ecological ethics?

3 Humanity: Rethinking Human Nature and the Natural

Detectives from the vice squad
with weary sadistic eyes
spotting fairies.

Degenerates,
some folks say.

But God, Nature,
or somebody
made them that way.

Police lady or Lesbian
over there?
Where?

—Langston Hughes, "Café: 3 AM"

The sources [of the ecological crisis] lie not in the nature that scientists study but in the human nature *and, especially, in the* human culture *that historians and other humanists have made their study.*

—Donald Worster, *The Wealth of Nature*

A central claim of *Gay and Gaia* is that societal power dynamics that shape and undergird interpretations of sexuality and human nature in ecological ethics must be exposed and critically examined. As Donald Worster reminds us in the quote above, the source of the ecological crisis is *not* nature, but rather *human nature* and *human culture*—all the patterns, concepts, and activities we have devised for interacting with each other and the rest of nature. Feminist scholars in recent years have paid close attention to how historical and contemporary understandings of human nature reflect socially constructed views about gender, race, and sexuality. Typically these constructs have been seen as "natural," rooted in nature, and have been used to maintain a hierarchical social order, itself seen as part of a "natural order." As we approach the end of the twentieth century, it is clear that this hierarchical social order has had deleterious effects on many of its members and on nature overall. An ethic of ecojustice needs a keen awareness of the socially constructed character of views of human nature, both because of the human injustice caused by social orders based on domination, and because of the adverse impact on the biosphere of basing this social order on the domination and exploitation of nature.

Because constructs of human nature and culture typically have excluded lesbians and gay men as "natural" members, much lesbian and gay theorizing has focused on the assumptions and values that make up socially constructed views of humanity. Awareness of the historical and contemporary ways interpretations of homosexuality and homoerotic relationships have been constructed illumines several aspects of the social order that are critical to an adequate ecological ethic. Christian models of ecological ethics commonly draw either explicitly or implicitly on the Bible and the natural and social sciences to derive their understanding of human nature. These in turn may have liberating or oppressive ramifications for socially marginalized persons, including lesbians and gay men. It is therefore important to assess the assumptions and understandings of humanity in models of ecological ethics for their awareness of the socially constructed nature of the views they hold. In this chapter I look at key insights of selected scholarship by lesbians and gay men who examine views of human nature in ancient and biblical literature, and in the contemporary sciences. From this overview we can derive several criteria for a liberationist ecological ethic that can be used later to assess the adequacy of current models of Christian ecological ethics.

SOCIAL CONSTRUCTIONS OF HUMAN NATURE: EXAMPLES FROM ROMAN AND PAULINE ANTHROPOLOGIES

In the midst of intense debate in the churches in recent years about the acceptance of lesbians and gay men, several scholars have focused renewed attention on biblical teachings on homosexual activity.[1] Their work has challenged long-accepted views about sexual identity that are rooted in the unchallenged assumptions of heterosexual naturalism by exposing the historical and cultural factors that shaped these ancient positions. Among these, New Testament scholar Bernadette Brooten is clearest in separating examination of female from male homoeroticism and bringing to her work a consistent attention to gender and power dynamics. Undergirding this is her conviction that sexuality is a social construct that is ultimately about power. "[It] is not simply a matter of romantic love, nor is it a changeless, purely biological phenomenon; rather, sexuality is determined by societal structures."[2] Her work is thus particularly helpful in drawing attention to how social relations of power can become embedded in—and perpetuated by—our commonly accepted views of sexuality.

Brooten has focused extensive attention on Paul's condemnation of female homoeroticism in Romans 1:26 because of what it reveals about Paul's anthropology and because of the impact it has had on all women and men—both in his time, and those influenced since by his teaching. She maintains that it is critical to locate Paul's thinking within the contemporary discussions about female sexuality in his own time. Behind Roman condemnation of female homoeroticism is concern for the social order—what it means to be female and male within society. Female homoeroticism in the Roman period was debated primarily as a political issue of public morality rather than as an individual matter of private morality, because it challenged the hierarchical relationships that undergirded the social order.

Similarly, current debates on homosexuality are much more than issues of private acts and morality; lesbian and gay relationships have critical social and public ramifications for how we conceive a healthy, moral society (a point recognized by both right-wing opponents and progressive advocates, but often overlooked by liberal, centrist positions).[3] Debates on sexuality also have important ramifications for ecological ethics. Because claims for naturalized orders of social and environmental domination are among the key culprits of the current ecological crisis, how societies deal with homoerotic relationships that threaten naturalized hierarchies may be an important indicator of their ability to shape a different ecological ethics.

The key construct informing Pauline and wider Roman cultural understandings of sexuality is that sexual relationships are essentially, that is, biologically, asymmetrical. Sexual relationships thus represent and constrain a broader human social hierarchy understood as both normative and rooted in nature. The most fundamental category for expressing the asymmetrical character of this hierarchy was active/passive roles. Every sexual pairing required one active and one passive partner in order to be considered moral and natural. These roles in turn were correlated with gender, so that active equals masculine and passive equals feminine, but active/passive superseded gender as the primary criterion for determining the moral and natural character of the relationship. Greater freedom in male roles allowed males under certain circumstances to be passive, but females were always supposed to be passive.[4] Brooten notes that while the division between active and passive is not biological, Roman writers of this period biologized it by describing women as born to be passive.

Understanding the gendered active/passive dualism allows us to comprehend why some forms of male homoeroticism were accepted, or at least debated, while virtually all female homoerotic behavior was condemned. Males who were lower on the social hierarchy, either through age (boys who had not yet reached manhood) or status (slaves) could be penetrated by males of higher rank without loss of status to either participant since the ordering of the hierarchy was not threatened. In accordance with this logic, however, only the active male was seen as masculine; the passive or penetrated males were defined as "effeminate."[5] Women in female homoerotic relations presented a different problem for these cultural constructions: how are female/female relations possible at all if all women are passive and all sexual pairings must involve an active and a passive participant (as well as a phallus)? Unable to fit female homoeroticism into a gendered social hierarchy with a phallus-centered view of sexuality, virtually all (male) writers of this period condemned it.[6]

Paul's views on sexuality must be placed within this widespread cultural context of asymmetrical sexual relations in order to help us better comprehend his condemnation of female homoerotic acts in Romans 1. Paul had a complex understanding of gender such that he could praise women for their labor and leadership in Christian communities while still calling for strict gender distinction in hairstyle, dress, and behavior. Within this framework, Paul's condemnation of female homoeroticism is best understood as based on the widely shared assumption in the Roman world that nature calls for men to be superordinate and active, and women to be passive and subordinate: "If Paul did not condemn sexual relations between women and between men from within the gendered framework of human beings, then he would have been unique in the ancient world."[7]

Paul's condemnation of female homoeroticism is also consistent with Jewish legal discussions of his time on homoeroticism. While Hebrew scriptures make no mention of female/female sexual relations, Paul extends the Levitical condemnation of male homosexual acts to include female homoeroticism as worthy of death (Roman law had no such prescription). While Paul lifted Jewish purity restrictions that set social boundaries such as dietary laws and circumcision, he does not set aside purity codes regulating sexual behavior. Brooten thus concludes:

> For Paul, same sex love in Romans 1:26–27 is a sin against the social order established by God at creation, and not just a private

> sin against a system of private morality. Paul envisages a social order in which Jew and Gentile are no longer relevant categories and attempts to break down the boundaries distinguishing them, such as circumcision and dietary laws. In contrast, within this same social order, Paul deems necessary a natural, and therefore immutable gender boundary, even though he respects women's work for the Gospel. Those who do not confine themselves within these demarcated areas break down the required gender polarity and gender distinctions thereby becoming impure. . . . Crossing the boundary of circumcision no longer threatens one's salvation, but crossing the gender boundary still makes one deserving of death.[8]

Recent scholarship on the relation of purity and property codes to sexual ethics in the Bible reinforces the importance of examining the socially constructed character of these views. In his groundbreaking book, *Dirt, Greed, and Sex: Sexual Ethics in the New Testament and Their Implications for Today*, New Testament scholar L. William Countryman argues that we cannot understand accurately biblical teaching about sexuality without placing these teachings within the cultural assumptions about purity and property that largely defined the patriarchal context of the biblical period. Purity refers to the way cultures symbolically draw social boundaries between acceptable and unacceptable groups and behaviors as a part of maintaining group identity and cohesiveness, often using body boundaries to represent these social boundaries.[9] Persons or behaviors that fall outside these boundaries are seen to be "dirty" by their very nature and thus to be excluded from the body politic. Cross-cultural anthropological studies demonstrate that while the form through which purity operates is similar between cultures, the content varies widely and reflects more the distinctive values and assumptions of each culture rather than empirical scientific evidence for what is actually hygienic or not. Similarly, personal and family relationships in biblical times largely were defined by patriarchal property codes that assigned each person a fixed place within the patriarchal family or clan. Violations of relationship such as rape or incest were treated as crimes that damaged the property of the patriarch, which then required compensation.

Purity and property assumptions and codes become relevant for Christian ethics because, early in its history, Israel associated purity—being set apart and different from its neighbors—with holiness and fidelity to God, and its patriarchal family patterns with God's will for ordering human society. Much early Christian debate centered on which if any of these codes were still relevant in the wake of Jesus's ministry and the gospel; Paul's arguments against circumcision and dietary laws as prerequisites for Gentile Christians are perhaps the clearest examples of rethinking purity requirements.[10] Christian communities ever since Paul have been notoriously inconsistent on which of the Levitical purity and property laws they still follow, which they reject and why; a critical step in good ethics is to understand what values and assumptions gave rise to purity and property teachings and then to ask if they are consistent with gospel values and commitments. Here again the tools of constructionist theory can be very helpful in understanding the histories and social relations that lie behind purity and property commitments.

Several implications of Brooten's and Countryman's work on sexuality and the biblical texts are relevant to the task of revisioning a just ecological ethic. Both expose the danger of grounding the social order in hierarchical assumptions of heterosexual naturalism and divine mandate. A social order that depends on women orienting their lives toward men must define lesbians as unnatural and unfeminine. Similarly, gay men who refuse to participate in relationships of compulsory heterosexuality must also be defined as unnatural and unmasculine.

There are also important implications here for our understanding of human "nature": "By living out a definition of femaleness contrary to the socially accepted norms of naturalness, lesbians expose nature as a social construct, rather than as a biological necessity."[11] Historical and current lesbian and gay existence and the tremendous diversity of social/sexual relationships cross-culturally and through time demonstrate that women's orientation toward men and men's orientation toward women is not required by nature, but rather is created and enforced by culture. We in turn must be critically aware of the ways contemporary constructs that inform our views of human nature become essentialized in nature. In social orders that assume heterosexual naturalism, maintaining total silence about the possibility of homosexuality or bisexuality is the most politically effective means of keeping all women oriented toward and subordinate to men, and gay men subordinate to heterosexual men. When that fails, defining lesbians and gay men as unnatural and monstrous can also keep their behavior in check and preserve the hierarchical social order as naturalized, and therefore unchangeable.

Two important ramifications for views of humanity in ecological ethics result. First, to expose and reshape assumptions of heterosexual naturalism that inform contemporary ecological ethics, it is important to challenge the invisibility and silence about lesbians and gay men that typify many ecological models. Such invisibility leaves unchallenged heterosexist structures destructive to human relations in society, and contributes to antiecological views of nature as female, passive, and open to domination and exploitation (explored further below). Second, a key component of ecological ethics is providing a compelling vision of relationships in an ecologically sustainable society. Gay and lesbian relationships have the potential to serve as positive resources within an ecological ethics to the extent that they provide models of egalitarian and mutual relationships outside the gendered social hierarchy.

SOCIAL CONSTRUCTIONS OF HUMAN NATURE: CONTEMPORARY EXAMPLES FROM SEXOLOGY AND THE NATURAL SCIENCES

Historical studies such as Brooten's, which are sensitive to constructed categories of gender and sexuality, help us to understand how we have inherited ways of thinking that shape the way we understand human nature. Similarly, lesbian and gay scholarship has drawn on constructionist theory to examine contemporary fields such as the social and natural sciences. The insights of these scholars provide an important contribution to the necessary revisioning of interpretations of human nature that inform ecological ethics. In this section I turn to two areas that have figured prominently in debates about sexuality, and particularly homosexuality as a natural or unnatural part of human nature: sexology, the scientific study of sex, and the biological sciences.

The work of the British gay scholar Jeffrey Weeks exemplifies a constructionist approach to developing a historical understanding of sexology. He examines the web of historical, theoretical, and political forces that shaped the emergence of sexology in the late nineteenth and early twentieth centuries and the discourses on sex and sexual identity it produced.[12] Key to the premise of *Gay and Gaia* is Weeks's thesis that the allegedly neutral and objective stance of sexology as a science actually reflects a complex interweaving of social values and assumptions about sex, sexuality, and nature. He challenges essentialist theories of sexuality by analyzing the "historical interactions between sexual theory and sexual politics over the past century in order to question the neutrality of sexual science and to challenge its hegemonic claims."[13] Particularly important are questions of epistemology, especially when truth claims about sexuality make appeals to the "essence" or "natural" human state. How have social relations and social controversy influenced what counts as scientific knowledge, and how do these processes continue today? How do these in turn affect our views of human nature and influence contemporary ecological ethics?

While often reformist and humanitarian in its impulse, the heritage of sexology has left behind essentialist understandings of sexuality that are highly problematic. They have contributed in the twentieth century to shaping and maintaining an elaborate system of social control of those who deviate from a procreative heterosexual norm. Essentialism here means "ways of thinking which reduce a phenomenon to a presupposed essence . . . which seeks to explain complex forms by means of an identifying inner force or truth."[14] Against this view, Weeks argues that "there is no *essence* of homosexuality whose historical unfolding can be illuminated.

There are only changing patterns in the organization of desire whose specific configuration can be decoded."[15] What behaviors sexologists claim are "natural" are actually social constructs that reflect the social mores of the status quo rather than what is unchanging or natural.

The historical context of the rise of sexology shaped its essentialist moorings. Reflecting its nineteenth-century European roots where the insights of Darwinian evolution led many scientific fields to a search for "laws" presumed to govern behavior in the natural world, the goals of sexology were to discover the laws of nature that determine human sexuality. Two linked processes shaped its early orientation. Paralleling the rise of liberal economic theory and philosophy, the nineteenth century saw an individualization of sex. This led to the discovery of "the sexual impulse," understood as a natural inner urging of sexuality, and to corresponding efforts to describe and classify pathological deviations from it. This corresponded well with the growing authority of biology (following Darwin and burgeoning interests in natural origins) and psychology (with its focus on pathologies) for insights into what is authentically (i.e., "naturally") human.

Paralleling similar trends in economic theory, emphasis in sexology was placed on the productive and reproductive aspects of nature. Alleged perversions of or deviations from this perceived norm in turn helped shape understandings of what was normal. Ironically, therefore, homosexuality as a concept was described before heterosexuality in order to name a nonprocreative pathology of same-sex attraction and behavior. The concept of heterosexuality was constructed later in order to describe what was presumed to be "natural."

Yet while creating these new concepts, sexology did not merely invent these behaviors. Early sexologists were both naming and describing what they saw around them in society, and in so doing they helped to create the conditions for new constructs of sexuality by giving them a name and identity. In the process, sexology functioned to translate concrete social problems of changing gender and sexuality roles that emerged during the rapid social change taking place in Western industrialization and urbanization into ahistorical theoretical terms by giving them a biological grounding. Hence "sexuality came to mean . . . both the study of the sexual impulse and of relations between the sexes, for ultimately they were seen as the same: sex, gender, sexuality were locked together as the biological imperative."[16]

This process of naturalizing cultural constructions of sex and sexuality became increasingly important in the twentieth century as sexuality has become a symbolic battleground for contested social and economic change. Increasingly, sex discourse was tied to socially authoritative discourses from the medical, psychological, and legal professions. While often the intent of these moves was progressive reform, one consequence was to legitimate the power of these institutions to regulate sex and place its control in the hands of experts. Hence Weeks observes, "Sexology has never been straightforwardly outside or against relations of power; it has frequently been deeply implicated in them."[17] Rather, given the dominant social mores of the time, sexology resulted in the *naturalizing* or *essentializing* of sexuality in general, the *privileging* of heterosexuality as normative and natural, and the *categorization* of other forms of sexuality, such as homosexuality, as *deviance* or *perversion*.

This illustrates a fundamental problem of essentialist categories and definitions in understanding human dynamics: by naturalizing and universalizing categories, they hide the historical roots that have shaped them and make alternatives seem impossible. In making constructions that benefit those in power seem natural and unchanging, they inevitably end up supporting the status quo.[18] Weeks's work, like much gay constructionist theory, suggests the importance of investigating all scientific understandings of human nature for the socially constructed values and assumptions they may reflect. This is especially true for ecological ethics where biological investigations of nature often serve to ground claims about human nature, which in turn often ground arguments about what forms of human behavior toward the rest of nature are "natural" or "unnatural" in formulating an ethic.

An excellent example of this has been the role of scientific investigations in debates about the naturalness or unnaturalness of homosexuality. The presence or absence of homosexual behavior among other creatures in the animal world has served as one important element informing issues of human sexuality. Increasingly, gay and lesbian scholars and scientists have become aware of the socially constructed nature of the scientific representations involved in these investigations. These studies in turn have important implications for the ways ecological ethics draw on the natural sciences.

Gay scientists John Kirsch and James Weinrich examine the nature of scientific assumptions in investigations of homosexual behavior in animals.[19] They survey numerous biological studies of same-sex behavior in the animal world to show the contradictory ways these studies have been used to address the question, "Is homosexuality biologically natural?" Several of their examples illustrate the importance of constructionist insights into the assumptions we make about nature, biology, humans, and other animals.

Noting that most discussions about homosexuality implicitly assume that naturalness is directly related to the presence or absence of homosexual behavior in nonhuman animals, Kirsch and Weinrich review scientific documentation of homosexual behavior in domestic and wild animals in a variety of environments. Homosexual acts in nonhuman animals appear to express different meanings in a variety of contexts, such as expressions of male dominance over other males, as continuous monogamous same-sex pair bonds, and as expressions of affection in primates. They conclude: "These homosexual patterns do not duplicate human homosexuality but do resemble certain aspects of human homosexual behavior. . . . Animal homosexuality covers a wide range of behaviors, just as human homosexuality does. There is scarcely any aspect of human same-sex behavior that does not have at least a moderately close parallel in some animal species."[20]

At first glance, therefore, it would seem logical to argue for the naturalness of human homosexuality based on its occurrence throughout the animal world. Yet the premise implicit in this is problematic because it assumes that the meaning of different homosexual behaviors among animals can be understood and then extrapolated to shed light on meanings of human homosexual behavior and orientation. To move into interpretations of meaning of homosexual behavior in animals and humans is to move into the realm of societal values and mores. These in turn reflect society's assumptions about what is natural. Depending on the content of these values, what is natural in different nonhuman animal behaviors may or may not be seen as natural in humans.

In an interview with *Playboy* magazine, former anti-gay activist Anita Bryant was asked how one explains homosexuality to a nine-year-old. She responded:

> I explained in simple terms to the little ones that some men try to do with other men what men and women do to produce babies; and that homosexuality is a perversion of a very natural thing that God said was good, and that it is a sin and very unnatural. I explained to the children that even barnyard animals don't do what homosexuals do.[21]

To the interviewer's response that her claim is untrue, that there is evidence animals do engage in homosexuality, Bryant replied, "That still doesn't make it right."

Bryant seems to ground the initial part of her argument in nature; because "even" animals don't have homosexual behavior, it is unnatural for human beings. She changes her logic, however, once her argument from nature is shown to be false. Clearly what is authoritative for Bryant is not actual empirical evidence from the behavior of animals other than humans, but rather a prior commitment to a culturally constructed view: homosexuality is wrong. Hence, even if it is natural for animals, that does not make it right for humans.

The following diagram shows four different logics that are possible from the premise of the naturalness or unnaturalness of homosexuality based on its presence or absence in animals.[22]

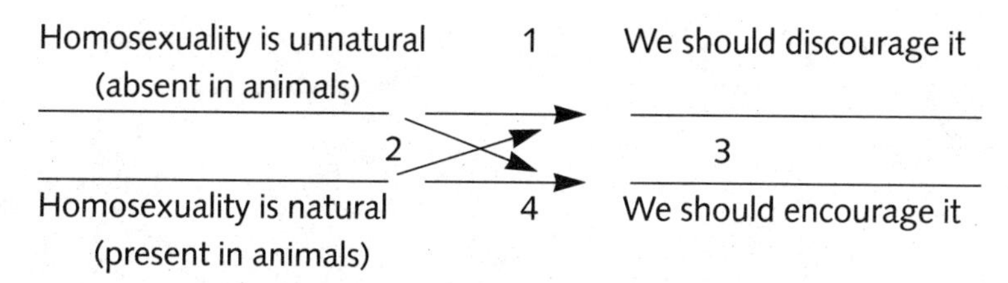

Bryant begins with the premise and logic of path 1: since homosexuality is unnatural, it should be discouraged. When her premise is contradicted, instead of keeping her logic parallel and following the logic of path 4 (since homosexuality is natural, it should be encouraged), she instead develops a different "logic" and follows path 3: even though homosexuality is natural, it should still be discouraged. The same shift in logic is possible between the premises informing paths 4 and 2. An example is the debates in ancient Greece over which was superior, heterosexual or homosexual love. Many of the advocates for the superiority of homosexuality agreed that it was not "natural." They argued on this basis for its preference: *because* homosexuality does not exist in animals and heterosexuality does, and *because* human beings have the ability to transcend "lower" nature, nonprocreative homosexuality is a purer, and hence superior, form of love.[23]

This exercise illustrates the problem of using studies of animal behavior to answer questions of human morality. Kirsch and Weinrich argue that the whole enterprise symbolized in the diagram above is wrong. "When one rises above the *scientific* level of truth or falsity of any one 'fact,' or the appropriateness of any one arrow, we have to ask whether *any* of these arguments are valid. We take a jaded view and answer the negative. When animals do something that people like, they call it 'natural.' When animals do something people do not like, they call it 'animalistic.'"[24] The important point is that facts about nature are never self-explanatory. Drawing on them to develop social policies and moral positions always involves cultural assumptions and societal mores, in addition to scientific observation.

This example also illustrates why investigating the claims and assumptions about human nature are important to ecological ethics. Claims about human nature nearly always contain implicit claims about nature in general and the appropriate relationship between humanity and nature. Note the different views of nature and humanity's relation to nature that inform each position. In Bryant's first argument (path 1), animals are inferior to humans, but are related enough to give guidance about what is appropriate human behavior. In her second argument (path 3), in order to keep her conclusion (human homosexuality is unnatural, wrong), Bryant severs the relationality between humans and nonhuman animals. For path 4 (homosexuality occurs naturally in animals, therefore we should encourage or tolerate it in humans), a parallel relation between human beings and the rest of nature is assumed. In contrast, for the ancient Greek advocates of the superiority of homosexual love, nature is separate from and inferior to human beings. In this case, when nature is viewed as lower than human beings, homosexuality is seen as superior precisely for being unnatural. Each view thus incorporates a different view of nature and what should be humanity's relationship with the rest of nature.

Other arguments against considering homosexuality to be a natural human behavior reveal the relation between social values and how science is conducted. For example, while many field scientists have observed homosexual interactions in the wild, virtually none of them have pursued or published their findings. In the absence of other factors, it appears that in these cases homophobia, not scientific observation and objectivity, has shaped what questions and observations are considered scientifically important. Similarly, negative attitudes toward homosexuality have affected assumptions that inform scientific interpretation itself. Consider the following quotation from a text on sexual behavior:

> There is also recent evidence that homosexuality may be related to certain biological pathologies: Research on animals and human beings . . . has shown that virilizing hormonal *imbalances* in the mother during gestation lay down important male patterns of receptivity in the developing neural structure of a fe-

> male fetus, who later on, in childhood and beyond, exhibits classically male patterns of play and aggressiveness. Comparable congenital *"errors,"* it is thought, may underlie some or much of the classically female behavior of some male homosexuals.[25]

On what grounds are the prenatal hormones judged to be "imbalances" or "errors"? Arguments that they are imbalanced because they produce a homosexual orientation are circular; they presuppose a negative judgment of homosexuality. Hence Kirsch and Weinrich ask, "How do we know that homosexuality is something having gone wrong, rather than a normal variation in which something went right?"[26] Similar to the findings of constructionist studies of sexology, close examination of scientific statements about homosexuality such as these often reveal as much about the cultural and moral presuppositions informing them as they do "objective" scientific assessment and interpretation.

Against the assumption that homosexuality is inherently unnatural because gay and lesbian people do not reproduce and thus pass on their genes, more open-minded scientists have considered mechanisms by which homosexuality could serve a constructive evolutionary purpose (and hence would not have been eliminated by natural selection in evolution).[27] Because of the role culture plays in human evolution, this also separates the question of the naturalness of human homosexuality from homosexual behavior in animals: "In suggesting how homosexual behavior may have contributed to the real evolutionary success and unique adaptations of human beings, [these evolutionary explanations] make homosexuality a very natural thing—regardless of whether its occurrence in animals is 'natural' or not."[28]

Kirsch and Weinrich thus conclude that homosexuality in humans is as biologically natural as human heterosexuality. This has important implications for public policy formation: "Its only social consequence is that biological arguments cannot be used to distinguish morally between homosexuality and heterosexuality. Like left- and right-handedness, the two are expressions of a single human nature that can be expressed differently in different individuals."[29] The fact that biological arguments are commonly used to make moral distinctions between homosexuality and heterosexuality again argues for careful consideration of how socially constructed views of human nature affect science and shape our assumptions about nature and what is natural. An ecological ethic that integrates justice concerns must develop sensitivity to these issues.

To summarize, the constructionist insights of these lesbian and gay scholars provide a critical guide and liberationist perspective from which to assess the views of humanity that inform models of ecological ethics. They contribute to both of our paired ecojustice norms of Gay and Gaia. With respect to Gay, lesbian and gay constructionist insights draw attention to the ways cultural values and social power relations along lines of sexuality, gender, and race inform our views of human nature either to perpetuate or destabilize social relations that directly influence questions of social and ecological justice. With respect to Gaia and ecological well-being, implicit in each view of human nature are attitudes toward nature and the humanity/nature relation that must be explored and assessed in a liberationist ecological ethic. The insights used to examine claims in science and theology about human nature can be summarized as follows.

- *Use of Constructionist Theory*: Lesbian and gay scholarship has shown the importance of paying attention to the ways societies construct categories such as gender, sexuality, race, and nature. Constructionist theory provides a critical tool for destabilizing categories and relations that are assumed to be rooted in nature in order to create more liberating positions, identities, and relations. It draws attention to relations of power and how these inform the assumptions we make about what is natural for human beings and how we relate to nature. An ecological ethics needs to have critical awareness of how the assumptions it makes either reproduce or destabilize the social relations rooted in these constructs.

Gay and Gaia

Ethics, Ecology, and the Erotic

About the Artist

Cover art: *Peaceable Kingdom*
© 1994 by John August Swanson
Serigraph 30" x 22.5"

Los Angeles artist John August Swanson is noted for his finely detailed, brilliantly colored biblical pieces. His works are found in the Smithsonian Institution's National Museum of American History, London's Tate Gallery, the Vatican Museum's Collection of Modern Religious Art, and the Bibliothèque Nationale, Paris.

Represented by Bergsma Gallery, Grand Rapids, Michigan (616-458-1776).

Full-color posters and cards of Swanson's work are available from the National Association for Hispanic Elderly. Benefits go to its programs for employment of seniors and for housing low-income seniors. For information, contact:

National Association of Hispanic Elderly
3325 Wiltshire Blvd., Suite 800
Los Angeles CA 90010
(213-487-1922)

- *Attention to Difference in Social Analysis*: At the heart of a liberationist ecological ethic must be attention to the multi-layered intersections of constructs of social difference—including but not limited to gender, race, sexuality, class, and nationality—for how they reflect or enable social and ecological relations of domination and subordination. An adequate social analysis forms an integral part of an adequate ecological analysis. Central to this is the awareness that human social relations built on inequality and domination are unjust and nonsustainable, having multiple deleterious effects not just on human beings, but also on human/nature relations and the biotic communities in which humans live and find sustenance.
- *Examination of Assumptions about What Is Natural/Unnatural*: An ecological ethic needs to incorporate historical/critical examination of societal and religious assumptions about what is considered natural and unnatural. Because ecological ethics is dependent on the sciences for ecological data, it needs to develop a critical understanding of the process of moving from scientific observation and its perceived neutrality and objectivity to formulating human values and social mores. In what ways does this process either facilitate or obstruct an ecological project that is also socially liberating for all members of society?
- *Attention to Social Location*: A liberationist ecological ethic must pay explicit attention to the role of social location in epistemology and theo-ethical analysis. In chapter 10 I explore the ramifications of expanding this into a notion of ecological location, but that must always include the role of social locations in the human community in generating knowledge and values. An ecological ethic must ask whose experience is reflected and whose is missing—including the perspective of creatures other than humans. It will give priority to those locations and standpoints that better generate critical knowledge and perspectives by paying attention

to the "subjugated knowledges" in its midst in order to gain a fuller understanding of social and ecological realities.

- *Attention to Communities of Accountability*: Liberationist ethics is a communal endeavor, and as such must be accountable to all members of the community that produces and is formed by it. A liberationist ecological ethic extends this accountability to all members of the biotic communities in which human communities are located. It shifts the primary location to the margins of society as sites of resistance and creativity in order to overcome center/periphery dynamics that maintain and reproduce hierarchical social relations.
- *Opposed to Heterosexism and Homophobia*: Lesbian and gay liberationist thinking pairs together the criteria of Gay and Gaia to signal an ethics of ecojustice grounded in and sensitive to lesbian and gay insights and experience. Any ecological ethics that perpetuates exclusion and invisibility of sexual minorities is inadequate as an ethics of ecojustice. Hence a liberationist ecological ethic must make explicit its inclusion of lesbian, gay, bisexual, and transgendered persons and community in the ecology of human culture and nature. It therefore explicitly opposes heterosexist epistemologies and understandings of nature and human nature.
- *Centrality of Sexuality Issues*: A liberationist ecological ethic must understand sexuality as an issue at the heart of ecojustice that exposes societal assumptions about right relation and the social order. Sexuality is not merely an issue of private morality, but of social power that is central to understandings of human ecology and how we relate to the rest of the natural world. Lesbian, gay, bisexual, and transgendered relationships can provide ecological ethics with positive resources for drawing on the erotic as a constructive moral resource in developing egalitarian and mutual relationships outside the gendered social hierarchy.

"The sources [of the ecological crisis] lie not in the nature that scientists study but in the *human nature* and, especially, in the *human culture* that historians and other humanists have made their study." Donald Worster's reminder that ethical analysis of the ecological crisis must begin with the constructs, values, and practices of human culture as its primary source has proved fruitful in examining some of the ways Western society has viewed human nature, especially with respect to lesbians and gay men. Equally important to an ecological ethics is examining the way society has constructed its views of nature, and the relationship of human culture to nature. It is to that question that we move in the next chapter.

4 Earth: Rethinking Nature and the Nature/Culture Split

At the root of the identification of women and animality with a lower form of human life lies the distinction between nature and culture fundamental to humanistic disciplines such as history, literature, and anthropology, which accept that distinction as an unquestioned assumption. Nature-culture dualism is a key factor in Western civilization's advance at the expense of nature. . . . If nature and women, Indians and blacks are to be liberated from the strictures of this ideology, a radical critique of the very categories nature and culture, as organizing concepts in all disciplines, must be undertaken.

—Carolyn Merchant, *The Death of Nature*

As environmental historian Carolyn Merchant argues in the quote above, liberationist approaches to ecological ethics maintain that the fate of both nature and subordinated human groups are tied inseparably together. Lesbian, gay, and feminist scholars have paid close attention to how cultural categories such as gender, race, and sexuality are constructed in attitudes toward nature and human nature to reflect social mores and values of particular cultures and societies. Implicit in much of Western thought has been a strong culture/nature dualism that has acknowledged categories on the culture side of the pairing to be culturally constructed while assuming categories on the nature side to be fixed and unchanging. Hence in the sex/gender pairing, gender has been seen as a sociological category and acknowledged to be a social construct, while sex was assumed to be biologically determined and therefore objectively verifiable.

As we have seen in the cases of Weeks and Brooten, in recent years pro-gay and profeminist scholars have turned their attention to questioning the allegedly fixed character of categories on the "natural" side of the pairing, such as sex and sexual orientation. This has been especially important for lesbians and gay men, for whom society has found ways to justify our exclusion from both sides of the pairing, as threats to both nature and culture. Yet increasingly scholars see the nature/culture dualism itself as a primary source of the ecological crisis as well. In its classic Cartesian forms, this dualism has served to separate humanity out of nature as the only active agent in history, and to view humanity-less nature as passive and unchanging, available to human beings as an endless resource for creating culture. Within this division, certain parts of the human community—particularly women, people of color, and homosexuals—have been more closely identified with nature and less with culture and hence less fit to engage in social and moral agency than educated, white, heterosexual men. Scholars are now paying close attention to how we have understood nature and seen it in relation to constructs of culture for the social and ecological dynamics they foster or constrain.

Implicit in any ecological ethics are assumptions about nature and the relation of humanity to nature. These have critical ramifications for the human and nonhuman members of biotic communities, and especially for lesbians and gay men with the ways assumptions about nature and human nature have been used against us—often in totally contradictory, but nevertheless damaging, ways. A critical examination of these assumptions lies at the heart of any revisioning of ecological ethics. To begin to develop the criteria that can guide an ethic of Gay and Gaia in revisioning the humanity/nature relation, the work of ecofeminist historian Carolyn Merchant is very helpful for what it reveals about Western constructs of nature and their ramifications for its human and nonhuman members. Other ecofeminists have built on Merchant's effort to rethink the nature/culture dualisms that inform assumptions about nature and culture. Yet this nature/culture dualism pervades not only the humanities and the social sciences, but also the natural sciences in the ways scientists have studied and theorized the world. Here the work of Donna Haraway in primatology provides a wealth of insights into how the nature/culture dualism has shaped the natural sciences and our views about what is possible and preferable in "nature" and in "culture." Each of these areas provide a necessary component for a critical liberationist ecological ethics.

DECONSTRUCTING WESTERN VIEWS OF NATURE

In 1980 Carolyn Merchant published *The Death of Nature: Women, Ecology, and the Scientific Revolution*, a book that revolutionized current thinking about nature and the nature/culture dualism and quickly became a standard in the field. At the heart is a gender analysis of the ways Western society has depicted and understood the natural world. Merchant seeks "to examine the values associated with the images of women and nature as they relate to the formulation of our modern world and their implications for our lives today."[1] Against the widespread view that women are inherently closer to nature—a view held by many feminists and those they oppose alike—Merchant argues that concepts of nature and women are both historical and social constructions: "There are no unchanging 'essential' characteristics of sex, gender, or nature. Individuals form concepts about nature and their own relationships to it that draw on the ideas and norms of the society into which they are born, socialized, and educated."[2]

Modern views of nature and women as culturally passive and subordinate must therefore be seen as social constructs rather than as physically or biologically rooted in unchanging evidence. This understanding is critical to making connections between social change for women and other oppressed groups and changing society's domination of nature. Examination of economic, cultural, and scientific changes in Europe during the sixteenth and seventeenth centuries reveals a dramatic transformation of the centuries-old image of an organic cosmos with a living female earth at its center into a mechanistic worldview of a dead and passive nature waiting to be dominated and controlled by humans. This Enlightenment view of the natural world has dominated the sciences and technologies that made possible the Industrial Revolution and the sweeping changes in the last three centuries of how humans across the globe relate to their natural environment.

A similar revolution of ideas with respect to nature, women, and other socially marginalized groups is under way today. Understanding how nature has been socially constructed in each of these periods contributes to new interpretations that can better represent reality and can undergird an ecologically and economically sustainable and just ethics for today.

Critical to the story Merchant uncovers is how images and metaphors that evolved to understand nature in one historical period can function very differently—and destructively—in another. Within the pre-Enlightenment European organic worldview, identifying nature as a nurturing mother who provided for human needs in an ordered universe had served as a cultural constraint on humankind's exploitation of the natural world: "One does not readily slay a mother, dig into her entrails for gold, or mutilate her body. As long as the earth was conceptualized as alive and sensitive, it could be considered a breach of human ethical behavior to carry out destructive acts against it."[3] The controlling images of a culture operate as ethical restraints or ethical sanctions, giving guidelines to the society as subtle "oughts" or "ought-nots."

In contrast, in the context of a rapidly changing social order, the Enlightenment reconceptualization of the world as a machine drew on these age-old associations of nature and women to justify the domination and exploitation of both. Nature as nurturing mother was not the only metaphor available to Enlightenment reconceptualizations. Nature had also been seen as a wild and uncontrollable female force that threatened "man's" survival. As the scientific revolution mechanized and rationalized its worldview, this image of nature as disorder became prominent and led to the emergence of a new modern idea, that of human power *over* nature. The cultural constraint of seeing nature as nurturing mother gave way to the cultural demand to tame and subdue "wild" nature: "Two new ideas, those of mechanism and of the domination and mastery of nature, became core concepts of the modern world. An organically oriented mentality in which female principles played an important role was undermined and replaced by a mechanically oriented mentality that either eliminated or used female principles in an exploitative manner."[4]

Like nature, images of woman were two-sided: she was seen as both virgin and witch. With the establishment of the mechanical logic of physical domination and rational control, women's historical association with nature linked femaleness, nature as disorder, and sexuality, and was used to justify their social subordination and exclusion.

> Symbolically associated with unruly nature was the dark side of woman. Although the Renaissance Platonic lover had embodied her with true beauty and the good, and the Virgin Mary had been worshipped as mother of the Savior, women were also seen as closer to nature than men, subordinate in the social hierarchy to the men of their class, and imbued with a far greater sexual passion. The upheavals of the Reformation and the witch trials of the sixteenth century heightened these perceptions. Like wild, chaotic nature, women needed to be subdued and kept in their place.[5]

This same linking of nature-as-disorder with sexuality has been used to justify oppression of women and men who engaged in homosexual acts.[6] The Enlightenment logic of control and domination of animalistic sexual desires underlies the sodomy debates of the seventeenth and eighteenth centuries and the medicalization and pathologization of homosexuality in the nineteenth century, which continue to inform much popular debate on homosexuality.[7] The heritage of this logic continues today where anti-gay discourse draws on the self-contradictory positions of condemning lesbians and gay men for being both "unnatural" (where natural sexuality is read as procreative heterosexuality) and "too close to nature" in the sense of homosexual behavior being "lower" or "animalistic," outside the boundaries of acceptable human culture.

This association of nature with certain segments of the human community as a means of control and exploitation is further complicated when the construct of race is considered in the nature/culture dichotomy. African American feminist Patricia Hill Collins, for example, has argued that white and black women are viewed differently in cultural stereotypes of womanhood, and assigned a different place in the relation between nature and culture: "Within the mind/body, culture/nature, male/female oppositional dichotomies in Western social thought, objects occupy an uncertain interim position. As objects white women become creations of culture—in this case, uncontrolled female sexuality. In contrast, as animals Black women receive no such redeeming dose of culture and remain open to the type of exploitation visited on nature overall. Race becomes the distinguishing feature in determining the type of objectification women will encounter."[8]

As attitudes toward women began to change in the United States in the nineteenth century, race continued to be the dominant marker that divided attitudes toward women. Women were still identified largely with nature, but with the Industrial Revolution, culture in the public sphere was increasingly seen as a negative arena that threatened woman's pure nature. In the United States, a unique combination of ongoing white supremacy with the emergence of the "cult of True Womanhood"—which used gender to define a public sphere/private sphere dichotomy to maintain that a woman's natural sphere of activity was the home where she could remain unsullied by the brutal masculine public sphere—resulted in very different views of white women and black women:

> The heated debate about black women tended to revolve around two diametrically opposed notions about the nature of black womanhood. Most whites, male and female, maligned the black woman on the grounds of racial make-up and question-

> able moral character, which resulted in her inevitable conceptual deviation from the True Womanhood ideal. Blacks, on the other hand, agreed that she departed from the ideal, but not because she was morally defective; rather she was the victim of sexual abuse and exploitation and therefore not to be blamed for circumstances beyond her control.[9]

Both white and blacks of this time associated women with a pure nature, but whites saw black women through the added filter of blackness, which automatically in white eyes placed black women in the realm of animals and therefore morally suspect "by nature." Blacks, by contrast, while agreeing that "True Womanhood" was the nature of woman, argued that degrading social conditions, not nature, prevented black women from attaining that ideal natural state.

What links all three of these constructs—gender, homosexuality, and race—is the association of women, homosexuals, and people of African origin with "nature as other" in order to justify their treatment as less than fully human. While the specific treatment of each of these groups has varied with the historical context, what remains constant is the underlying assumption that the natural realm is somehow other than—inferior to, and hence a moral threat to—human culture. This mechanism of "othering" creates categories of persons seen to be less than fully human and closer to nature; what is associated with nature therefore must be controlled to protect that which is fully human, civilized, and moral. It is important to recognize that this same practice of othering, based on an assumed culture/nature dichotomy, justifies and facilitates both social and ecological domination and exploitation. Hence, deconstructing not only the gendered, heterosexist, and racist identifications of nature but also the culture/nature dichotomy itself is an important task for a liberationist ecological ethics.

As these examples suggest, an adequate ecological ethic requires reexamination of not only constructions of nature alone but also humanity's relation to nature in its multiple forms. We return to the opening quote by Merchant to see how Western constructions of the nature/culture relation lie beneath many contemporary ecological and social issues:

> At the root of the identification of women and animality with a lower form of human life lies the distinction between nature and culture fundamental to humanistic disciplines such as history, literature, and anthropology, which accept that distinction as an unquestioned assumption. Nature/culture dualism is a key factor in Western civilization's advance at the expense of nature. . . . As the unifying bonds of the older hierarchical cosmos were severed, European culture increasingly set itself above and apart from all that was symbolized by nature. . . . If nature and women, Indians and blacks are to be liberated from the strictures of this ideology, a radical critique of the very categories *nature* and *culture*, as organizing concepts in all disciplines, must be undertaken.[10]

Many feminists, and particularly ecofeminists grounded in social constructionist theory, have taken up this challenge, extending their critique of Western epistemological dualisms to include the dichotomy separating nature and human culture.[11] In examining the debate in feminism of whether women are closer to nature than men are, Joan Griscom noted early on the problematic patriarchal construct informing the discussion that allows the separation of culture from nature: "The question itself is flawed. Only the nature/history split allows us even to formulate the question of whether women are closer to nature than men. The very idea of one group of persons being 'closer to nature' than another is a 'construct of culture.'"[12] Whatever the reality of "nature," our experience of nature is always mediated through cultural categories such as gender, and so there is no experience of nature apart from culture: "If gender is shaped by culture, ideology, and history, and how one experiences nature is culturally mediated, then gender conditioning would tend to shape our experience of nature."[13] In examining such gendered images as "virgin forest" and "rape of the land," Greta Gaard argues that linking women to nature by feminizing nature also sexualizes nature, with important ramifications for how we view culture and the humanity/nature relation: "When nature is feminized and therefore sexualized in such constructions, culture is masculinized, and the human/nature relationship becomes one of compulsory heterosexuality."[14] Because compulsory heterosexuality is so closely identified with masculine control of women (and nature), is it any wonder that men who show environmental concern often are dismissed as "soft," "tree-huggers," somehow less than "real men"?

As in an ethic of Gay and Gaia, social constructionist ecofeminism aims its critique of the nature/culture dualism at both social justice and environmental movements. Stephanie Lahar, for example, argues that ecofeminism has a unique contribution to make in that it works to deconstruct the nature/culture dualism from both sides. Ecofeminism is critical of progressive social movements that ignore the natural environment or see it as undifferentiated and passive, a resource available for the work of justice. Similarly, however, ecofeminism critiques other groups within the environmental movement for an inadequate social analysis of the human community. Thus Lahar notes that while deep ecology overcomes the nature/culture dualism by including humanity fully within nature (environmental degradation is thus a symptom of our alienation from our naturally "wild" selves), deep ecologists obscure unequal power dynamics *within* the human community in the process: "In using a universal 'we' that is powerful, privileged, and historically alienated from natural processes, [deep ecology] fails to see human diversity (including diversity in human environmental relations) and abuses of power played out in ethnocentric, classist, and sexist acts and institutions."[15]

The critique of both sides of the nature/culture dualism is critical to lesbian and gay liberationist thinking. Lesbians and gay men are nearly always invisible or excluded from the universal "we" critiqued by Lahar (note our invisibility as well in Lahar's critique). Because lesbian and gay oppression and exclusion has been justified from both sides of the nature/culture dualism, both sides must be deconstructed. Acknowledging the "natural" and "cultural" reality of lesbians and gay men in the face of constructs that exclude us from both nature and culture serves to problematize the nature/culture boundary itself. Lesbian and gay liberationist thinking insists that lesbian and gay people are fully part of the natural world (as is all of humanity) and form an essential and inherently valuable part of humanly constructed "culture." In addition, underlying attitudes of nature as "other" must be identified and critiqued to move away from othering mechanisms that justify the exploitation of nature and, by association, women, people of color, and lesbians and gay men. Hence critique of both sides of the nature/culture dualism corresponds closely to the double ecojustice criteria of "Gay" and "Gaia" that forms the basis of this ethics.

DECONSTRUCTING SCIENCE AND THE NATURE/CULTURE BOUNDARY

Because of the important role the natural and human sciences play in questions of both ecology and sexuality, the work of feminist historian of science Donna Haraway in examining and deconstructing the nature/culture boundary in the sciences makes an especially critical contribution to a liberationist ecological ethics. Haraway examines how the narrative frameworks of science set the boundaries for beliefs about nature and culture in order to reconstruct a more healthy (or at least less hostile) relationship for all members of the earth community.[16] Because I will draw extensively on Haraway's insights in order to formulate criteria for an ethics of Gay and Gaia, it is worth looking closely at the structure of her argument here.

The central theme of Haraway's work critical to our thesis here is the claim that themes of race, gender, sexuality, nation, family, and class have been written into how nature has been understood in the Western life sciences since the eighteenth century, and that nature continues to be a critically important and deeply contested myth and reality. Primatology provides one key filter by which to understand how Western constructs of nature have taken shape. For Western people, nonhuman primates have held a privileged position in helping to define what is human because they are seen to constantly straddle the boundaries of what is understood as nature or culture. Seen as "almost human," primates and the history of their study can reveal what counts for both nature and culture, how that boundary has shifted and in what ways, and what is at stake in the boundary constructions and shifts.

This becomes all the more critical because the stories written about nonhuman primates are often used to explain human reality: "Especially western people produce stories about primates while simultaneously telling stories about the relations of nature and culture, animal and human, body and mind, origin and future."[17] Despite changing interpretations over time, a constant feature of Western knowledge about primates has been its use in grounding the human story in nature, producing the "naturalization of the human story." The ideological constructs preserved in these naturalized human stories often serve to buttress the claims and practices of the dominant cultural ideologies to the detriment of socially marginalized persons and nature itself, and therefore need to be examined critically.

In examining the claims of science, Haraway pays close attention to the relation of the practice of science to the surrounding culture. She avoids reductionist analyses of any kind, whether from scientific realism (the position that what science portrays is objectively true) or from social constructionism (what science portrays is merely a reflection of the dominant societal power structure), because the production of scientific knowledge is more complex than either of these perspectives alone allows. Hence while social, economic, and cultural structures always enable and constrain the claims produced by scientific discourse, they do not directly produce them. Yet neither are they left out of or unimportant to the production of scientific texts. The point is that the objects of scientific study are always mediated to us through language and culture, and their meanings have important material as well as cultural implications.

An example of this is the impact on primatology of the newly developed communications and control theories and technologies that followed World War II and how these in turn changed our views of the natural world. With the emergence of the fields of ergonomics, cybernetics, and semiotics, with their emphasis on communication and control within a system of community, ideas about ecological communities began to shift and animals came to be seen "as biotic components in technological communications systems."[18] Changed social relations and technology provided the conditions to "know" nature differently than before. Hence Haraway argues: "Meanings are applications; how meanings are constituted is the essence of politics. No one can constitute meanings by wishing them into existence; discourse is a material practice. The meanings of cybernetic communications systems include *particular* structurings of objects of knowledge—not as ideology, but as that which can be known in a particular time and place, called Nature."[19] This relation between science and the surrounding cultural matrix in which it is practiced means that our culturally constructed epistemological categories both enable us to know new things "in" nature as well as constrain what can be known "about" nature.

A similar relation between social location and scientific knowledge is revealed in the kinds of metaphors employed by scientists. In her studies of the mother/infant relationship in baboons, for example, feminist primatologist Jeanne Altmann uses the metaphor of "dual career mothering" to describe the division of work of female baboons, and employs the key terms "juggling" and "budget" to describe how female baboons organize and respond to a hierarchy of demands. Haraway notes the silent implications of both race and class in metaphors that best reflect the reality of white middle-class professional women. Yet her point in locating the metaphors that science uses is not to discard them as "polluted," but to demonstrate how they are enabled by the cultural context, and in turn enable specific kinds of accounts of nature, *and* what they may hide in the process. Altmann's writing both produces new insights about the life patterns of baboons while at the same time it reflects the constraints of the concepts and logic of early 1970s white middle-class feminism.[20]

This relation between the social and epistemological constraints of science becomes important to ecological and liberationist ethics because of the complex relations between cultural values, scientific knowledge, and criteria for what counts as moral behavior. Haraway shows, for example, how concern during the Progressive Era among the dominant elites in United States society over the perceived threat of decadence and decay of civilization led to three linked public practices designed to preserve a threatened manhood: (1) preservation and exhibition of ideal animal types in museums (and the paradox that the animals must be killed to be preserved from decay), (2) racial eugenics (preservation of a pure white race), and (3) conservation (preservation of pure wilderness from decay). These activities, seemingly contradictory today under the logic that governs much of the contemporary environmental movement, were a logical outcome of the perceived need of many in the Progressive Era to fight decadence, "the dread disease of imperialist, capitalist, white culture."[21]

An example of this is the work of primatologist Robert Yerkes on chimpanzees during this period, which was enabled and constrained by this constellation of social values. Yerkes believed that the purpose of science was to enhance human possibilities of control: "Man's curiosity and desire to control his world impel him to study living things."[22] To do so, Yerkes drew on the new principles of human engineering to study chimpanzees.[23] Laboratory studies were critical for this. Yerkes saw the laboratory animal as engineered and standardized, as well as being a natural object; both were needed to respond to human queries to yield "objective" knowledge. Critical to this enterprise were underlying assumptions about nature and culture: "Conceptually, both culture and human engineering assumed a plastic human nature, where underlying biological mechanisms supplied the raw materials for social manipulation."[24] Yerkes saw chimpanzees as pure units of personality undistorted by culture, the ideal "raw material" with which to separate what belongs to nature and what to culture, and thus what can be reshaped among people through human engineering.

Throughout her study of Yerkes, Haraway weaves together the influences of contemporary scientific developments (particularly in reproductive and nervous physiology) and of contemporary social developments (particularly scientific management of society and liberal corporate reform) to demonstrate the constraints and influences of the narrative web in which Yerkes's primate studies take shape. Again, the political context is shaped by the assumptions and values of liberal Progressive reform:

> There was no conspiracy to place science in service of white, male-dominant capital involved in this remarkable concatenation of organizations and people. Operating from the "unmarked category," where the social conditions of one's existence are invisible, these men saw their role to be human service and believed strongly in democracy, individual rights, and scientific freedom. They saw themselves to be self-sacrificing reformers who believed in and practiced science. They helped define what could count as rational, making to this day the act of marking the privileged category with modifiers look merely ideological. Under this banner of rational reform, Yerkes created a chimpanzee community for human service.[25]

Yet one result of Yerkes's understanding of the nature/culture boundary was the development of a naturalistic ethics of evolutionary adaptation where natural function was made a moral criterion. In the social conclusions drawn from Yerkes's study of primates, morality, biology, and scientific service became inseparable and were ruled by an obligatory reproductive heterosexuality. It is critical for the revisioning of ecological ethics to understand how naturalistic ethics that ground views of humanity and culture in "objective" nature in fact are grounded in and reproduce social values and relations. Yerkes's work is infused with a logic of domination, a mix of cooperation and control that moves from instinct through personality to culture and finally to human engineering. Hence at a critical historical and cultural time of reformulating understandings of nature and culture, primatology served as a mediator between the life and human sciences, straddling the boundaries of both, and setting guidelines for the moral in its views of the natural.

A different recoding of science, gender, and race took place in the 1960s as nature was reimaged as the healer of a "mankind" burdened with stress from the modern postcolonial world. From *National Geographic* and Gulf Oil's representation of Jane Goodall as the lonely young white woman returning to an African Eden to bridge the symbolic nature/origins gap with her gorillas and chimpanzees in the African wilderness, to the chimpanzee astronauts ("cyborg neonates") Enos and Ham carrying the hopes of humanity's future in their Mercury launches into outer space, to the ongoing fascination with chimpanzees and apes learning to "talk" like humans, the "primitive" Third World provided the setting and the primates for First World audiences to reinvent nature within postcolonial, multinational capitalism. Haraway observes that each of these representations involving primates makes use of fundamental transgressions or boundary crossings: the origins of man are represented by a woman; using a masculinist reproductive "organism/machine" chimpanzees are "birthed" into space and into the future; and primates learn speech through sign language, violating one more human/animal boundary. What is at stake in the recoding of these representations for understanding the nature/cultural dichotomy and its shifting boundaries?

Among several narrative possibilities, it is revealing to note which boundaries are *not* violated, and which dimensions are made invisible by these representations. Together they present complex pictures of nature with both radical differences and continuities with those in earlier decades. For example, despite the multiplicity of surrogates that cross boundaries, such as species barriers, machine/organism barriers, language barriers, and earth/space barriers, the injunction to be heterosexually fruitful and multiply—that is, a compulsory reproductive politics—is never relaxed.[26] History also is conspicuously absent. The narratives place the animals just at or over the line into "culture," and people are placed at or over the line into nature, but the stories emphasize their communication across nature/culture boundaries, not history. The narratives are mediated by a triple code of gender, science, and race. Woman serves as mediator between scientific man and nature. Her whiteness shows that it is not any man that is fully human, but white, Western man. Left out of the story are people of color and their histories of struggle that both enabled the expeditions to retrieve and study these African primates, and that contest the Western discourses of meaning. Haraway thus concludes:

> In all of these stories humans from scientific cultures are placed in "nature" in gestures that absolve the reader and viewer of unspoken transgressions, that relieve anxieties of separation and solitary isolation on a threatened planet and for a culture threatened by the consequences of its own history. But the films and articles rigorously exclude the contextualizing politics of decolonization and exploitation of the emergent Third World, obligatory and normative heterosexuality, masculine dominance of a progressively war-based scientific enterprise in industrial civilization, and the racial symbolic and institutional organization of scientific research. Instead, the dramas of communication, origins, extinction, and reproduction are played out in a nature that seems innocent of history. If history is what hurts, nature is what heals.[27]

The problem with universalizing narratives such as these is that they hide the history of exploitation and colonization that makes them possible and present science as an ahistorical practice that takes place in a "pure" nature without historical and social consequences. Haraway examines Jane Goodall's chimpanzee research to expose a reality just the opposite of *National Geographic*'s picture of the young woman alone discovering the secrets of a fixed nature. Rather, Goodall's work emerges from a complex historical/cultural web shaped by decolonization in Africa, the interactions of chimpanzees, Africans, and North Americans, and a series of contested discourses within primatology. It is precisely the contested nature of "nature" that the *National Geographic* representations so skillfully hide; the implicit and explicit message is that the world is given and "discovered," not made.[28]

Tracing the origins of claims about human nature, nature, and culture such as these are critical to a liberationist ethic. Lesbians and gay men all too often have faced demands that we change ourselves into heterosexuals to conform to dominant views of what is natural, buttressed by scientific claims of "a plastic human nature," a "fixed and given nature" (which places lesbians and gay men outside), "natural [evolutionary] function as a moral criterion," and "obligatory reproductive heterosexuality." The universalizing narratives constructed from these claims hide the social practices and history that have produced them, and which justify exploitation and oppression of lesbians, gay men, and nature. In addition, they hide the historical dimension of nature, that nature is always a process and product of the interaction of its members, including its human members, rather than a static or given entity.

The writings of feminist primatologists give insight into the importance of deconstructing the sex/gender boundary as part of overcoming a reified nature/culture dualism. Each seeks to problematize *gender* by contesting what can count as *sex*; that is, they attempt to reconstruct understandings of nature as well as culture. To do so, however, these scientists first had to contest and revision what may count as scientific knowledge. The narrative field constraining scientific practice that women primatologists entered in the 1960s and 1970s was already set by what Haraway terms the "Law of the Father": the dominant scientific masculinist epistemology, which follows Aristotelian logic that active male agency and paternity are the keys to humanity. Disconnection from "nature" is seen as essential to man's "natural" place in culture and self-realization through transcendence. This understanding of "unmarked man" is universalized to represent fully human status; anything that deviates from it through markings such as female gender or nonwhite race by definition is less than universally human (i.e., particular) and closer to nature (and thus further from fully realized human culture and status). Man's destiny is culture and activities defining culture; women's destiny is nature and "natural" activities.[29]

By their gender, these women primatologists were already "marked," that is "biased," and the only way within this narrative field to produce unbiased and therefore scientifically acceptable work was to operate as "man." Yet in the changing global context that culminated in the 1975–85 United Nations Decade for Women, these women did begin to destabilize that narrative field of primatology by both working within and simultaneously challenging its constraints. Hence by 1985 several women primatologists were explicitly using feminist concepts of power-charged difference and contradiction rather than the masculinist search for universal origins as their starting point of analysis.

One critical result for a liberationist ecological ethics of the feminist primatologists' deconstruction of the nature/culture boundary is the recognition of the *active historical agency* of *both* the animals and people who are the objects and subjects of primate scientific knowledge. The result of this is an epistemological shift which critiques the "productionist" bias in much evolutionary and anthropological theory that perceives animals, sex, and nature as passive resources to be appropriated through a logic of domination for the production of science, gender, and culture. The effects of such a shift on the ways nature is understood are dramatic. "When biology is practiced as a radically situational discourse and animals are experienced/constructed as active, non-unitary subjects in complex relation to each other and to writers and observers, the gaps between discourses on nature and culture seem very narrow indeed."[30] The nature/culture boundary appears increasingly permeable and constantly shifting in these works.

With this thorough deconstruction of a fixed nature/culture boundary, what is left of nature? Haraway proposes provocative alternatives for reconstructing our understandings of nature that can ground an ecologically friendly justice ethics. Positing nature as "something we cannot do without, but can never 'have,'" she begins by recognizing our need to find another relationship to nature besides reification and possession.[31] The challenge is to turn decisively both from Enlightenment-derived modern and postmodern premises about nature and culture, and logics of productionism and its corollary, humanism, that define "man" as the only actor in a world where all else is reduced to resource. In the process, several critical shifts occur in the way we understand what may count as nature and how we conceive acting in the world. Here I focus on some of these key shifts with an eye to the implications for a liberationist ecological ethic.

Nature as a Matrix of Actors

The key epistemological shift is recognizing nonhuman agency in nature. Nature is made, rather than an ahistorical entity that "exists," but it is made as a co-construction of humans and nonhumans. Nature is made as both fiction and fact by a matrix of actors that includes humans, machines, and nonhuman organisms. Coining the term "material-semiotic actor" to emphasize that the object of knowledge is also an active partner in its production, Haraway defines nature as "a commonplace and a powerful discursive construction, effected in the interactions among material-semiotic actors, human and not."[32] Thus, while nature cannot pre-exist as such, neither is it merely an ideological construct, solely the product of human practices. Rather the boundaries in nature and between its entities, such as organisms, materialize in the social interaction among humans and nonhumans; this interaction generates the discourses that constrain what may count as nature.

An Amodern Rather Than Modern or Postmodern Perspective

Haraway suggests that we need a perspective on science and history that is neither modern nor postmodern, but "amodern." What is required as an alternative to the modern understanding of history as progress is not postmodern deconstruction that relativizes and discounts different disciplines as simply discourses about power, but an *a*modern view of the history of science as culture that refuses beginnings (universalizing origin narratives that hide the social history that produced them), enlightenments, and endings (especially apocalyptic scenarios about the end-times from which we derive our current identity and meaning): "[The] world has always been in the middle of things, in unruly and practical conversation, full of action and structured by a startling array of actants and of networking and unequal collectives. . . . The shape of amodern history will have a different geometry, not of progress, but of permanent and multi-patterned interaction through which lives and worlds get built, human and unhuman."[33]

The upshot of this perspective is that we must accept responsibility for our place among other actors in the social and ecological relations that construct a future that is anything but determined. Apocalyptic narratives that predict the return of "the sacred image of the same" are inadequate because the system is not closed and the world is not full. "It's not a 'happy ending' we need, but a non-ending."[34]

Articulation Rather Than Representation

Recognizing nonhuman agency in nature means progressive ethical theory and praxis need to be shaped through a politics of articulation where each actor articulates her or his positions and commitments, rather than a politics of representation where some speak for or represent others. To illustrate this difference, Haraway gives the example of current debates over the future of the Amazon region. Who should decide the fate of that region? Historical analysis of the Amazon reveals a history of hundreds of years of the Amazon's "social nature," a region and history co-inhabited and co-constituted by land, humans, and other organisms.[35] Such analysis debunks the modern image of the Amazon as "pure" wilderness, "Eden under glass," in order to insist on locations of responsibility and empowerment among all those involved in the current conservation struggles.

Hence, against a politics of "saving nature" that disregards the histories and lives of the people already living there, Haraway argues for a politics of "social nature" where the practice of justice leads to a different organization of land and people and in the process restructures our concept of nature. She describes the efforts of the resident indigenous peoples of the Amazon working together with the petty resource extractors of mixed ancestry to defend the rain forest and secure their joint future in it. Their authority as defenders of the Amazon is relational, derived "*not* from the power to represent from a distance, *nor* from an ontological natural status, but from a constitutive social relationality in which the forest is an integral partner, part of natural/social embodiment."[36] Instead of claiming authority through their ability to represent the rain forest, the people of the Amazon claim authority through articulating a social collective entity that includes humans, other organisms, and other kinds of nonhuman actors in an ecologically and socially sustainable whole.

Why is this distinction important? Haraway illustrates this by posing the simple question that some environmentalists have raised in opposition to a politics of social nature: "Who speaks for the jaguar?" Haraway compares this to the question asked by some pro-life groups in the abortion debate: "Who speaks for the fetus?" The problem with both questions is that they rely on a politics of representation that requires the services of a "ventriloquist" to represent the permanently speechless. Through the distancing process of representation, nature and the unborn fetus are disengaged from the surrounding and constituting discursive environment and relocated in the domain of the representative, in the process gaining greater status than the subjugated human adults (rain forest people or pregnant women) whose agency now is called into question as "biased." Everything that used to surround and sustain the represented object, such as local people and pregnant women, simply disappears or reappears in the drama as the "environment" that threatens the represented. Finally, only the spokesperson, the one who represents, is left onstage as the unbiased actor. "In the liberal logic of representation, the fetus and the jaguar must be protected precisely from those closest to them, from their 'surround.' The power of life and death must be delegated to the epistemologically most disinterested ventriloquist, and it is crucial to remember that all of this *is* about the power of life and death."[37]

In a world structured inequitably along lines of power, "Who speaks for nature?" is not an innocent question. Following a modernist logic, the "detached, objective" scientist has most often been granted the right to represent nature, that is, the permanently speechless world of objects. The constructionist Bruno Latour has sketched the double structure of representation through which scientists establish the objective authority of their knowledge. Scientists first shape new objects through operations Latour calls inscription devices that give status or definition to the object. Then scientists speak as if they were the mouthpieces for the speechless objects they have just shaped. In this double move, authorship rests with the representor who claims independent object status for the represented.[38] The ambiguity of this process of representation is striking: "First a chain of substitutions, operating through inscription devices, relocates power and action in 'objects' divorced from polluting contextualizations and named by formal abstractions ('the fetus'). Then, the reader of inscriptions speaks for his docile constituencies, the objects. This is not a very lively world, and it does not finally offer much to jaguars, in whose interests the whole apparatus supposedly operates."[39]

At its core, a logic of representation depends on the possession of a passive resource, the silent object, now stripped of its status as actant. More helpful and less open to distortion is a clear articulation of the world, from each person's point of view (not represented views of the object) through partial, locatable "situated knowledges."[40] It is precisely the world that gets lost (not represented) in doctrines of representation and scientific objectivity. A politics of articulation is not innocent conversation; it pays attention to the dynamics of power inherent in all social interactions. It is precisely in having all the people who care, who are committed and engaged (including scientists and North Americans), articulate their positions in a field constrained by the new collective entity that the most promising narratives and practices may develop. Articulation as politics must remain open to new actions, interventions, and shifting boundaries within the collective entity.

Affinity Rather Than Identity as a Basis for Praxis

One ramification of stressing articulation over representation is a praxis based on affinity rather than identity. In learning to articulate individual and community positions and practices from locations that are accountable rather than essential, naturalized identities, we recognize both that all subjectivities are multidimensional (and so, therefore, is vision), and that there is no way to be simultaneously in all of the "epistemologically privileged" (i.e., subjugated) positions structured by critical position such as gender, race, class, and sexual orientation.[41] It also forces us to recognize that consciousness of these contradictions is not a result of "being" some category on the nature side of the nature/culture dichotomy (such as "being female" or "being gay"), but rather it has been forced on us by the historical experience of the contradictory social realities of patriarchy, colonialism, racism, heterosexism, and capitalism. Emphasizing a fixed sense of identity can lead to efforts to replicate sameness rather than opening us to new, unimagined possibilities. It results in a series of taxonomies of oppressed groups and liberation theories, each looking for the essential standpoint from which to incorporate and understand all the other contradictions, and often ends up producing epistemologies used to police deviation from "politically correct" experience.

Stressing the affinity or relatedness we have to others in the struggle rather than identity opens up the search for effective ways to build political unity to confront fluid and constantly shifting forms of oppression. Committed to building a political form "that actually manages to hold together witches, engineers, elders, perverts, Christians, mothers, and Leninists long enough to disarm the state,"[42] affinity is about power-sensitive conversation and practice among actors and actants who articulate perspectives grounded in locatable particularity in order to build effective political unity. Examining coalitions like conservation groups working to defend the Amazon and the AIDS Coalition to Unleash Power (ACT UP), Haraway stresses that not all the actors are equal. Each coalition has an "animating center," such as the indigenous people of the Amazon or persons with AIDS, with whom other actors must engage.[43] The collective effort is to visualize a "social nature" out of a heterogeneous body rather than narrowing vision to "saving nature" by expelling invaders to return to an original unspoiled garden. Articulation is done in community; moral questions are not grounded in "Who am I?"—the question of always unrealizable identity—but "Who are we?"—a question open to newly contingent, friction-generating, and possibility-producing articulations and coalitions. As Haraway concludes, "We articulate, therefore we are."[44]

Nature as Coyote: Playing with the Trickster

Revisioning nature as made from a matrix of human and nonhuman actors allows us to recognize the world itself as witty agent and actor, always full of surprises and new possibilities. By giving up our futile attempts to be in charge of the world, we open ourselves to new awareness, including the somewhat unsettling possibility that the world may have an independent "sense of humor." The coyote of southwest Native American stories suggests a situation where "we give up mastery but keep searching for fidelity, knowing all the while that we will be hoodwinked."[45] The ironic paradox of Western experience is that just as we have gathered impressive technology and concentrated more power in human agency than was imaginable in any other time in human history, we are being forced to realize that we are not the only actors on the stage. The world resists being reduced to resource and emerges as coyote—always problematic, yet always potent with possibilities for negotiating the tie between meaning and bodies. Here one sees a connection with scientific theories of Gaia which hypothesize that the earth itself is a living self-regulating organism in which humans play a vital part, but whose activities must be healthy for and accountable to the sustainable well-being of the whole planet.

From this overview of constructionist insights of Western understandings of nature, the nature/culture dichotomy, and the humanity/nature relation, criteria for developing an adequate understanding of nature can be grouped under the following rubrics: general methodological guidelines, views of an interactive nature, and methodological ramifications of these views.

Methodological Guidelines in Understanding Nature

- *Deconstructing the Nature/Culture Dualism*: The experience of lesbians and gay men and others in being excluded from both sides of the nature/culture dichotomy (either unnatural or too natural) requires that both sides of the relation be examined for the ways they have been constructed and understood. Any rigid separation of human culture from the rest of nature must be rejected. A liberationist ecological ethic must understand humanity and human culture to exist fully within nature, although with distinctive characteristics that affect human interaction with other parts of the natural world.

- *Rejection of Biological Essentialism*: While insisting that human beings exist fully within nature, a liberationist ecological ethic rejects biological essentialism as a means for understanding human nature and the humanity/nature relationship. It does not deny that there are biological, genetic, and ecological influences and constraints on human behavior, but also focuses attention on the ways scientific information is culturally mediated and naturalized to reflect social values and mores.
- *Attention to the Relation of Science to Ethics*: In light of a rejection of biological essentialism and awareness of how scientific information is culturally mediated, a liberationist ecological ethic must incorporate a critical perspective on the role science and scientific knowledge play in ethics. Particular attention must be paid to (conscious and unconscious) heterosexist values and assumptions that inform scientific knowledge and methods.
- *Unmarked and Marked Categories and the Nature/Culture Dichotomy*: A liberationist ecological ethic must be aware of how social categories that are marked or unmarked by gender, race, and sexuality function in assumptions about the nature/culture dichotomy. Enlightenment presuppositions too often interpret unmarked categories that implicitly mean male, white, or heterosexual as unbiased, objective, and closer to culture, while marked categories such as female, black, or homosexual are biased, closer to nature, and thus a threat to culture and objectivity. A liberationist ethics must de-

construct both the assumptions informing the nature/culture dichotomy and the relation of other societal constructs to it.

- *Attention to the Mechanism of "Othering"*: A particularly important component of this ethic is attention to the ways associating certain groups or categories of persons with a "lower" or "animalistic" nature serves to devalue both persons and nature as somehow less important than "fully human culture and human nature." Both sides of this practice must be deconstructed: the assumption that nature is less valuable or morally threatening to culture, and that certain groups of society are somehow closer to nature.

Views of an Interactive Nature

- *Agency/Subjectivity of Other Creatures*: A liberationist ecological understanding of nature must include recognition of the historical agency and subjectivity of other creatures in the ways we construct nature. This also implies a shift away from utilitarian and productionist biases that view nature primarily as a resource for creating human culture and meeting human needs—biases that are reflected within the human community through attitudes and practices such as compulsory heterosexuality that tie one's worth to one's procreative capacity.
- *Nature as a Relational Matrix of Actors*: A liberationist ecological ethic will recognize that nature is both made up of and made by a matrix of interacting humans, machines, and other organisms. Such recognition compels us to pay attention to the historical dimension of nature, emphasizing the idea of social nature in which human beings play a part rather than ahistorical constructs of nature as pure or innocent, contaminated by human involvement.
- *The Agency of Nature*: The world has an element of agency as actor in itself, apart from the sum of its con-

stituent parts. Metaphors of nature such as Gaia or the Native American use of coyote emphasize that nature cannot be controlled, and is always full of surprises and new possibilities. A liberationist ecological ethics recognizes this and so aims for fidelity with the earth rather than mastery over it.

Methodological Ramifications of an Interactive Nature

- *Articulation Rather Than Representation*: A liberationist ecological ethic should avoid the logic of representation that assumes the passivity of nature, other groups or persons, or other resources. The alternative of trying to articulate accountable, locatable, and partial perspectives is a more promising way to include the voices of different constituents of nature in ethical discourse. Rather than neutral or unbiased representation, articulation recognizes the importance of power and social location that shape different human relations with parts of the natural world and includes these elements as critical components of efforts to articulate stances and perspectives.
- *A Praxis of Affinity Rather Than Identity*: Paying attention to the social dynamics involved in articulating positions from locations that are understood as historically constructed rather than naturally fixed means the politics of a liberationist ecological ethic will be based on shifting coalitions, connections, and affinities between different groups rather than on fixed essential identities. Because all locations and subjectivities are both multidimensional and shifting, so are the ways we see and interpret the world, and so must be our political responses. An emphasis on affinity recognizes this ecological dimension—the changing, interrelational character of where we are located—while not ignoring power differences critical to a liberationist approach.
- *Critical Assessment of Metaphors*: A liberationist ecological ethic must pay close attention to the language it uses

in describing nature and our relation to it, particularly how its metaphors either enable or constrain different values, knowledges, and practices. The goal is not to discard some images as polluted or incorrect, but to understand where they come from, what material practices shaped them, and what they enable and constrain in the contemporary context. An examination of metaphors includes understanding how they function differently in different contexts and historical periods.[46]

- *Critical Examination of Universalizing Narratives*: Characteristic of narratives that universalize, such as origin or ending stories, is that they both hide the material practices that produced them and shift the treatment of their subjects to "totalistic" or universalizing categories. Now hidden from view, the original social relations may then be replicated uncritically in the story's retelling and reenacting. Appropriated uncritically or nonrelationally they can limit our imagination for possible futures as well as constrain or fix the meanings we have of ourselves and nature in the present. A liberationist ecological ethic engages these narratives critically, open to the new and constructive meanings they may help to generate while attuned to ways they may replicate destructive views and social relations.
- *Openness to the Future*: A liberationist ecological ethic must foster human acceptance of our responsibility as one actor among others in constructing a world whose future is open (within ecological constraints). This implies critical engagement with eschatological teachings that see the future as predetermined, undercutting human agency and responsibility today.

With this exploration of the ways we conceive the relationship of humanity to the rest of nature, we may now turn to the third issue critical to a Christian ecological ethics: the presence of the sacred in our ecological world. Where do we find and discern the sacred in our efforts to build a liberationist ecological praxis? What attributes of the divine can point to a more ecologically friendly and sustainable relationship with each other and the earth? And what contributions can the efforts of lesbians and gay men to name the presence of the sacred in our lives, joys, and struggles make to this ecojustice ethics? It is to these questions that we turn in the next chapter.

5 God: Sensing the Divine in Right Relation

Walking with my dogs, I become aware in a fresh way of how marvelous it is to be a member of a "group"—women—which, along with blacks and animals, has been associated closely in western history with "nature." Not that wind and water and fire and earth are simply or always benign, because they are not. Nature can be cruel and deadly—human nature especially. And each of us dies in our own embodied way, which is seldom easy. But keeping slow pace with my older dog, Teraph, I know what I have in common with the trees' gnarled roots at the water's edge, the windchill whipping my cheeks, the pile of dog shit I step in, the crows harping from the fence, the joggers and other walkers, some smiling and nodding, others preoccupied and aloof. I know them all, the people and the trees. I do not know their names, but I know that our sensuality, our shared embodied participation in forming and sustaining the relational matrix that is our home on this planet, is our most common link, and that our sensuality can be trusted.

If we learn to trust our senses, our capacities to touch, taste, smell, hear, see, and thereby know, they can teach us what is good and what is bad, what is real and what is false, for us in relation to one another and to the earth and to the cosmos. I say to myself, as I return to campus from my outing with my dogs, that sensuality is a foundation for our authority.

—Carter Heyward, *Touching Our Strength*

How do we know? How do we know one another, how do we know the world around us, how do we know the experience of the sacred in our midst? And if, as Carter Heyward suggests, all knowing is sensual, mediated through our bodies, how is this reflected in the ways we name God and God's relation to us and all of creation?

Lesbian and gay theologians working within a liberationist framework typically pay close attention to these questions by first asking how socially constructed categories and presuppositions grounded in normative heteromale experience have shaped claims about divine/human relations. We agree with feminist theologians that the divine, while distinctive and distinguishable from humanity, can only be known by human beings through the filter of human experience. Attempts to name and understand God therefore inevitably reflect human self-understandings with all our contradictions related to differences in experience and power. It becomes critical for us to ask how the forced invisibility and exclusion of lesbian and gay experience from human self-understanding has shaped and distorted views about the divine, and how heterosexist claims about God in turn serve to justify lesbian and gay oppression.

A liberationist ecological ethic grounded in lesbian and gay experience will ask how these images contribute to exploitative relationships in society and with the rest of the natural world. Hence this deconstructive step is a critical starting point for an ethic of Gay and Gaia that we have explored in the previous two chapters: what can constructionist insights attuned to lesbian and gay male experiences and other liberationist insights tell us about ways of naming, imaging, and acting that block connection and relation? Yet an ethic of Gay and Gaia goes on to ask about the constructive step: what is it that *fosters* connection and relation? What ways of naming, imaging, and acting can lead to deeper social and ecological ways of connecting and sustaining our communities? And if, as this ethic claims, sensuality, embodiment, and erotic connection in right relation are necessary prerequisites to empower an ecological moral agency, what does this suggest for where and how we experience the sacred in our lives?

It is to this second step, the theological revisioning rooted in lesbian and gay male experiences needed to ground a liberationist ecological ethics, that we turn in this chapter. Donna Haraway's attention to the function of images and metaphors is useful in this theo-ethical task. How we name and understand our experiences of the divine and the divine/creation relation both enables and constrains our intrahuman and human/nonhuman nature relations. For example, how does perpetuating the procreative masculine image of "Father" as an indispensable part of the Godhead constrain lesbian and gay efforts to overcome the norm of compulsory heterosexuality that shapes human relations in Western society?[1] Or how might drawing instead on lesbian and gay male-rooted experiences of deep connection found in relationship with our friends and lovers reveal different aspects of divine presence and activity?

In an effort to move beyond patriarchal understandings of the divine that ground ontology in relations of socially constructed hierarchical dualisms, lesbian and gay theologians along with feminists have focused on justice understood as *right relation* as the best grounding for our theo-ethical praxis.[2] Originally referring primarily to divine/human and intrahuman relations, right relation increasingly is seen as an ecological category that grounds relations between divinity, humanity, and all of nature. Integral to this is reconceiving the divine in ways that emphasize mutuality and relationality and move away from hierarchy in understanding the divine/creation relation.

Two themes in lesbian and gay theological reflection on right relation have critical contributions to make toward developing an ecological understanding of the divine for a liberationist ecological ethics. Both are developed in contradistinction to socially constructed norms of relationship and love that reflect heterosexist or solely masculine experience. To counter the exclusivist relational norm of heterosexual marriage, lesbian feminist Mary Hunt examines women's friendships to propose an ethical norm of friendship. This in turn evokes a suggestive reimaging of the divine as friend. Reconceiving our relational norm and divine image as based in friendship has important ecological implications for how we reimage the web of divine/nature/humanity relations.

Following the groundbreaking work of the African American lesbian feminist writer Audre Lorde, Carter Heyward and J. Michael Clark have emphasized love and power as rooted in eros and the erotic in reshaping understandings of the divine. Critiquing the normative view of divine love as agape—selfless and sacrificial—as reflecting constructions rooted in masculine and hierarchical experiences of reality, they reimage love and power as erotic, rooted in mutuality and right relation.

In this chapter I examine some of the insights of Hunt, Lorde, Heyward, and Clark for the contributions they make to an ecological understanding of the divine, building on the constructionist insights of the previous two chapters. From there we can assess some contemporary Christian models of ecojustice for their contributions to formulating a liberationist ecological ethics.

FRIENDSHIP: RECONCEIVING THE DIVINITY/HUMANITY/NATURE WEB OF RELATIONS

In her book *Fierce Tenderness: A Feminist Theology of Friendship*, Mary Hunt contends that when people live in right relationship, new and more just paradigms of the holy emerge.[3] She focuses on human friendship—understood especially, but not exclusively, in the light of friendship between women—as a helpful paradigm of right relation for the whole of creation. Despite the fact that friendship is a universally shared human experience with numerous positive ethical ramifications, little attention has been paid theologically to friendship in the Christian tradition. Instead, with a central sacred narrative focused so strongly on the death and resurrection of Jesus of Nazareth, primary attention has been given to themes of divine and human suffering, such as in the various atonement theories, theologies of the cross, and more recently Alfred North Whitehead's process metaphor for God as "fellow-sufferer."

This raises the question of whose experience and interests are represented by a focus on suffering: "While it is true that suffering is as common a human happening as friendship . . . my sense is that women would rarely choose to lift it up as paradigmatic of anything positive."[4] In ethical efforts to shift toward right relationship, Hunt finds a focus on suffering as primary to be problematic for two reasons. First, suffering tends to divide the sufferer from those who do not suffer and in this sense may be seen as the antithesis of relationship. More importantly for women and other oppressed persons, however, is that focusing on suffering tends to emphasize that area over which we have no or limited control, as opposed to choices that evidence our religious and moral agency. While an ecological ethics needs to recover an appropriate sense of ecological limits to human control, this cannot come at the expense of recent efforts to reclaim agency by women, lesbians and gay men, and other oppressed persons.

Friendship provides an alternative, constructive foundational way in which to orient our lives. Hunt sees friendship as a shared, communal praxis that overcomes patriarchal dichotomizing of the personal and the political, of love and power, of theory and practice, by uniting and integrating each of these in relationships of "fierce tenderness." They are fierce because of the intensity of attention—to self, to others, to social injustice—and they are tender because they draw on and generate a quality of care and nurture that only friends create and share.

A focus on friendship offers a number of ecological qualities important for the task of revisioning ecological ethics. First, as part of human (and some nonhuman) culture, friendship is a vital part of creation with important potential for constructive good: "The potential for friendship and the willingness and conditions to actualize it are simply part of creation. Women loving women are a wonderful part of creation that has been passed over socially as well as in theological work that shapes so much of Western culture. As we lift up that reality to public expression, as we sacramentalize it like any other good, its true value is revealed."[5]

Friendship is also deeply ecological. While friendship includes both joy and loss, nothing is ever wasted or lost. Each experience of friendship is recycled into new experiences, adding a strand to the fabric of our friendship histories that give meaning and continuity to our lives. Friendship provides an ethical grounding of actively befriending creation as our normative stance. In an age of potential and actual ecological and nuclear devastation, "cultivating a friendly stance toward the whole of creation . . . may spell the difference between a friendly environment and no environment for our descendants."[6]

Women's experience of friendship suggests constructive ways to overcome antiecological competitive and hierarchical dualisms endemic to Western patriarchal cultures. Hunt argues that women experience friendship as a plural experience that includes other human beings, ourselves, the divine, and other parts of nature in a web of relationships. In this the personal/communal split is meaningless, for women experience multiple friendships as simultaneous rather than competing. The nature/culture line is less rigid as well, and women's experience of friendship with animals and with the earth reveals something about the oneness and connectedness of the cosmos that adds to an ecological ethics: "The gradual blurring of lines between human and animal life means increasing friendliness to both."[7]

This ecological understanding of friendship also critiques heterosexist norms for relationships:

> A friendship norm implies new patterns of relating that reflect values of love and justice lived out not two by heterosexual two, but in many combinations of genders and in threes, fours, and dozens as well. This is what is meant to call for a new relational ethic based on friendships, lots of them. It assumes that we will be just as rigorous in our evaluation of friendships as others have been dogmatic in their condemnation of anything that did not approximate their coupling. Such rigor, I believe, serves to inform us about the subtleties of love and caring, and the difference these make in our search for a just, participatory society.[8]

This is a richly suggestive metaphor and grounding for a liberationist ecological ethics grounded in and sensitive to the experience and insights of lesbians and gay men.

Four components critical to women's experience of friendship provide a model for the rigorous attention friendships merit as referents for right relation: love, power, embodiment, and spirituality.[9] Hunt understands *love* to be an orientation toward the world where friends are more united than separated, and more at one among the many than separated and alone. Love entails commitment to recognize, foster, and deepen this drive toward unity and connection without losing the uniqueness of individuals.

Power in women's friendships is the ability to make choices for their selves, for their dependent children, and with the community. Power is always found at the intersection of social/structural power and personal/individual power, and becomes distorted when either matrix is ignored.[10] Friendship as right relation is where empowerment and relinquishment are most likely to take place in healthy, sustainable, and mutual ways.

Embodiment includes sexuality, but is a broader recognition that everything we do and all that we are is mediated by our bodies. Embodiment in an ecological paradigm reflects our relation to each other and the rest of the earth, as well as our access to the goods of the earth such as healthful nutrition, labor, and rest.

Spirituality grounds the ability to make choices about the quality of life for oneself and one's community. This implies that the spiritual impulse toward meaning and value in women's friendships is expressed in very concrete ways: "Spirituality is part of an intentional, accountable process of making choices that affects the whole community."[11]

Theological reflection on friendships and their components in turn reveals something about the divine. It also reveals how other images of God may serve to hide or distort important parts of the divine and devalue or ignore altogether the important ethical value of women's friendships:

> Irrational reactions to inclusive language about God/ess are a clue to what is being revealed about our friendship with the divine. When She is our Mother instead of our Father, common assumptions about the nature, relationship, and function of the

> divine are called into question. When He is Father and never Mother we sense the imbalance. . . . Friendship with the divine, whether he, she, or it, is inspired by human friendship, and vice versa. This does not trivialize the divine nor elevate the human. It simply names friendship as the most adequate relational referent. As women begin to value friendships with women, the referent for divine-human friendship is given new content. . . . It inspires new depths of friendship in both arenas.[12]

An ecological ethics needs to be alert to the dangers of anthropomorphizing in calling the divine (and nonhuman nature) "friend." Hunt sees it as a step in the right direction, but fraught with the same problems of projecting human attributes onto the divine found in other metaphors such as "Lord" and "comforter." All images both enable and limit our understanding of the divine, and they must emerge from a liberating praxis in each community. Instead of being a final answer, imaging the divine as friend points to one attribute that illumines a critical ethical dimension of divine/human relations. It provides a needed corrective and generative image that emerges from women's experience for this time and context.

Beginning with friendship as ontological grounding provides a constructive and empowering context in which then to consider suffering and the tragic dimension in life.[13] Affirming women's (and all human) agency at the outset enables us to explore its limits and the suffering that occurs through loss. Loss is real and painful, both within friendships and the loss of friendships themselves. Loss of friends reminds us that no matter how communally oriented we are, each person is at one level radically alone. Important to a liberationist ethics, it reminds us how important it is to befriend ourselves.

Yet loss also teaches us that, although it can be a severely painful and distorting experience, it is nevertheless one that we can survive. Survival can lead to empowerment as we discover resources within ourselves and among those who befriend us along the way who were unknown to us. It also teaches us something about divinity and our relation to the divine. Thus Hunt concludes:

> Theologically a divinity that overarches all that is, remains static even as people change, is unnecessary. To the contrary, loss of friends teaches that being alone is inevitable, that even the divine abandons people, or so it seems. Befriending ourselves and making new friends teach that being alone, while real, is temporary in an interconnected universe. Who wants a divinity that does not measure up to reality? Loss shows that being radically alone, even without a divine friend, is an experience one can survive. A God who is absent is a venerable yet vulnerable part of the Christian tradition.[14]

Imaging the divine as friend provides an empowering and generative way of ontologically grounding right relation. Friendship covers the broadest array of emotions and relations, and is therefore a more accurate reflection of our experience of the divine: "As in all women's friendships, the potential for nurture and nastiness, comfort and challenge resides in all divine-human relationships."[15] Yet more importantly, more than other divine images such as Father and Mother, spirit or force, friend has the advantage of being widely available as a positive relationship. As an image it is personal without being intrusive and powerful without being mystical. It can ground our relations with the divine in a plurality of justice and love-producing ways.

Three themes common to women's friendships—attention, generativity, and community—tell us important things about the divine. These three in turn converge on justice, an attribute at the heart of any liberationist experience of the divine. How might these help us to image the divine in ways that are faithful to the demands of ecojustice?

Imaging the divine as an attentive friend testifies to the abiding presence that religious people experience. "This friend is waiting and cooperating creatively in the unfolding of history. This friend thinks of the smallest detail and permits the largest indiscretion without dominating or breaking the friendship."[16] Imaging the divine as a generative friend testifies to the creative side of God that in turn generates creativity and novelty through us and all creation. We can align ourselves with this energy leading us creatively into the future, trusting that a divine friend is leading the way. Community flows from divine inspiration. As the boundaries between culture and nature become creatively blurred, a holistic sense of community extends outward to include our biotic communities. Central to this is recognizing common roots as friends of the One who nurtures all. "This imagery is born out in the history of common suffering and promise that religious traditions have chronicled. Likewise it is useful for breaking down the particularities that set up hierarchies of privilege and dominance."[17]

Finally, friendships built on attention, generativity, and community converge on justice-seeking activities by justice-seeking friends. Authentic friendship is not an alternative to justice, but takes place within and fosters just relationship. The divine as just friend is the One who stands on the side of the oppressed and with her people in the face of danger. "Justice stands with open arms and ample bosom ready to embrace and to nurture as necessary, to propel and encourage as appropriate."[18]

Grounding right relation in the norm of friendship—friendship that includes but is not limited to sexual intimacy—offers many possibilities for a liberationist ecological ethic. It fosters attention to both sides of ecojustice. It provides images of the divine and the divine/human relation that are fundamentally ecological in their interwoven relationality, while also liberating in the sense that they are available to all. It provides a way of understanding the distinctiveness of human agency, best expressed in friendship, while suggesting that this reflects the ontological grounding of creation and defines an appropriate stance toward creation. It overcomes many of the antiecological dualisms of the Western tradition in a praxis that integrates spiritual, social, and psychological/personal dimensions. And it maintains a commitment to justice by providing an ethical norm that incorporates at the center a thoroughgoing critique of sexism and heterosexism.

THE DIVINE AS EROTIC: KNOWING GOD IN SENSUOUS, EMBODIED RELATION

Related to but distinct from efforts to ground right relation in friendship has been building on women's, especially lesbians', and gay men's experience of the erotic in our relationships to reshape our understandings of divine power and relationship with the divine. In this effort lesbian and gay theologians have followed the groundbreaking work of Audre Lorde, the late African American lesbian poet who laid the foundation for reclaiming *eros* and the erotic as the critical grounding for women connecting to themselves and to each other.[19] Gay men have also found in her words ways to understand how our experiences have been denied and our power thwarted through societal rejection of the validity of gay male eroticism and relationality.

Lorde speaks of the erotic as an assertion of the life-force of women that serves as a deep, nonrational source of knowledge that women can reclaim to transform their lives and their communities. Western societies have taught women and men to fear and suspect the erotic as untrustworthy, thereby cutting ourselves off from one of our deepest sources of self-knowledge and empowerment for change. But the erotic is the power that comes from connecting. This "self-connection shared" means connection with oneself, with other women and with other men, and with the world around us.[20] In this connection power comes from sharing deeply with oneself and with others, and opens one to a deep capacity for joy.

The power of the erotic, of connecting with other women and marginalized men, in turn leads to the empowerment of these women and men. Hence it has had to be repressed in patriarchal society. Yet when women and marginalized men let the erotic permeate and integrate all they are and do, its energy empowers by heightening, sensitizing, and strengthening their experience. Western society has socialized especially women to respond to external (male) authorities, deeply alienating themselves from their own experience, feelings, and intuitions. In contrast, the erotic empowers them to live from within outward, taking responsibility for themselves in the deepest sense.

Hence the erotic provides the grounding, a primary source for power and energy in the struggle against oppression. It frees women and marginalized men from externally imposed definitions and authorities to pay attention to the yearnings for connection and life within. The erotic urges us on, requiring excellence while not demanding the impossible, from ourselves or from others. The power and capacity for joy enable and demand a response that moves beyond patriarchal constructs for answering these desires: "That deep and irreplaceable knowledge of my capacity for joy comes to demand from all of my life that it be lived within the knowledge that such satisfaction is possible, and does not have to be called *marriage*, nor *god*, nor *an afterlife*."[21]

The erotic can help to break down false dichotomies. The truth of its knowledge comes from integrating action and feeling: "the erotic is not a question only of what we do; it is a question of how acutely and fully we can feel in the doing."[22] Hence "it feels right to me" is more than a superficial phrase; it is an indicator of listening to the urgings of the erotic within. The erotic connects the spiritual and the political: "the bridge that connects them is formed by the erotic—the sensual—those physical, emotional, and psychic expressions of what is deepest and strongest and richest within each of us, being shared: the passions of love, in its deepest meanings."[23]

Lesbian and gay theologians working from a liberationist commitment have found in reclaiming the erotic a generative means of grounding truth claims about right relation and the divine rooted in the lived experiences of lesbians and gay men. One sees this particularly in the development in Carter Heyward's early work of a theology of mutuality and right relation and her more recent full grounding of theology in the erotic. J. Michael Clark has drawn on the work of Heyward, other feminists, gay male, and Jewish writers to ground a theological understanding of the erotic in the lives of gay men and lesbians. Four areas of their revisioning the divine out of lesbian and gay experiences of the erotic have important implications for a liberationist ecological ethics: starting with lesbian and gay experience; understanding the divine in mutuality and right relation; the relation of the divine to human agency; and the experience of limits in the divine.

The starting point for both Heyward and Clark is an exploration of the sacred and divine through the sexual and erotic dimensions of women's and gay men's lives. Heyward defines *sexual* as "our embodied relational response to sacred/erotic power" and the *erotic* as "our embodied yearning for mutuality. . . . To speak of the erotic or of God is to speak of *power in right relation*."[24] Clark notes the ecological or interwoven dimension of the erotic in gay and lesbian experience that in turn reveals the divine: "As a community knit together by erotic and sexual relationships which are fundamentally good, we discover God/ess-with-us in our erotic energies and in our communal empowerment to seek and to create justice and liberation."[25]

Reclaiming sexual imagery for the divine is an important component of dissolving hierarchical dualisms that have separated spirituality from sexuality and the body and justified the exploitation and marginalization of women, gay men, and the natural world. Within the context of mutuality, sex can be an expression of commitment to right relation, and hence to justice. In this sense, lovemaking is a form of justicemaking, and its necessary grounding. It connects us with our deepest selves and with others, simultaneously turning us both into ourselves and beyond ourselves. The praxis of lovemaking as justicemaking, grounding our efforts toward justice in the erotic, reintegrates images of the divine as love(r) and justice.[26]

Within this context, the act of coming out is an act of relational integrity that contributes to the praxis of justice. For lesbians and gay men, coming out is a relational process that can be considered both justice-oriented and ecological. Through coming out we reclaim the integration of our private and public selves by reclaiming as positive and whole the erotic and sexual dimensions of our being and acting. In the context of heterosexism and homophobia, coming out must be seen as an act of resistance to wrong relation, that is, to injustice. It is a constructive, political act toward building justice as right relation. When viewed holistically, insofar as ecology is understood as the natural web of right relation, coming out can be an act of ecojustice. It contributes to dismantling social structures of wrong relation that violate ecological right relation.

Coming out also reveals aspects of the divine. It is never a one-time act, but an ongoing process that involves deepening relationship with oneself and with others. So is our journey with the divine. Coming out always involves a tension between moments of revelation and concealment. Lesbians and gay men learn to trust our intuition and our relationships with others to guide which moment is necessary and appropriate in different contexts and times, in order to both survive and thrive in a homophobic society. There are similar tensions in our relation with the sacred. While the divine is in our midst constantly, we do not always notice God: "The wisdom of God is not always evident, not always revealed, because we are not always ready to see what is happening."[27] The divine presence is often hidden from us until we are ready to help generate the conditions for its revelation. Like coming out, it is a relational process that involves all of who we are (not just our genital activity) and opens us to the deepest parts of who we are in relation.

Coming out can generate solidarity and compassion. It can help reveal the divine as both just and indignant as well as tender and nurturing. In coming out we can learn with others to express anger and active indignation with injustice while gaining a sense of patience for struggle for the long haul and for other lesbian and gay persons who may be more frightened and disempowered than we are. God, too, relates to us in these ways.

Coming out paradoxically often moves us further to the margins of society while revealing those margins to be the creative edge of life and living.[28] Hence the social location of lesbians and gay men at the fringes may be the best location from which to explore theologically the cosmos and our relation to the divine.[29] Precisely from being excluded, lesbians and gay men often are able to perceive and experience aspects of the divine not recognizable to dominant sectors. Heyward describes this "special vocation" as follows:

> It occurs to me that it may be the special privilege of lesbians and gay men to take very seriously, and very actively, what it means to love. As lesbians and gay men, we have had to fall back on the category of lover in order to speak of our most intimate, and often most meaningful, relationships. . . . Deprived of civil and religious trappings of romantic love, we may well be those who are most compelled to plumb the depths of what it really means to love. Our deprivation becomes an opportunity and a vocation: to become conscious of the things we have not seen, and to make others conscious of these same things.[30]

Many lesbian and gay theologians join other liberationists in grounding ontological understandings of God in a view of justice as mutuality and right relation. A liberationist ecological ethics maintains that an ethics of right relation must be extended to all things.[31] Heyward uses the image of relational matrix as the common home of all earth creatures to expand on her claim that God is our power in mutual relation: "God is our relational matrix (or womb). God is born in our relational matrix. God is becoming our relational matrix."[32] Mutuality in relation does not necessarily imply equality as the sameness of position or status, but rather is experienced as a dynamic relational movement that signals relational growth and change. In mutuality as the power of God, we call each other forth into our most liberating and creative possibilities. Understood ecologically, mutuality in right relation recognizes that our relational matrix extends to include the whole earth community.

It is the erotic as the divine within us that grounds and empowers our yearning for mutuality, for right relatedness. This claim has important implications for theological epistemology. It argues for the authority of the sensual, of the erotic, for knowing the divine, maintaining that "our bodiliness and our sexuality [are] the locus of our knowing and our touching the divine."[33] Our bodies mediate all our knowledge, including our spiritual insights. The divine is both personal and inclusive, and thus is best understood as pangendered and pansexual, "embracing but not limited to *all* possible gender expressions and *all* possible sexual orientations."[34]

This erotic epistemology is also ecological: "God reveals herself through our relationships not only to other people but also to other creatures and nature."[35] Clark adopts a process perspective to argue for a panentheistic view: the divine integrates physical and spiritual energy to infuse all of the cosmos. Hence the divine is radically immanent throughout the natural world. From this perspective it is extreme hubris to assume that only people are created in the imago Dei. Instead, opening ourselves to the earthly embodied immanence of the divine in all things leads to an appreciation of "ecological communion," a celebration of the interconnectedness and interdependence of all that is. As an alternative to an ethics rooted in a human-centered imago Dei, Clark draws on Carol Christ's work to call for an ethics grounded in the desire to enhance the life possibilities of all creatures, human and nonhuman.[36]

These efforts to reconceive our understanding of the divine by reflecting theologically on our experiences of the erotic have important implications for human and nonhuman agency in an ecological ethics. Congruent with Mary Hunt's work, both Heyward and Clark see the erotic as a grounding for friendship as paradigmatic of human agency in right relation.[37] Hence to love humanity and the earth is to befriend God: "The human act of love, befriending, making justice is our act of making God in the world"; and, "To generate friendship—embodied/incarnate mutuality—is the purpose of a sexual theology and ethics, just as it is the heart of a liberating God."[38]

Grounding ontology in mutuality in right relation means that divine agency in the divine/human relation is related to and depends in part on human agency. Hence Clark maintains, "in our erotic community we affirm that God/ess is so intimately with us in our lives, in our sociopolitical struggles for justice, and in our loving and passionate sexual interactions that the divine depends on us even as he/she nurtures us into being."[39] Doing good and undoing evil are human acts that incarnate God as power in relation.

Sin therefore is the denial of our power of relation, that is, the denial of God.

Yet the reality of sin points to the limits of human agency. Agency becomes alienating rather than liberating when it breaks down the mutuality of right relation. Human hubris too often results in our tendency to fail to set or heed limits. In portions of the gay male community one aspect of this has been the drive toward alienating sex with multiple partners, which Clark sees as a distortion of the erotic in our lives: "Insofar as we have misconstrued gay *pride* as nothing more than an arrogant and defiant insistence upon promiscuous sex as the meaning of our liberation, we have allowed ourselves to become imprisoned in a sexual underworld. . . . Genuine liberation must include freeing ourselves from such self-alienating sexual acting out."[40] Within an ecojustice framework of Gay and Gaia, another way of understanding this is that affirming human agency, or Gay, cannot mean ignoring or violating the ecological emphasis of Gaia on appropriate limits and interdependent relationality.

These insights on agency have important ramifications for ecological ethics. They emphasize the interrelatedness of all things, and the need to pay attention to power in relation. Distinguishing between mutuality and equality in relationship can help to affirm nonhuman agency in nature while recognizing power differences in intrahuman and human/nonhuman relations. If mutuality in right relation is our ontological grounding, our understanding of the divine, then we must recognize the agency and subjectivity of nonhuman creatures and creation in our relationality with them.

Finally, the devastation of AIDS in the current social location and reality of the gay male community forces us to reflect on theodicy, the nature of divine presence in the face of human (and other) suffering. In contrast to human-caused suffering, Clark understands AIDS to be an example of natural or nonmoral evil in the world.[41] He argues that in the face of nonmoral evil and tragedy such as that of AIDS, the notion of a God benevolent in nature is meaningless. We experience limits to God's benevolence in the natural world as well as from distortions in human agency.

To try to find theological meaning in the senseless suffering of AIDS, Clark considers and rejects as inadequate several traditional answers to theodicy.[42] Drawing on insights of process theology and post-Holocaust Jewish reflection, he concludes that we must abandon traditional claims about God's omnipotence and omnipresence in order to reconceptualize the divine as limited, a co-suffering companion and helpmate.

Creation and life emerge not ex nihilo, but out of the ongoing and endless divine ordering of chaos. An element of chaos and randomness thus pervades creation and constantly threatens life. Our bodies are subject to randomness, mixtures of possibility and risk. Randomness has a moral dimension when it enters human reality: "'Randomness' or 'chaos' is evil for humankind because of our awareness of pain, our capacity for suffering, and the ways in which these combine to alienate us from God's goodness, co-suffering, and empowered presence. The inbreaking of random chaos, humanly experienced as tragedy, simply happens."[43]

As Hunt also notes, an adequate understanding of the divine therefore must include humanly experienced limits to God's ability to "rescue" us from suffering and evil. Yet the divine may still be experienced as co-sufferer and companion amidst suffering, which in turn guides our agency when we confront the suffering of others. The experience of God's presence in God's absence—the awareness that God is not a divine rescuer—enables us to recognize and accept our responsibility as moral agents. We must face the enormity of human and nonhuman suffering such as found in the threat of death in AIDS, without, however, resorting to understandings of the divine constructed from a fear of mortality. Whether these views result in traditional appeals to life after death, reincarnation, or God as divine rescuer, too often the effect is to undermine the radical immanence of this life "as the only reality which we actually know and experience."[44] Critical to an ecological ethics, this reconceptualization of the divine emphasizes the need to take seriously this world as the only locus of our reality, while affirming human agency and responsibility in the face of suffering and wrong relation.

Each of these four areas where lesbian and gay theologians are revisioning the divine from lesbian and gay experience has important contributions to make to a liberationist ecological ethics. Grounding theological reflection in the experience of the erotic in the lives of lesbians and gay men addresses the heterosexism and invisibility of sexual minorities in ecological ethics while highlighting the importance of sensual embodiment for all experiences of the divine. Lifting up the erotic as the ontological grounding of reality provides a suggestive way of drawing attention to the ecological insight that all things are interrelated, while highlighting the source of human motivation for connection and justice with each other and the biospheric community. Seeing agency grounded in the mutuality of right relation that pays attention to differences in power acknowledges the distinctive nature of human agency while drawing attention to the reality of subjectivity in other creatures. And the experience of absence and limits of the divine that reimage God as co-suffering companion *and* source of empowerment/relation point to human responsibility within an ecological understanding of limits.

Focusing on the ways lesbian, gay, and feminist theologians draw on constructionist insights to examine how we name and describe the divine in turn suggests several criteria for theo-ethical reflection. Liberationist writers have drawn attention to the relation between images of God and religious and moral agency: our understandings of the divine and divine/creation relations influence how we relate to each other and can serve both to maintain or subvert unjust social and ecological relations. The guidelines proposed here can be grouped in two rubrics: methodological groundings for imaging the divine; and selected liberationist attributes of the divine.

Methodological Groundings for Imaging the Divine

- *Right relation*: Liberationist perspectives name justice as right relation as the grounding for knowing and naming the divine. The praxis of right relation can generate new and more just paradigms of the holy, emphasizing mutuality and relationality and moving away from hierarchical models that reproduce hierarchy in the social realm. A liberationist ecological ethic expands this to provide an ecological grounding of right relation with and among all creation as the context for knowing the divine.

- *Friendship and Positive Experiences of the Erotic—Enjoyment as Basic Values*: A liberationist ecological ethic highlights friendship and positive experiences of the erotic as paradigms for knowing the divine in right relation. Friendship gives insight into both the ontological grounding of creation as right relation and befriending as an ethical stance toward creation. The divine in right relation is grounded in and experienced most deeply through the erotic, through the ecstasy of an integrated sensual/spiritual connection with the deepest parts of ourselves and others.
- *The Divine and Moral Agency*: A liberationist ecological ethic must pay attention to how imaging the divine fosters or impedes moral agency. Images such as friend, lover, right relation, and the erotic provide positive moral norms and empower agency that in turn provide a liberating context and resources to engage and find meaning in painful experiences such as suffering. They can help to reintegrate love(making) and justice(making) in our religious and moral agency.
- *Limits to Naming the Divine*: A liberationist ethic takes seriously the ancient Hebrew concern with naming God, thereby limiting God or reproducing our own image as idols. At the same time it recognizes that we must risk naming at least partial images that emerge from and further foster a liberating agency. A liberationist ecological ethic therefore needs a critical understanding of the limits to anthropomorphizing and projecting images onto the divine: all images both reveal and obscure aspects of God. What are needed are partial, changeable, and locatable images of God that empower us in right relation while remaining open to change and new images.

Attributes of the Divine

Attention to the experience of God in right relation suggests several attributes of the divine and the relationships of God/humanity/creation important for a liberationist ecological ethics. Many are classical attributes given a new ecological and liberating grounding:

- *Love*: Experienced through justice as right relation at all levels of creation, we are more united than separated—with each other and with nonhuman nature, thus calling into question a rigid nature/culture dichotomy;
- *Power*: Found in mutual empowerment, the ability to make choices in relation, within ecological limits;
- *Embodiment*: Experience of the divine and each other is mediated bodily, rooted in and connected to the earth;
- *Spirituality*: Embodied, connected to quality of life for one's self and community, understood now to include all of creation;
- *Agency*: Relational in focus, maintains the dialectic of individual/relation/community. Recognizes the different elements and interweaving of divine and human agency, and agency/subjectivity among other creatures in nature.
- *Relational*: The divine is experienced as a relational matrix, where hierarchical patterns shift within and between subjects in relation. Fixed or rigid hierarchies reveal differences in social power and location, rather than divine ontology which constantly calls into question power relations that subordinate and marginalize.

Friendship and the Divine: Attention to the dynamics of right relation in friendship suggest additional components or attributes of the divine to incorporate into an ecological ethics:

- *Attention*: God's abiding presence is like the deep attention to relation found in true friendship;
- *Generativity/Creativity*: The experience of God in relation generates creativity and novelty, new possibilities;

- *Community*: Experience of the divine creates community as the locus of relationship; attention to community in the broadest ecological sense reveals important aspects of God;
- *Justice*: Each of the other attributes converge on justice as right relation, experienced in many facets through solidarity and compassion. God is experienced as both just and indignant as well as tender and nurturing.

Limits of Divinity—God's Absence and Presence: An ecological ethics must incorporate an understanding of experienced limits of the divine as God's presence and absence. God is experienced both as co-suffering companion and as the source of empowerment who guides our agency, but the divine does not violate the ecological parameters of the earth as a living system or intervene to rescue us from the destructive consequences of our actions. God's presence often is revealed most clearly at the margins of human society, which can serve as a locus of creativity and resistance to unjust and nonecological ways of relating.

We have now outlined the primary criteria for an ecological ethics of Gay and Gaia that emerges from the insights of lesbian and gay constructionist approaches and theological hermeneutics. How might these insights help us to evaluate the adequacy of some of the current approaches to Christian ecological ethics for the contributions they might make to a liberationist ethics? In chapters 6 and 7 I look at two contemporary efforts to reform Christian approaches to ecological ethics by drawing primarily on the legacy of biblical and orthodox theological doctrine: Douglas John Hall's "stewardship ethics," and James Nash's "ethics of ecological integrity and Christian responsibility."Then in chapters 8 and 9 I look at two efforts to rethink Christian ecological theo-ethics through a more radical reformulation that is highly critical, yet still appreciative, of the normative Christian biblical and theological tradition: John Cobb Jr.'s "process theological ethics of justice and sustainability" and Rosemary Radford Ruether's "ecofeminist theology of earth-healing." After outlining the main assumptions and approaches of each position with respect to the questions examined in the previous chapters—hermeneutics, humanity, nature, and God—I assess the contributions and problems of each from the perspective of a liberationist ecological ethics rooted in the experiences and insights of lesbians and gay men.

Part 2

Conversations with Companions on the Road

6 Biblical Theology: An Ethic of Prophetic Stewardship

Stewardship constitutes a special and unique charism of North American Christianity.... The stewardship tradition, which is rooted in biblical tradition and has of necessity been retained in the experience of North American Christianity, has come of age.

—Douglas John Hall, *The Steward*

The biblical image and metaphor of stewardship, grounded in the Genesis accounts of creation and assumed throughout much of the Bible, historically has been the dominant theological influence shaping Judeo-Christian and Western understanding of humanity's relation to the rest of nature. With the birth of the environmental movement in the 1960s, the Judeo-Christian ethic of stewardship has come under increased attack by some who argue that it is a primary source of the current ecological crisis.[1] Critics have argued that its anthropology holds humanity apart from and above nature. Others have responded that this is a misreading both of Judeo-Christian understandings of stewardship and its contribution to ecological destruction.[2]

Because stewardship has been the dominant theo-ethical stance of Christianity toward nature and continues to be the ethic to which many ecologically aware congregations aspire, it is important to look carefully at its content and presuppositions from a liberationist perspective. In his widely read book *The Steward: A Biblical Symbol Come of Age*, Canadian theologian Douglas John

Hall seeks to reclaim the biblical metaphor of stewardship as both central to Christian identity and the appropriate stance for human beings to be able to confront current global crises in justice, environmental, and peace/war issues.[3]

Within the framework of the theology of the cross, Hall turns to the biblical emphasis on stewardship and its implications for ethics to derive a cultural and theological analysis that can address ecological issues. In the face of the global ecological crisis, non-Christians as well as Christians are taking a new look at an ethic of responsible human stewardship in their search for a new understanding of human nature and our relationship to nonhuman nature. Drawing on Paul Tillich's understanding of symbol as a part of the received heritage that comes to life and gains meaning through the context of the community and the struggles it faces, Hall argues that the biblical metaphor of stewardship has achieved such a status of symbol in our time: "The only adequate response to the great physical and spiritual problems of our historical moment is for the human inhabitants of the planet to acquire, somehow, a new way of imagining themselves. Just in that connection, 'the steward' is one of the most provocative as well as historically accessible concepts to contemplate for anyone who cares about the destiny of our civilization."[4]

Hall sees the purpose of *The Steward* to be an "exercise in theological praxis" that grounds a theology of stewardship in a critical reappraisal of the historical North American church experience of stewardship. This will allow an understanding of stewardship as a holistic image of human and Christian vocation that is universally applicable.[5] Hall's primary thesis is that the biblical metaphor of stewardship is central to the Christian gospel; it can be understood, in fact, as the gospel in miniature. It should therefore be the central stance and identity of the Christian churches, which need to reclaim and revision stewardship for the times and context in light of the biblical witness. Hall argues that we need a holistic vision that enables us to reclaim the basic orientation of stewardship to the world—that is, we need a holistic theology of stewardship. Stewardship is a symbol rooted deep within the Western, Judeo-

Christian tradition whose time has come to be reappropriated by the churches to provide ethical guidance in confronting the social and ecological issues of our time.

Hall grounds his ethical reflection in a dialogue between the insights of biblical theology and Christian tradition, and questions posed by the contemporary world.[6] He seeks to explore, "in the light of contemporary circumstances, the depths of a biblical metaphor which already in its own historic context is potentially more than metaphoric."[7] Reflecting his own social location as a Canadian, middle-stratum, highly educated white man, the primary focus of praxis in Hall's stewardship ethics is affluent First World societies, particularly the mainline Protestant churches. He argues that the primary task of First World Christians is to change their own society, whose overly consumptive lifestyle is responsible for the bulk of problems in both First and Third World societies: "All have sinned. But scripturally as well as practically, the sin of the rich is the thing that must be dealt with first and most stringently because it is both the source of the oppressive conditions under which the poor exist and the reality which, if it is confronted and changed, could alter the situation of the oppressed as well."[8] Hall recognizes that there are problems of poverty and oppression within North America as well as the Third World, but most of his analysis and praxis address the wealth and affluence of North American society gained at the expense of the Third World and the environment.

The focus of this praxis is therefore on reversing the hubris of self-proclaimed human sovereignty to the appropriate vocation of servant and steward. Hall directs this praxis primarily at mainline Protestant denominations and congregations, working particularly at the level of the World Council of Churches and denominational programs.[9] Yet Hall also intends his work to aid Christian praxis at the congregational and individual level, as evidenced by its accessible writing style and the fact that at least two study guides for congregational use have been written to accompany *The Steward.*[10]

Hall begins his examination of stewardship ethics with the Bible both due to respect for its authority in Christian theology as well as its location as the source of the steward symbol. He looks first to the Genesis 1–2 creation accounts to ground human identity as fundamentally relational in a threefold manner: "This being has its being—no, receives its being—as it stands in relationship with God and with its own kind and with 'otherkind' (i.e., non- or extra-human creatures)."[11] Within creation, humanity has a special place and vocation as God's covenant partners in the sustaining and enhancing of all life. Exploring other Hebrew scripture usage of this image, Hall lists the qualities of stewardship as humbleness of spirit, lack of pretension or ostentation, and parental behavior toward those for whose welfare the steward has responsibility. The steward operates between two poles: closeness to the master who gives authority to the steward to act as the master's representative, and keen awareness that the steward is not the master and therefore all authority and responsibility are provisional and contingent upon faithful and adequate fulfillment of duties.[12]

New Testament accounts of stewardship add a Christological dimension: the risen Christ is seen as the ultimate steward, and therefore discipleship means stewardship in the Christian community. Stewardship means creaturely solidarity: we are all bound up with one another, and even Jesus as God's steward is accountable within a "chain of mutuality." Sovereignty is God's alone, which "puts an enormous question mark over all human pretension."[13] Theological motifs in New Testament accounts reveal further dimensions. The Christological emphasis grounds stewardship in grace: ethically we do not attain status as stewards through works or imitation of Christ but we are brought into stewardship by God's grace already enacted in God's chief steward. Ecclesiastically, the church becomes a stewarding community that exists for others and undercuts all its own efforts at power. Anthropologically, stewardship applies not just to Christians, but to humanity as a whole: "An even more important observation in this instance is that what is described and prescribed in the Scriptures of the New Testament as the appropriate life of the followers of Christ is at the same time the authors' way of discussing humanity in God's intention. Christians are not aberrations or superhuman beings; they are essentially persons who are in process of becoming truly human. . . . Stewardship is a human calling. It applies not only to those who are called into the life of explicit discipleship, but to the human species as such."[14] Finally, eschatologically, the New Testament symbol of the faithful steward implies accountability to God in the care for each other and the earth to the end.

With this biblical survey, Hall concludes that the symbol of the steward is rich enough and important enough to the biblical message to justify its use as a vehicle of transmission of the Christian gospel. He moves on to analysis of contemporary North American society, asking what it is in our present context that makes stewardship a ripe idea, a symbol come of age. Focusing primarily on cultural analysis, he posits two pervasive, problematic, and seemingly opposite attitudes that define our age. The first he calls the "technocratic mind-set," the collective mentality that produces machines and makes technique the "reigning mythology of our epoch."[15] It results in both a loss of meaning and a quest for mastery in First World societies. The combination of these lies at the root of most of our physical crises. The second he describes as "programmed indifference," the loss of meaning at the level of public and personal due to the technocratic, mechanistic mind-set, which has led to widespread apathy that finds expression in new forms of privatism, escapism, narcissism, and individualism. As a result, "We find ourselves at the end of a process that began with the Renaissance and expressed itself mightily in the Industrial Revolution: the great thrust forward, the bid for sovereignty over nature and history. But just as such sovereignty seems within reach the whole vision has ceased to charm us. It has in fact turned sour."[16]

Hall analyzes these attitudes theologically, naming as "sin" the distortion of being that is their consequence. Drawing on traditional understandings of sin, he sees the technocratic mind-set as a result of pride, and apathy and indifference as examples of sloth: "If pride implies reaching too high, sloth means sinking too low, settling for something less than real humanity."[17] Both are a distortion of the human vocation of stewardship, which challenges human beings to take the proactive posture of those who serve. Here Hall makes explicit in a series of questions the biblical anthropology that informs a stewardship stance:

> When this biblical metaphor [of stewardship] is brought to life by the Spirit of God blowing through the psychic desolations, the spiritual emptiness and future shock of our corporate life, it has the effect of direct address: "Will you at last assume your rightful role in this creation, child of Adam and Eve? Can you find it in yourself to take responsibility without being carried away by your own cleverness and power? Can you act the servant without groveling and demeaning yourself? You have it in your power to love and to change the world! Can you at last take hold of that vocation, without thinking it either too high or too low a calling?"[18]

Hall explores further those aspects of North American culture and Christianity that impede freeing up the symbol of the steward to transform our praxis. He begins by noting that the North American social location on the global stage is of those who have taken more than our fair share and who have oppressed others in the process. He asks, if liberation theology is a theology for those who have fallen among thieves, how do we develop a theology for the thieves? Making this more difficult is a basic impediment that he identifies as "a certain abiding ambiguity in our Christian attitude towards This World."[19] Christianity has become split along different lines of attitude toward the world. What Hall calls *the theocentric way* gives clear priority to God, usually (in practice, at least, if not in theology) to the neglect or devaluation of the natural world. A strong spirit/nature dualism makes this form of God-consciousness antithetical to creation consciousness. *Christian humanism,* on the other hand, with its grounding in nineteenth-century liberalism's unabashedly anthropocentric form of Christianity, is equally problematic. With its overly optimistic anthropology, it provides little room for challenging the plunder of nature by modern society.

Hall is sympathetic to the spirit of humanism in its passionate commitment to humanity, but he reminds the reader that the biblical God is both anthropocentric *and* geocentric. He checks liberalism's optimistic anthropology by grounding theology and human identity in the theology of the cross:

> I remain convinced that the direction of the Christian gospel is human-ward and earthward. For that reason I am able gladly to join hands with every sort of humanism. But as a Christian my own reason for taking up the cause of humankind and earth will not be found in any humanism, altruism, socialism, or ethical culturalism. It will be found in a faith whose God is the God of Golgotha—that is, a faith that is more honest about the darkness of the world than the most determined realist; a faith whose God inhabits the darkness and is "not overcome" (John 1:5). . . . Only a God who can be understood to participate in a world that is courting "catastrophe" (von Weizsäcker)—only a "crucified God" (Luther)—could give us the courage to believe that such a world is nevertheless worth caring about, perhaps even dying for.[20]

The theology of the cross for Hall is a whole theological and faith posture. It takes as its point of departure the brokenness of the human spirit and the human community, and places its hope in God's transformative solidarity with fallen creation.[21] It thus provides a radical this-worldliness that tempers the antiworld tendency of theocentric forms of Christianity, while avoiding falling prey to the human optimism and arrogance that result from liberal anthropologies. Stewardship theology recognizes this dialectic of a human creature called into and capable of responsibility, yet prone to turning this responsibility into unconditional management and mastery.

At this point Hall moves more explicitly into ethics by describing five "principles of contemporary stewardship praxis" that serve as "middle axioms" to bridge stewardship theology with practice. Middle axioms, as John Bennett and others have described them, serve to speak about principles fundamental to the Christian faith by rewording them in language and concepts accessible to others outside the faith.[22] They therefore are critical in making Christian ethics and praxis intelligible and accessible to others.

The first principle, *globalization*, claims that the whole earth is the steward's responsibility. While claiming holism and wholeness as the primary frame of reference, it nevertheless recognizes the importance of particularity: we can only love and care universally if we have loved and cared for particular persons and other creatures. This principle is critical to developing an ecological stewardship ethics: "Today the physical interrelatedness of our habitat has become the primary spiritual truth by which all our faiths and ideologies and values are tested and judged. Any system of meaning, religious or secular, which lacks the capacity to see things whole is quite simply irrelevant—and possibly pernicious. The tradition of Jerusalem in its biblical and best traditional expressions not only has this capacity, but places its concern for wholeness of vision, reconciliation of the parts, and the breaking down of 'dividing walls of hostility' at the very centre of its message."[23]

Communalization implies that personal stewardship is always participation in the stewardship of a community. To combat the negative consequences of North American individualism, Hall draws on the biblical dialectic between authentic individuality and community to call for a concerted effort to communalize the theology of stewardship.

Ecologization moves ethics beyond the human community to include the stewardship of the earth and its creatures. It means the recognition of the complex state of mutual dependency between human and extrahuman creatures and their and our environments. Development of this principle means, negatively, the rejection of all religion that in the name of the divine devalues nature, and positively, openness to the mystery of all created life—including the willingness sometimes to sacrifice human well-being for the sake of the nonhuman.

Hall's fourth principle, *politicization*, implies both moving stewardship out of the realm of sentimentality and private morality to address the difficult political realities of the world, as well as extricating stewardship from its association with economic capitalism.

This means distinguishing stewardship from philanthropy and willingness to prophetically critique the political economy of capitalism.

Finally, *futurization* means that stewardship must be exercised in the light not only of the immediate context but also the near and immediate future. The alternative to the current hubris of attempted human mastery is not passivity but stewardship, which combines hard work and planning with the humility of recognizing that we are not God.

Hall is very aware of the resistance such an ethics of stewardship will encounter in contemporary North American society, noting that each principle implies criticism and threat to the status quo:

> *globalization* can occur only against the entrenched spirit of a narrow nationalism, provincialism, and localism;
>
> *communalization* can take place only by confronting head-on the persistent individualism of both society and religion;
>
> *ecologization* runs headlong into the spirit of a rampant technology and the continuing bid for absolute sovereignty over nature;
>
> *politicization* along any lines but those of the market economy immediately creates enemies in our society, and amongst influential segments within all of our churches as well;
>
> *futurization* flies in the face of every private and institutional desire to secure 'the good life' here and now *and* the fatalized sense that in any case we can do nothing to alter the prospect of catastrophe.[24]

Yet for Hall this is precisely what it means to be the church—and to be fully human.

Drawing on the principles of globalization and communalization, the first task of First World citizens is to learn how to think corporately in order to be able to confront the corporate sin and guilt of the rich. Radical truth-telling will be required to break through our habits of constructing images of the world that omit the experience and reality of the poor and hide systemic structures of injustice. Taking seriously an ethics of stewardship would mean attacking directly First World worldviews of sovereignty and ownership, particularly the institution of private property. The second strategy of Christian stewardship is taking care of the earth's greatest victims, whom Hall sees as the poor—particularly poor women, and the earth itself.

Finally, Hall argues that humanity's proper vocation with respect to nature is to have dominion—but dominion not in Western understandings of domination, but rather the biblical perspective where God exercises a dominion of love through Jesus Christ. Thus human dominion with respect to nature in a stewardship ethics is understood as a stance of servant, keeper, and priest who represent nature to God in gratitude. It is only in accepting our creatureliness that humanity can represent other creatures—in fact, it is only in accepting our creaturehood that redemption is found: "Not the overcoming of our creaturehood, but its joyful acceptance: this is salvation!"[25] The image of the steward best represents the dialectical relationship of humanity with nature: fully creature yet with a unique place and vocation.

How does Hall's stewardship ethic compare to the goals and assumptions of a liberationist ecological ethic? To answer this, it is helpful here to look more closely at the methodological approach of biblical theology on which Hall draws and the resulting views of humanity, nature, God and their interrelation, and then assess these in light of the criteria for an ethic of Gay and Gaia outlined in the previous chapters.

Because Hall's stewardship ethic is grounded in biblical theology, his primary hermeneutics is shaped by his understanding of the relationship between Christian tradition, particularly the Bible, and contemporary experience. Christian theological and ethical reflection for Hall is rooted in praxis: "It is a type of reflection upon our concrete involvement in existence, undertaken from the perspective of a message about God's involvement in our existence. This message transcends any of its specific articulations, including the biblical one, because it is always addressed to the specific realities of our 'here and now'—our context. That is to say, it is a living Word of the living God intended for living human beings."[26] For Hall this takes place in the dialogue between tradition and experience: "Two interfacing points of concentration or orientation are necessary, then, to the witnessing community: an ongoing struggle to comprehend the Scriptures and the long tradition of those who in the past have done the same thing; and a continuous attempt to decipher the worldly context in which the community is to make its witness. Theology is what happens when these two orientations intersect—that is, when the biblical testimony to the Word encounters the spirit of the age, and when the Word encounters that spirit (Zeitgeist)."[27] What gives Hall's stewardship ethic its distinctive shape is grounding his hermeneutics in the theology of the cross. This hermeneutics both makes the ethic radically this-worldly, rooted in the suffering and brokenness of the world, and serves radically to relativize all human effort at mastery, authority, or superiority.

As is clear from his writing, the Bible is Hall's primary source in developing a stewardship ethic. He follows the Protestant principle of *sola Scriptura* in seeing the biblical witness to this living Word as unique and normative for Christians. With Luther and other reformers, Hall maintains that scripture can become the Word of God for us when it bears witness to the presence and ultimacy of the *living* Word in our context. The Bible for Christians is the witness to a sequence of historical events that culminate in the event of Christ; the biblical record thus consists of both event plus interpretation. For the Christian community in its efforts to live out faithful discipleship, the Bible serves as the primary source of information and inspiration. Hall, while not minimizing the importance of the Bible as source of information, of "original revelation," nevertheless places primary emphasis on the inspirational role of scripture. For the disciple community, the Bible provides imagination and courage.[28]

One can see this double function of information and inspiration in the way Hall develops his stewardship ethic. The Genesis and other Hebrew scripture accounts provide the raw material for fashioning an understanding of the basic outline of stewardship relationships, but it is the total biblical witness to "faith in God as the one acting in and through history for the world's redemption"[29] that inspires Hall to rethink what this ancient ethic might offer humanity today. He is thus free to draw on both the biblical witness and contemporary experience to develop his five ethical principles to guide our current praxis.

A strength of Hall's approach to the Bible is his refusal to equate God's word directly with the biblical text because God's word is always addressed to the specific realities of our context and cannot be contained by any one historical reality, including the biblical reality. Problems with methodologies of liberal biblical approaches such as Hall's arise primarily in their avoidance of the initial deconstructive phase—appropriating insights critically from the biblical tradition and science—which is so central to liberationist thinking and praxis. Hence, while Hall's stewardship ethic turns to the biblical tradition for recoverable resources to help us confront contemporary issues and crises, his noncritical appropriation of these symbols and resources reproduces many problematic assumptions that need to be lifted up and examined, not simply assumed. While Hall argues for a dialogue between Scripture and contemporary experience, the biblical witness clearly predominates. Yet when the deconstructive, critical phase is left out of appropriation from the Bible, the result often is an uncritical reproduction of problematic social relations that have shaped the text and tradition.

Hall's discussion of humanity's relation to nature is indicative of the influence of his Neo-Orthodox training for the way he uses the Bible as his primary source in developing this theological ethics. He draws on neither the social nor natural sciences as primary sources for understanding either human nature or the relationship of humanity to nature. Instead, he uses the Bible as the primary source in his theo-ethical analysis, and draws on science as a secondary source to corroborate or supplement his views. This approach of biblical theology is problematic in its assumptions that the canonical biblical witness is both adequate and authoritative for ecological ethics. This in turn affects the way biblical materials are used to engage other sources such as science and human experience. While Hall has a critical stance toward science, valuing it for its contributions while critiquing its elements of mastery and mechanism, he does not extend this same critical stance toward the biblical tradition of stewardship as the appropriate and normative human vocation. Hall draws on science primarily to inform the contemporary context, rather than also letting it engage the biblical materials to expose and examine the historically particular assumptions they contain about the natural world. Hence the biblical understanding of such critical areas for an ecological ethics as human nature and the humanity/nature relation ends up basically immune to critique of science, left unengaged and uninformed. For example, nowhere do insights of evolution or physical anthropology explicitly inform his discussion of human origins and nature. For this task biblical and classical Protestant perspectives remain his primary sources.

Hall's use of cultural criticism and the experiences of others to inform the contemporary context is similarly problematic. He draws heavily on recent literature and cultural analysis to help discern the state of contemporary North American society. Yet the majority of writers upon whom he draws are from social locations similar to his own: liberal to progressive, white, educated, North American and European men.[30] The fact that he draws on virtually no voices of women or racial, ethnic, or sexual minorities in developing his cultural and social analysis is indicative of the lack of attention to differences in power and location in the sources he uses. This results in an analysis of free-floating ideas, blind to the societal dynamics and relations that produced them.

One result of this is the tendency to develop an ahistorical, universalized understanding of human nature that ignores the influence of particular historical and contextual experience. In the process, socially and culturally constructed aspects of human identity built along lines of social and environmental difference drop out or are rendered invisible. While Hall clearly is aware of recent theological developments and social movements among historically marginalized communities such as feminist, black, lesbian, and gay, and Third World liberation theologies, they rarely inform his analysis, at least not directly.[31] With the absence of these voices and attention to power in his cultural analysis, and the lack of engagement of this analysis with the biblical materials he cites, it easy to see why he does not incorporate such perspectives as current feminist critiques of a stewardship metaphor and model.

Finally, while Hall pays close attention to justice issues, his vantage point is clearly from the perspective of the powerful and affluent in Western societies. While he does privilege a minority Christian tradition that often has expressed solidarity with the suffering—the theology of the cross—there is no epistemological priority given to hearing from marginalized perspectives themselves in the way he develops his ethic, either in which voices in the Bible and tradition are drawn on as authoritative or in listening to marginalized perspectives in contemporary experience that may not be present or easily discernible in the tradition (such as lesbian and gay voices). This reveals the weakness of an ethic that does not include an explicit commitment to the praxis of solidarity with the oppressed, taking seriously learning from and contact with their locations. It is a reflection of the location of Hall's praxis with and among the affluent: even though his sentiments and commitments are on the side of justice for all, they are filtered through and reflect the dominant (and primarily white, male, and heterosexual) location of his sources. The result is to perpetuate the invisibility of socially marginalized persons such as lesbians and gay men, *and* the social mechanisms that both marginalize these social groups and justify the exploitation of nature.

These methodological issues are reflected in the views of humanity, nature, God, and their interrelation that Hall develops for a stewardship ethic, and these in turn are problematic for a lesbian and gay liberationist ecological ethic. Let us briefly synthesize his views in these areas and then reflect on both their promising and problematic characteristics for an ethics of Gay and Gaia.

Hall reflects his roots in Neo-Orthodoxy and Christian realism in that his stewardship ethic is explicitly theocentric and anthropocentric. Although he maintains a strong awareness of human capacity for evil and in places argues for inclusion of a geocentric perspective, for Hall the biblical witness is clear that God has given humanity the primary vocation of agency through a stewardship relation to the rest of nature. What results, then, is an understanding of creation that is thoroughly interrelated and relational, yet also hierarchically ordered. Hall cites Paul in 1 Corinthians to illustrate creation as a "chain of mutuality" that includes even Jesus Christ and sees all bound up with one another—no one can claim independent dignity, authority, or worth, for all are related and ultimately accountable as stewards to God.[32]

To reflect on human nature, Hall turns to an ancient source, the "Jerusalem tradition" found in the Bible, particularly the Genesis 1–3 accounts. Just as creation is interrelated and relational, human beings are fundamentally relational beings. According to Hall, humanity cannot be understood in isolation, but only within a threefold relational matrix: "Human identity is bound up with God on the one hand, and all the rest of creation on the other. For the biblical way of defining human being is from first to last relational.... Following this scriptural lead, then, one is not justified in separating humanity off from the others and asking about the character of the human being in isolation. This being has its being—no, receives its being—as it stands in relationship with God and with its own kind and with 'otherkind.'"[33] For Hall, the steward is a particularly apt metaphor to demonstrate this threefold relationality, for it shows humanity's *accountability* to God and *responsibility* for its fellow creatures.

Yet, Hall maintains that two things set humanity apart from the rest of creation. First, while human beings are creatures who "feel kinship with the dust as well as with the many beings who do not apparently ask about their being," the human creature "cannot find comfort in its own sheer animality, for it feels within itself a yearning and a potential that makes it 'restless' (Augustine) and never content with the joys of the body alone."[34] Second, sin leads to an "anthropological distortion" that leaves human existence flawed to the core. Hence, according to the classical Christian doctrine which Hall affirms, fallenness in creation is consequent upon this anthropological distortion, affirmed in the Genesis account as the fall of Adam and Eve. Only an anthropology that affirms both sides of this dialectic—affirming our full creatureliness yet unique self-consciousness as well as our true nature inevitably distorted by sin—can avoid the distortions in views of human nature found historically in romantic (fully animal) or liberal (separate from animal) views.

With respect to the sacred, for Hall, the biblical understanding of God is never explored apart from an understanding of humanity: "Who is God?" and "Who is the human creature?" form two inseparable parts of one question.[35] He argues that while the emphasis is on the divine/human relationship, all biblical theocentrism is both immediately anthropocentric and geocentric. Yet Hall also follows the classical, Reformation, and Neo-Orthodox emphasis on a strong distinction between creator and creature; there is no hint of pantheism here. Sovereignty is God's alone, which radically relativizes all human authority with respect to the rest of the world. Drawing on the patriarchal images of Yahweh that infuse the Hebrew scriptures, Hall emphasizes this point as he elaborates his understanding of God vis-à-vis the stewardship metaphor:

> What is established here [in the stewardship relationship], once and for all, is that *ownership, mastery, ultimacy of authority, and sovereignty* are attributable to God alone. . . . When this idea is combined with its immediate anthropological implications, as it must be, it has a critical theological and ethical connotation: namely, it puts an enormous question mark over all human presumption, not only in relation to material realities (including the alleged possession of properties), but also with respect to more nebulous realities such as authority. As soon as God is pictured as *the owner and sovereign of everything* in relation to which human beings can be at most stewards, institutions such as the holding of property, the hierarchic distribution of authority, the technocratic mastery of the natural world, and the like are thrown into a critical perspective.[36]

For Hall, God is radically connected and committed to this world not through pantheism, the divine *in* nature, but through the cross. It is at the point of the cross that God demonstrates divine commitment to the world, particularly at its points of suffering and brokenness, while not glossing over the distortion and alienation of God's creation that results from human sin: "the cross symbolizes the impossibility of the human condition; but as the point of God's deepest solidarity with us it also stands for the grace that overcomes our impossibility, the negation of that which negates."[37]

Because of the cross, Hall's understanding of God is thoroughly Christological. In addition to the offices of prophet, priest, and king, Christ's work as second person of the Trinity is fulfilled through the office of the steward. Jesus is the preeminent and authentic steward who models divine participation in stewardship and invites us through grace into obedient discipleship as stewardship. In the cross Jesus is the faithful steward who desires nothing for himself, but lays down his life for his God and his friends. Through the redemptive work of Jesus in the cross—redemption that comes not from overcoming human creaturehood but in its joyful acceptance—God makes human beings covenant partners in enhancing and sustaining life. For Hall this is what prevents the classical affirmation of *imago Dei* from being reduced to a crass human superiority over nature. Being in the image of God refers to the *unique relationship* of accountability and responsibility in grace that humanity has vis-à-vis the Creator, not to the possession of unique and/or superior qualities. It leaves no room for human arrogance, but rather humble gratitude in response.

With a stewardship metaphor mediating God's relationship to humanity and humanity's relationship to extrahuman nature, how does this ethic understand nature? Somewhat surprisingly, there is relatively little mention of or focus on nature in Hall's work. While in several places he describes both the Bible and stewardship ethics as "anthropocentric and geocentric," a clear tension exists between these two perspectives throughout the book. On the geocentric side, nature as created by God has inherent worth, and human beings as stewards are responsible for the well-being of "otherkind" both for their own sake as well as for humanity's. On the other side, the anthropocentric focus of stewardship ethics dominates. Nature frequently is described as a "resource" given by God for the well-being of both human and extrahuman creatures. Nature is essentially passive. Although all creation is seen as relational and interrelated (Hall affirms the ecological principle of mutual dependency), there is little or no emphasis on the agency of nonhuman creatures or nature. Rather, agency lies in the human vocation of stewardship, which makes humanity responsible for representing nature. Hall considers three ways of formulating "the relation between humankind and otherkind": (1) humanity above nature; (2) humanity in nature; (3) humanity with nature.[38]

Hall sees the humanity-above-nature position—and particularly humanity as master of nature—as the dominant view of Western modernity since the Enlightenment. It is the perspective that informs the mechanistic, managerial, and mastery paradigms that Hall critiques so rigorously. Yet, equating this view with a Judeo-Christian perspective is too simplistic, Hall argues, for it ignores the Bible's association of the human bid for mastery with disobedience rather than the positive will of God. One must still ask, however, why Christians at the beginning of the modern era permitted their sacred texts and faith to be used to justify rampant exploitation of nature. Citing a number of historical factors, Hall locates the heart of the problem precisely in the understanding of the relation between human and extrahuman:

> If one thinks of the human creature as a possessor of superior endowments such as intellect and will, and locates the relationship between this creature and the others on a scale or hierarchy of being which is determined by such gifts, one is bound to end up with the humanity above nature alternative. But if the image of God does not refer to a quality that we possess (making us superior to other creatures), but to a relationship in which we stand vis-à-vis our Creator, and a vocation to which

we are called within the creation, a very different conception of the humankind/otherkind relation follows.[39]

The second position, humanity in nature, sees human beings as one species among others, having no more to offer nor more right to life than other forms of life. Hall sees this naturalizing of humanity merely as a romantic (and modern) reaction to the excesses of the humanity-above-nature position. He finds this second position equally problematic. It can lead to the logical conclusion that if humankind is simply one part of nature, and yet is the primary problem in the biosphere, would not the world be better off without humanity's participation? In response, Hall maintains that, "It is not only a counsel of despair but patent nonsense, in my opinion, to assume either that humanity is or that it ought to be simply 'in' nature."[40] Both the biblical perspective and common sense demonstrate that humankind is qualitatively distinct from the rest of nature.[41]

Instead, drawing on the biblical view of reality as ontologically relational, Hall argues that humanity *with* nature is the most accurate depiction of the human/extrahuman relation. He notes that throughout the Bible the preposition "with" is employed to connect two subjects in relationship to emphasize both sameness and difference: husbands with wives, friends with friends, Jesus with the disciples, even Emmanuel as "God with us." Using "with" to name these relationships highlights a dialectical tension between participation and individuality, commonality and uniqueness between the subjects. "Being" in the Bible is therefore really "being-with." Hence "the distortion of our being (sin) is nothing more nor less than our alienation from all that we are created to be with."[42]

For Hall this is not merely a matter of semantics, but is key to understanding the biblical ontology of relatedness and love. The ontology of relatedness means that there is both a strong sense of the interconnectedness of everything that exists as well as an emphasis on the uniqueness of each creature. It is this uniqueness or difference that enables love, for "love means difference: I am not you, you are not me. If we love each other, it does not mean . . . that we simply merge into an ontic unity. . . . Love means that I am *with* you and you are *with* me in a special sense. We evoke and support each other's individuality even as we discover our fundamental solidarity."[43]

Affirming both our relatedness to as well as our uniqueness from nature permits humanity to love nature through our vocation as stewards. Stewardship as vocation implies many facets of humanity's relationship to nature. First and foremost we are *responsible* (to God) for the well-being of both humanity and the rest of creation. Second, as "the speaking creature," the place where creation becomes reflective about itself, humanity *represents* nature: "When we are true to our own essence, to the ontology of communion and community, we speak for all our fellow creatures, for the totality." Third, humanity's rightful role is to "have dominion" in the sense of exercising a unique responsibility and answerability for other creatures. But quite apart from modern understandings of dominion as domination, it is the biblical testimony to God's dominion as love—made concrete in Jesus' life and death on the cross—that defines human dominion.[44] Finally, citing the biologist Charles Birch, Hall argues that what stewardship as dominion means today is *solidarity* with nature, issuing in a life of service and sacrifice.[45]

Building on this, Hall's understanding of the threefold relationship between God, humanity, and nature could perhaps best be described as stewardship relationships grounded in an anthropocentric and geocentric theology of the cross. The cross underscores God's fundamental commitment to the world. Humanity figures prominently in this relationship as accountable to God and responsible for the well-being of the rest of nature.[46] Seeing creation as a chain of mutuality, including God in Jesus, Hall concludes his book with a call for an ethic of caring as a way of being. Because God cares deeply for the world—made most evident in the cross—we are freed by grace to care deeply and live and act accordingly.

While Hall's exploration of the stewardship metaphor reveals several rich theological insights, it also reveals the problematic nature of a liberal biblical theological approach for a liberationist ethics. First, in contrast to feminist scholars such as Brooten and Ruether, Hall ignores constructionist approaches and thus shows little awareness of how views of humanity and nature constructed in ancient patriarchal cultures shaped biblical and traditional understandings of stewardship. His uncritical appropriation of stewardship leaves in place the hierarchical ordering of relationships patriarchal cultures assumed was natural and therefore normative. Hall modifies this by stressing relationality of all creation in a "chain of mutuality" that extends to include Jesus Christ, but the hierarchically ordered pattern of relations remains. Stewardship thus means exercising a benevolent paternalism toward the rest of nature: humanity's vocation is accountability to God and responsibility for nature, exercised by representing nature to God. While Hall rules out any mechanistic or managerial expression of stewardship that either denies humanity's full creatureliness or denigrates other creatures, retaining a hierarchically ordered framework that restricts agency to human beings is highly problematic for a liberationist ecological ethics.

This is complicated by a second weakness in Hall's approach, his ignorance of the natural sciences in making claims about humanity. This shortcoming is illustrated in his views on dominion. Here he ignores the evolutionary history of nature and place of humanity within this, and instead places human beings in a redeemer role vis-à-vis nature as representatives of Christ, a relation that exactly reverses the true dynamics of the ecological crisis where human beings have been the primary cause. Viewed ecologically, the rest of nature more likely needs to be redeemed *from* rather than *through* humanity as Christ's representative.[47] The other models we will examine here incorporate the scientific evidence for humanity's late arrival on the evolutionary scene in their reinterpretation of the doctrine of dominion as best understood today as a realistic description of the current positioning of human beings in nature; whether we like it or not, human technological power in our current social/historical context is such that the fate of much of nature does lie in human hands as we near the twenty-first century. Hall, however, ignores this, and continues to maintain a theological view that the dominion relation is itself inherent to nature and human nature, one created by God. This is no longer credible today, given scientific evidence for over three billion years of plant and animal life that precede the evolution of human beings. Ignoring the lessons of science—even in a highly theological treatise such as Hall's—risks perpetuating untenable views of humanity and nature.

Hall does not advocate ignoring science in an ethic of stewardship; it plays an important role in giving content to his middle axioms, such as ecologization. But drawing on science as a second step after fundamental attitudes toward humanity and nature have already been set leaves these views critically unexamined and functionally normative. As Brooten, Weeks, and others have argued, it functions to leave in place assumptions about humanity as "natural," which are actually socially constructed and need to be examined to see how they play out in different contexts.

With respect to nature and the humanity/nature relationship, Hall's strengths include his consistent emphasis that creation is fully relational, that human beings are fully creatures, and that human salvation lies in accepting our creaturely condition. All these affirmations are critical to a liberationist ecological ethic.

Within these affirmations, however, the stewardship ethic proposed by Hall contains several problematic assumptions. While being critical of Enlightenment paradigms of mastery and mechanism, Hall nevertheless perpetuates a strong nature/culture dualism in his understanding of the humanity/nature relation. Reflecting an interpretation of the Bible influenced by his Christian realist mentors, for Hall agency belongs to culture and lies exclusively in humanity's God-given vocation of stewardship. Nonhuman nature has inherent worth as a good creation of God, but needs humanity to represent it to God. Noncreaturely nature, when it is mentioned at all, is imaged primarily as passive, a resource given by God to be used wisely and prudently. There is no recognition of the active agency of nonhuman nature itself in constructing the world, either in the agency of other creatures who construct nature as in the image of nature as a matrix of actors, or in recognizing the agency of nature as a whole, as in the image of nature as Gaia or coyote.

Instead Hall's stewardship ethic maintains the biblical focus on the God/humanity relation that highlights humanity's distinctiveness from the rest of nature. Rather than emphasize that all creatures, including human beings, fill unique ecological niches and have unique combinations of abilities and qualities, Hall focuses on humanity's uniqueness. He affirms human beings' full creatureliness, but argues that because of humanity's unique vocation of stewardship, the humanity/nature relation should be understood as humanity *with* nature rather than humanity *in* nature. Human beings maintain a realm of freedom from the constraints of nature through culture, where the uniquely human vocation of stewardship is exercised. Rather than engaging in a creative partnership with nonhuman nature that recognizes the unique gifts and agency of *all* participants, the stewardship ethics locates agency on one side and maintains a pattern and logic of paternalism where humanity is responsible for nonhuman nature.

To do this, Hall draws on the logic of representation that Haraway critiques so devastatingly, and argues that humanity must represent nature before God. In the process he renders nature basically passive and undifferentiated—as well as conflating all human difference in humanity's relation to nonhuman nature. What emerges is a hierarchically ordered relationship of creation where differences within categories, such as within humanity or within nature, are glossed over, while boundaries between categories are reinforced. Yet as Haraway and many ecofeminists have argued, it is precisely in paying attention to difference and how it is reflected or obscured in socially constructed boundaries and categories that we open up new possibilities for more ecologically sustainable and just relations. At a point in history where the Western emphasis on humanity's distinctiveness from the rest of nature has brought disastrous results, any discussion of humanity's uniqueness must be framed by our continuity with and full participation in nature. Human culture exists *in* nature, not apart from it. While there are helpful insights in the stewardship ethic, such as its emphasis on relatedness, love, and accountability for our actions, a liberationist ecological ethic must move beyond it as a grounding metaphor.

Finally, the view of God informing Hall's stewardship ethics is both promising and problematic for a liberationist ecological ethics. Among the constructive elements is Hall's emphasis on God's relationality with and radical commitment to the world. Hall grounds this ethics in what he calls the first ontological principle of biblical theology: God is love.[48] There are points of contact in Hall's understanding with our emphasis here on befriending and friendship as our ontological and ecological grounding. For Hall this grounding is best understood through the theology of the cross: the power of God's love paradoxically is revealed in the weakness of love, seen in divine commitment in solidarity with humanity on the cross. Numerous values and insights with theologies of the cross resonate with those articulated above, including solidarity, compassion, and the notion that the divine presence is found at the margins among those who suffer. Yet feminists have drawn attention to the negative social consequences that emphasis on the cross, as the central metaphor for divine/human relationality, has had for women and other marginalized persons.[49] While the cross can serve to expose and highlight certain power dynamics in society, it risks obscuring or hiding others. When theological reflection on biblical themes and metaphors is divorced from the lived experiences of socially marginalized persons, it can function to further social oppression.

Hall is not blind to the importance of contextuality and consequences in theology; in fact, this is a central element in his theological method.[50] Yet there is little awareness or willingness to use the consequences of context to question the socially constructed nature of biblical theological claims themselves. An example of this in Hall's stewardship ethics is his emphasis on self-sacrificial love as imaging the divine in the incarnation with no reference to the social consequences of this teaching for historically marginalized groups, particularly women and people of color. Thus he writes: "Jesus in fact defines and fulfills the office of the steward. Because he is a just and faithful steward, unlike the unjust steward of Luke 16, he desires nothing for himself. Because he is obedient to the one he represents, he is not concerned about saving his own life, but lays it down for his friends. He does not think in terms of possession, not even the possession of his own life."[51]

However, feminist, womanist, lesbian, and gay scholars have awakened our attention to the destructive consequences of just such an ahistorical biblical theology that reflects a self/other dualism. As Hunt points out, befriending others begins by befriending ourselves. We befriend ourselves when we take our embodied selves and needs seriously, not by desiring nothing for ourselves. Community is built by valuing all our lives, not by dualistically sacrificing some for the sake of others. This is not to deny the importance of sacrifice as an element in friendship and community, but it is sacrifice for the good of all, including ourselves in relation. In community, the lives of others are interdependent with our own, and a minimal level of "possessions" is necessary for the life and health of each individual and the interdependent whole. Thus, arguing as Hall does that God is connected to the world not through pantheism but through the cross is problematic for an ecological ethic grounded in lesbian, gay, and feminist experience and constructionist insights. Hunt articulates an alternative saving metaphor by emphasizing incarnational reality in solidarity through friendship and community:

> Women's friendships challenge the adequacy of Christianity's central metaphor, the death and resurrection of a man who laid down his life for his friends. . . . Women friends surely would have seen to a woman's survival and not to her death. Some friends Jesus must have had! Granted friendship does not stop death, but the mystique of giving up life in order to find it again does not appeal as readily to those who are struggling to survive as it might to those who in a "survival of the fittest" contest consider themselves the fittest. A more apt metaphor from women's experience would be the triumph of a group of women over injustice without losing anyone.[52]

This example illustrates another problem with a stewardship ethic based on biblical theology divorced from constructionist insights: the ease with which patriarchal masculine images are attributed to the divine and return to humanity via the *imago Dei*. Note the italicized attributes of God that ground a stewardship ethic in the quote cited earlier: *"ownership, mastery, ultimacy of authority, and sovereignty* are attributable to God alone. . . . As soon as God is pictured as *the owner and sovereign of everything."*[53] While Hall is emphatic that what is reflected in the *imago Dei is* responsible stewardship as vocation rather than attributes of owning, mastery, authority, and sovereignty, the historical legacy of linking patriarchal images of God with the *imago Dei* is far more ambiguous, if not primarily negative. It reveals the inadequacy of drawing on the metaphors and themes of biblical theology divorced from a constructionist analysis of their source and historical and contemporary function.[54]

In summary, while Hall's stewardship ethic is promising in its emphasis on a number of liberationist themes—the centrality of active praxis, a critique of human power and attempts at mastery, God's solidarity with the suffering and marginalized, extended now to include the whole earth community, and the potential of a theology of the cross for grounding an ontology of relatedness and an ethics of caring—from a liberationist perspective, it also reveals the inadequacy of basing ecological ethics on biblical theology alone. Assuming a priori the adequacy and authority of biblical perspectives, biblical theological approaches too rarely include engagement with critical disciplines, including the natural and social sciences, as indispensable to the task of biblical exegesis and hermeneutics. Rather, when engagement with insights from these disciplines occurs, it takes place as a second step in formulating ethical guidelines. Too often, however, this approach merely reproduces problematic assumptions and social relations found in the biblical texts themselves, leaving them closed to critique and critical assessment from other perspectives.

The reappropriation of the stewardship metaphor itself, without attention to the social relations of patriarchal cultures that enabled and constrained it, risks reproducing these power relations in our own context. While Hall emphasizes stewardship as solidarity with creation, his mechanism of representation replaces a logic of justice with a logic of benevolent paternalism that results in Christian stewardship as "the steward's care of earth's greatest victims"—especially poor women and the earth.[55] The ethic continues to draw on patriarchal attributes such as mastery and ownership to name God and leaves in place a hierarchically ordered pattern of relations with nonhuman nature in a subordinate place.

Finally, while sensitive to listening to the voices of the poor, Hall too often conflates human differences both within the North American context and in examining humanity's relation to the rest of nature. This is a direct result of his inadequate social and cultural analysis, which ignores the perspectives of women, people of color, and sexual minorities. It may be the inevitable result of a biblical theological approach that does not start from praxis rooted in interactive solidarity with oppressed communities or that does not grant explicit epistemological privilege to the perspectives that emerge from these groups. The result is an ethic that perpetuates injustice by reproducing the invisibility and exclusion of socially marginalized groups from the ecology of human and wider biotic communities.

7 Christian Liberalism: An Ethic of Realism and Responsibility

Christian ecological ethics is not simply the values and norms of social ethics applied to an ecological content that is conceived as an area of intrahuman responsibilities. It is also a relevant revision and extension of these values and norms, applied to an ecological context that is conceived as moral problems of human relations with all other beings and elements in the ecosphere.

—James A. Nash, *Loving Nature*

In addition to the biblical theology movement, Christian realism has been the most influential liberal Christian movement of the mid-twentieth century. Tracing its heritage to the thought and writing of Reinhold Niebuhr, Christian realism is practical theology that combines a liberal social analysis with a Christian theological and ethical tradition to provide ethical guidelines for Christian advocacy and participation in public policy formation. A pragmatic theology, realism makes the case for the responsibility of Christians to participate fully in public life. Christian realism in general, and Niebuhr's work in particular, provided the major influence shaping the practical theology of Christian social activists in North America until his death in 1971 and the subsequent emergence of various forms of liberation theology.[1]

In *Loving Nature: Ecological Integrity and Christian Responsibility*, James A. Nash extends the framework and heritage of Christian realism to include maintenance of the health and integrity of the earth's biosphere as a critical part of contemporary Christian responsibility. As executive director of the Churches' Center for Theology and Public Policy in Washington, D.C., Nash has spent years bridging the worlds of North American churches, at both the denominational and congregational levels, with the arena of public policy formation on environmental and ecological issues. At the simplest level, *Loving Nature* is the synthesis of the theological and ethical reflection that informs his advocacy in ecological issues. In addition, however, Nash's work is a constructive piece in theological ethics that creatively extends classical and contemporary Christian insights and doctrine into the ecological and biophysical realm. In the short time since *Loving Nature* was published in 1991, it has received wide critical acclaim and represents the best of liberal Christian ethical reflection and policy recommendations on ecological issues. Nash's work addresses some of the shortcomings of Hall's biblical stewardship ethic while raising additional issues that need to be addressed by a liberationist ethic.

Nash begins his reflections with an overview of the seriousness of the ecological crisis facing the globe. He considers his book to be an exposition on the meaning of ecological integrity and the current crisis that threatens to disrupt and destroy it. Yet for Christians, the ecological crisis is more than a biophysical challenge. Because the crisis is rooted in part "in philosophical, theological, and ethical convictions about the rights and powers of humankind in relation to the rest of the biophysical world," it also represents "a theo-ethical challenge" that the Christian faith must address.[2] This will mean a significant effort to redefine moral responsibilities and relationships in light of the teachings of the Christian faith and the implications of the ecological crisis.

Nash seeks to address several groups on both left and right in his writings. On the left, *Loving Nature* is an apologetic against secular and radical religious critics of Christianity who charge the Christian faith with primary culpability for and inability to respond adequately to the contemporary global ecological crisis. On the right, Nash follows Christian realism in arguing against conservative, spiritualized forms of Christianity that do not see the political and ecological implications and responsibility of the faith. He deliberately charts a middle ground that he hopes will have broader appeal and impact among practitioners of mainstream Christianity.

Nash's primary thesis informing his analysis and constructive efforts is that Christian theology can be reformed to provide an ethical foundation for the praxis of ecological integrity:

> Ecological responsibility does not require the abandonment or replacement of Christianity's main theological themes. "New" or "radical" or "imported" theologies are not necessary. What is required, however, are reinterpretations, extensions, and revisions, as well as cast-offs of cultural corruptions, in ways that preserve the historic identity of the relevant Christian doctrines and yet integrate ethical insights and ecological data. . . . The important point, however, is that Christian theology can remain loyal to the intentions of the faith in the historical affirmations of the church while developing a genuinely ecological theology.[3]

For Nash the key issue relating Christian theology to ethics is love (and justice as a dimension of love). Hence any effort to relate ecological integrity to the Christian faith must relate ecological ethics to the Christian tradition of love and justice: "Loving nature is the key issue, it seems to me, in Christian ecological ethics and politics."[4] Only with such a grounding can ecological ethics maintain its Christian identity while offering guidelines to Christians and others on how to maintain ecological integrity.

Nash is explicitly interdisciplinary in his approach, incorporating history, theology, biblical studies, philosophical ethics, ecology, and the social sciences. He is aware of the hazards of drawing on so broad a range of disciplines, yet argues that to be authentic and relevant, Christian ethics demands dialogue with relevant disciplines and responsiveness to empirical problems. He is aware, as well, that frequently these disciplines also make implicit ethical assumptions and assessments in their own fields, only to hide them under the guise of value neutrality.

As noted in the comment at the beginning of the chapter, Nash argues that ecological ethics cannot be understood as simply a different application of Christian social ethics. It incorporates much from social ethics, but because of the much broader context of ecological ethics, some of the standard values and norms for social ethics must be reshaped or discarded. One implication of this is that if an ecological context becomes the normative framework for Christian ethics, there will need to be significant redefinitions of moral responsibilities and relationships within all branches of Christian ethics. Here Nash shows a far greater awareness than Hall that ecological ethics must shift from an anthropocentric to a primarily ecocentric framework.

Like Hall, Nash sees Christian theology and ethics as closely interdependent. Theology provides Christian ethics with its grounding and critiques ethics on its consistency with theological formulations. Christian ethics contributes to theology by critically evaluating the ethical implications of theological formulations. In addition to being interactive, theology and ethics should be coherent, internally consistent, adequately comprehensive, and integrative of experiential data.[5] Nash emphasizes reform over radical revisioning of Christian ethics. Citing the need for "wider palatability" and "the desire to minimize the risks of overstating the case," in his theological and ethical reflections Nash eschews radical approaches, putting emphasis instead on reform and pragmatism. Informed by the empirical reality of the ecological crisis, the movement in Nash's work is from identifying theological principles and moral norms to advocating their implementation in public policy.[6]

A strength of Nash's book is his acknowledgment of the role and importance of social location in ethics and theology. Claiming that "biography shapes ideology," he begins his book with an "ecological autobiography." In it he illustrates an early awareness of the ecology/economics dilemma—the tension between preserving nature and exploiting it to meet human economic needs—by weaving together the ecological and socioeconomic history and themes that informed his childhood growing up in the shadow of the western Pennsylvania steel mills. Nash uses this discussion of his early social location to ground concretely his ongoing concern for the ecology/economics dilemma that lies at the heart of the ecological crisis as well as to alert readers to the "pre-ethical prejudices" that shape his thinking.

Yet Nash explicitly opposes any forms of moral determinism that leave us enslaved to the original sources of our moral perspectives. The point, rather, is to become critically aware of them in order to transform those that are no longer adequate: "Moral formation should be followed by ethical justification and transformation. The challenge of ethics is to examine our inherited values rationally, to revise or eliminate these values coherently in accord with empirical data and ultimate commitments, and to make choices that are consistent with our transformed values."[7] Aware of the influences of his own social location, Nash seeks to follow this process of ethical reflection critically informed by a Christian perspective.

Implicit in the methodological discussion above are a number of presuppositions informing Nash's work that also are important for a liberationist ethic. Primary among these is that ecological problems are also moral problems: "They are fundamentally moral issues, because they are human-created and soluble problems that adversely affect the good of humans and otherkind in our relationships."[8] Ecological perspectives, dispositions, and actions assume moral values that can be evaluated as morally right or wrong.

Second, Nash argues for and assumes the logical coherence in most cases for the ethical extension of human ethics to the biosphere to include other creatures. In claims as diverse as extending the love of neighbor in the story of the Compassionate Samaritan to include all creatures, to extending human legal rights to nonhuman creatures, Nash makes a consistent case for extending several ethical concepts developed in the human social sphere to operate in the global biosphere.

Third, informing his work from the opening pages is the assumption of the inseparability of ecological and economic justice. Nash argues on both moral and practical grounds that ecological and economic considerations must always be held together to see how each affects the other.

Finally, Nash signals his sensitivity to anthropocentric versus ecocentric and biocentric assumptions in preferring *ecological* over *environmental* to describe the ethics he proposes: "The latter often seems to have anthropocentric connotations, suggesting moral concern only for the *human* environment, rather than for the context of all life."[9] Nash considers all three perspectives critical to develop an adequate Christian ecological ethics and finds the term *ecological* more inclusive of a broad range of perspectives.

With this overview of Nash's approach, let us turn now to the structure and content of his argument. Nash moves broadly through five sections from identifying and analyzing key moral problems present in the empirical dimensions of the ecological crisis to an examination of Christian traditional and contemporary theological and ethical responses in order to explore the implications of these theo-ethical foundations for Christian advocacy in public policy. In this summation of his work I pay particular attention to those arguments that serve to highlight the theological and ethical assumptions informing his ecological ethics.

Working from his assumption that ecological problems are moral problems, Nash focuses initially on the major moral implications of the facts of the ecological crisis, which he groups under the rubrics of *pollution* (toxic pollution, global warming, and ozone depletion) and *exceeding nature's limits* (resource exhaustion, overpopulation, maldistribution, reductions and extinctions of species, and genetic engineering).

Consistent with the approach of ethical extension, Nash argues that pollution is a violation of the ethical principle of nonmaleficence, the moral obligation to prevent or minimize harm of neighbor. He explicitly defines neighbor to include both human and otherkind. "Poisoning our neighbors and wasting common commodities are not matters of privacy or free marketeering or national sovereignty; they are serious moral offenses against others that demand public regulations and prohibitions."[10] The question of justice also means that resolving the problem of poverty is a critical part of any responsible solution to the problem of pollution, as the poor in both First and Third worlds typically are the most adversely affected and have the least options to avoid the toxic affects of pollution.

The problems of global warming and ozone depletion raise critical moral lifestyle issues as many of the products that contribute to depletion of the ozone layer come from products that satisfy human "wants," such as aerosol propellants and automobile air conditioners, rather than essential human needs. A basic ethical issue involved here is responsibility to future generations, both human and otherkind, that are imperiled by human overconsumption today. Taking an ecological paraphrase of John F. Kennedy's famous challenge, Nash closes this section with the imperative "Ask what you can do *without* for your country—and the peoples and other populations of your planet."[11]

In looking at the problems that result from exceeding ecological limits, a key analytical assumption is that on a self-contained planet there are no infinite boundaries or inexhaustible resources—everything is limited. Hence a primary moral response is *ecological prudence*: the "adaptation to the forces and restraints of nature that cannot be changed, no matter how sophisticated our technology becomes."[12] Overuse of nonrenewable resources and nonsustainable depletion of renewable resources raise again questions about responsibility to future generations: how should such resources be apportioned and how far ahead should moral consideration be extended?[13] Taking seriously the ethical responsibility for future generations means that economic systems that stress sustainability and frugality are essential.

In looking at the environmental effects of population growth and maldistribution of wealth and resources, Nash makes explicit the linkage between economic injustice and ecological degradation. Hence, consistent with a liberationist approach, he argues that strategic responses to these problems must be linked: making a preferential option for the poor must also mean a preferential option for ecological integrity, and vice versa. Key to this moral analysis is distinguishing between pollution from poverty and pollution from prosperity. While the two are often linked—the excessive use of the world's finite resources by the wealthy contributes to the desperate lifestyle and resulting environmental degradation of many of the world's poor—pollution from prosperity often also reflects moral inequities.

Finally, in looking at the rapidly accelerated and now massive rate of human-induced extinctions and reductions of species, Nash understands the key moral question to be, why protect biodiversity? He notes that many of the most commonly heard answers are anthropocentric, reflecting solely instrumental assumptions about the nonhuman world: biodiversity, the argument goes, should be protected because it has large, untapped potential for human welfare. Nash acknowledges that these approaches are valid as far as they go, but he argues by themselves they are insufficient. As unique works of God and/or evolutionary process, other creatures have intrinsic value in themselves apart from their instrumental value to humankind.

Nash concludes the first section with a list of nine values or "ecological virtues" that can help to make ecological integrity a reality. Nash recognizes they are based on a mix of anthropocentric and ecocentric assumptions, yet sees all as both necessary and interactive: (1) *sustainability* demonstrates concern for long-range intergenerational equity; (2) *adaptability* implies living within ecological limits; (3) *relationality* is the acute awareness that all things are interconnected and have consequences for everything else; (4) *frugality* applies primarily to the prosperous and is the key to sustainability, equity, and biodiversity; (5) *equity* is an anthropocentric value synonymous with justice in the distribution of the world's goods and services; (6) *solidarity* today means global solidarity—seeking to institutionalize economically and politically global responses to the ecological crisis; (7) *biodiversity* is an ecocentric value that seeks to save other creatures for their own sake, extending solidarity to the whole biosphere; (8) *sufficiency* means political and lifestyle solutions proportionate to the intricacies and magnitude of the problems; and (9) *humility* is the guiding norm for all the other virtues, as it recognizes the limitations on human knowledge, technological ingenuity, moral character, and biological status.[14] Nash concludes with a critical question that introduces the following chapters: Can Christian theology and ethics support and nurture these ecological virtues?

Before outlining his ethic, Nash examines the classical Christian tradition to see whether the theological resources are there to undergird an ethics of ecological integrity. He devotes a chapter to responding to the commonly voiced critique, first popularized in Lynn White's 1967 essay "The Historical Roots of Our Ecologic Crisis," that the Christian faith itself is one of the main culprits in the ecological crisis.[15] Nash then reexamines Christian doctrine in light of his claim that the Christian faith contains all things necessary for ecological integrity. He grounds his reflections in contemporary biblical exegesis of selected portions of Scripture and in classical expressions of traditional Christian doctrines, and finds that they do indeed provide "a firm theological foundation" for an ecojustice ethics.[16]

From the doctrine of creation, Nash turns to the classical creeds' affirmation of God as Maker of heaven and earth. Despite widespread Christian practice to the contrary, he argues that the doctrine does not permit a nature/grace dichotomy. Rather, the whole of nature is a manifestation and beneficiary of God's grace. This in turn endows all of nature with an intrinsic moral significance. The radical monotheism of Christian faith means that only the Creator is worthy of worship, but nonetheless all creatures are worthy of moral consideration. By virtue of having been created by God and pronounced as good, nature is still sacred. The Genesis accounts testify to the kinship of all creation: all are created by God. Against the charges of White and others, he observes that divine valuations of creation in Scripture are primarily cosmocentric and biocentric rather than anthropocentric. An important task of Christian ecological ethics, therefore, is to "determine the moral responsibilities entailed by the reality of theocentric and ecological kinship."[17]

In linking creation with redemption, Christian faith affirms that the goodness of creation is an interim goodness still in need of liberation and reconciliation:

> For the Christian faith, however, the affirmation of the goodness of creation is also an expression of ultimate confidence in the goodness of God. The world now has an interim goodness. It is not to be despised or rejected or transcended; it is to be appreciated and valued as an expression of the goodness of God. . . . Yet it is also a world of systemic alienation, in which all life is temporary and destructive of other life. Empirically speaking, the classical theological propositions that the Creator and the creation are "very good" are virtually indefensible, in my view, apart from an eschatological expectation. The creation needs liberation and reconciliation. Thus the Christian church has always linked creation and redemption, though in most of its historical forms, it has strangely excluded otherkind from the realm of redemption.[18]

From this Nash argues that the eschatological expectation that all creation will be brought to fulfillment by a good God combined with Christian covenantal responsibility to God in turn further enhances Christian responsibility for ecological protection and integrity.

Nash interprets the Noachic covenant in Genesis 9 as an "ecological covenant of relationality" that demonstrates a cosmic covenant built into the earthly ecosphere. While the story of the flood is morally ambiguous, it does testify to a rational order of interdependence in creation that demands respectful adaptability from moral agents. In addition, in the Noachic pledge, God covenants to preserve all species and their environments. This divine promise entails human obligation because faithfulness to God means loyalty to God's covenants. Ecologically this means assuming responsibility to future generations, human and otherkind, as well as respect for relationality, since all creatures are kin.

Nash develops the theme of representation by exploring the incarnation of Christ as cosmic representation as portrayed in several of the Pauline epistles. Consistent with his principle of ethical extension, Christ as the representative of humanity is also the representative of the biosphere—indeed, the entire universe. The ecological implications are clear: "The Incarnation confers dignity not only on humankind, but on everything and everyone, past and present, with which humankind is united in interdependence. . . . It sanctifies the biophysical world, making all things and kinds meaningful and worthy and valuable in the divine scheme."[19]

The incarnation as the representative of the entire cosmos also argues against any human arrogance of exclusivity that might be derived from the doctrine of *imago Dei*. Like Hall, Nash places emphasis on what the doctrine says about relationality rather than ontology. Humanity's vocation rather than status results from the *imago Dei*. Nash argues against the anthropocentric tradition that sees the dignity of the biophysical world as derived from humanity's dignity. Rather, "it is more accurate to say that any dignity ascribed to humanity is derived from our natural history" because all creaturely dignity comes from having been created by the Creator.[20]

Nash examines the doctrine of the Spirit to understand God's relation to the natural world. Drawing on Scripture and the Nicene Creed, he affirms that God exists *in* Creation as the Holy Spirit. The indwelling of the Spirit makes the biophysical world both sacramental and sacred, but it does not divinize it. Rather, "nature is sacred by association, as the bearer of the sacred." As opposed to pantheism, which identifies the divine with nature, sacramentality de-divinizes nature. Yet as the Spirit's sacred habitation, the cosmos demands moral respect and responsibility, both for its own sake and for the sake of humanity's physical and spiritual well-being.

In looking at the doctrine of sin, Nash notes the perennial problem in Christianity of defining the meaning of sin too narrowly, often to the point of triviality. Yet sin is the most appropriate context for understanding the ecological crisis and the range of human actions that have induced it. Given the current context, therefore, our understanding of sin must be extended to cover ecological misdeeds and the human condition underlying them. Here Nash follows the theological analysis of Reinhold Niebuhr who saw the root sin as human pride or egoism that leads human beings to declare autonomy from God. In the current context, sin may be understood as alienation between humans and God and all the creatures of God. Ecologically, then, sin is "the refusal to act in the image of God, as responsible representatives who value and love the host of interdependent creatures in their ecosystems, which the Creator values and loves."[21] The answer to awareness of sin in resolving the ecological crisis is not some romantic utopia that ignores the flawed character of humanity, but perpetual vigilance and sufficient and ongoing reforms.

In examining the doctrine of redemption, Nash draws on the Christian tradition of consummation as cosmic redemption to argue against dualistic understandings of redemption that understand human redemption as release from nature. Historically, a minority Christian tradition has argued that the promise of Christ is not redemption *from* the world, but the redemption *of* the world. It argues that the divine purposes are cosmocentric and biocentric, rather than simply anthropocentric. Against postmodern and process thinkers whom Nash sees substituting present abundant life for the promise of eternal life, Nash argues that eschatological hope is central to Christian faith and ecological ethics. An eschatological perspective both protects the intrinsic value of creation against present human instrumental valuations as well as preserves the moral character of God in the face of all-to-real present injustices. "The key [moral] issue in the hope for cosmic redemption is the moral character of God in relation to God's valuation of creatures as ends in themselves."[22] The moral and ecological responsibility of humans is to work to approximate the eschatological vision of the new creation to the fullest extent possible under the constricted conditions of the creation.

Nash concludes with an eleven-point summation of what a "reasonably reformed Christian theology" offers as a foundation for ecological integrity:

1. All creation is valued by God; Christians are called to "practice biophilia."
2. Christian faith de-divinizes but also sacralizes nature.

3. Christian faith affirms ecological relationality which shapes our moral responsibilities.
4. Humans have both rights and responsibilities governing the use of biophysical goods. This implies practicing prudence, frugality, and fairness "in respect for our coevolving kin."
5. The biophysical world has *interim* goodness in experience and *ultimate* goodness in hope.
6. Christian faith counsels *human humility* in the face of divine and biophysical reality.
7. Human dominion means responsible representation as a dimension of the image of God. It is judgment on exploitation and abuse of nature.
8. Ecological negligence is an expression of sin.
9. God's ecological judgments against ecological sin call for ecological repentance.
10. Ecological responsibility is an inherent part of the ministry of the church.
11. Christian faith offers solid support for all the nine ecological virtues outlined at the end of the first section.[23]

Having established the adequacy of Christian doctrine for a foundation for ecological ethics, Nash turns now to the task of interpreting the central values of the Christian faith—love and justice—in an ecological context. Nash considers love to be the integrating center of Christian faith and ethics; therefore any Christian ecological ethics is seriously deficient if it is not grounded in love. What makes this challenging and difficult, Nash argues, is "the tragic condition of existence in a predatorial biosphere." The nature of ecological interdependence is that every species must feed on and struggle against other species in order to survive. It is no different for human beings: "As ecological predators and exploiters—as well as prey and hosts—humans must kill and use other life-forms and destroy their habitats if we are to satisfy basic human needs and exercise our peculiar endowments for cultural creativity."[24] This tragic condition can and must be morally restricted, Nash argues, but it cannot be avoided.

Turning to the nature of Christian love, Nash asserts that if the value of all beings is grounded in love, all creatures have intrinsic value and therefore deserve to be treated not merely as means to human ends, but as ends in themselves. To be in God's image means to reflect the ultimate lover, to love all that God loves. Hence in rooting an ecological ethics in Christian faith it is reasonable to extend the logic of love to the whole creation so as to represent God's love and care. Nash argues for a moderate definition of love, seeing it as a relational concept and initiative that seeks to establish connections and build caring relationships.[25]

Several dilemmas result, however, from grounding ecological ethics in love. Aside from the fact that virtually all previous debate on the nature of Christian ethics rooted in love has assumed divine/human and interhuman relationships, two additional factors make ecological ethics complex: (1) damaging and killing other creatures and their habitats are biological necessities, and (2) human relations with other creatures are between unequals. Acknowledging this, Nash follows a Christian realist starting point in observing that in a biophysical world of predation and unequal relationships, "love is always compromised."[26] Natural or nonmoral evil is built into the ecosphere bringing an inherent tragic dimension to creation quite apart from any human moral blame.

Against those who would romanticize nature, Nash maintains that the world is a morally ambiguous reality. Predation is a primary condition of human and other creaturely existence. Recognizing this is critical to ecological ethics, for "we are not a special creation, a species segregated from nature. This is bad biology which leads to bad theology and ethics."[27] Instead, the ethical problem involves making discriminate judgments to discern the most love-embodying acts and principles under natural conditions circumscribed by necessary evil. To act in such a morally constricted context, Nash proposes that humans act as *altruistic predators:* "beings who seek to minimize ecological harm that we inevitably cause and who consume caringly and frugally to retain and restore the integrity of the ecosphere."[28]

Nash describes three qualifications of Christian love in an ecological context. First, it cannot be considered "equal regard" because that assumes ontological equality between humans and other creatures which Nash explicitly rejects.[29] Yet the intrinsic value of other creatures as created by God and coevolved with humanity leads to a high degree of regard for otherkind. They are never to be seen or treated merely instrumentally. Second, in a context of ecological inequality, ethics cannot restrict Christian love to sacrificial love. Rather love is relational; it is a means whose intent is to create and enhance caring and sharing relationships. Nevertheless, Nash argues that an element of self-sacrifice is inherent in every form and context of love. In the context of ecological crisis, this leads to the unavoidable question of what human interests and benefits must be sacrificed in order to serve the needs of other creatures and protect the health of the biotic community? Third, some forms of Christian love that presume interaction between moral agents (such as forgiveness) are inapplicable in an ecological context. A result of these three qualifications is that Christian love in an ecological context necessarily will be less rigorous than in human social relations—yet it still makes serious ethical demands on human beings in ecological interactions.

From his exploration of the nature of Christian love, Nash ends this chapter with a list of six ecological dimensions of love.[30]

1. *Beneficence:* Love as Beneficence ecologically means being protectors of the biosphere.
2. *Other-Esteem* is an expression of love as eros, and helps to combat anthropocentrism in an ecological context. It rejects managerial images of humanity as gardener or zookeeper of the biosphere.[31]
3. *Receptivity* moves beyond other-esteem to express yearning for relationship to both give and receive from others. In contrast to modern notions of "self-sufficiency," receptivity recognizes and celebrates the interdependent and kinship relationship between humanity and the rest of the earth.

4. *Humility:* Love as humility rejects both self-aggrandizement and self-deprecation of humanity with respect to otherkind. Recognizing our origins and destiny in humus and our connection to all creation, it regards all creatures as worthy of moral consideration.
5. *Understanding* means loving God, and therefore God's creation, with whole mind. It recognizes that sound knowledge about ecological dynamics is critical for ecological love.
6. *Communion* is the full extension of love as other-esteem and receptivity. "Communion not only wants the loved ones to be in their distinctiveness; it wants them to be our loved ones in fully reconciled relationships."[32]

Nash devotes a chapter to a more thorough exploration of the meaning of love as justice in an ecological context. He starts by asking whether ethical extension into an ecological context is possible with Christian and societal understandings of justice: "Are concepts of justice applicable to beetles?" Hence the framing question for the chapter becomes, "What is love as justice, and what, if anything, does it involve for human and biotic (or organic) rights as well as human ecological responsibilities?"[33]

Consistent with his commitment to ground ethical norms in scripture, Nash maintains that Christian responses to ecological problems must be developed in the light of biblical commitments to justice. He notes that most biblical statements dealing explicitly with justice apply only to divine/human and intrahuman relationships, yet sees no inherent reason why they cannot be extended to the rest of nature: "The Lover of Justice sets no boundaries on justice. The gospel we are called to incarnate relates to *all* creatures in *all* situations."[34]

Turning to the tradition, Nash draws on Reinhold Niebuhr's analysis of love as agape (sacrificial love), realized only imperfectly under limitations of historical ambiguity through mutual love, including justice. Nash concludes that while love and justice are distinguishable, they are inseparable; love demands more than justice but also demands no less than justice. By justice Nash means primarily distributive justice, "the proper apportionment or allocation of relational benefits and burdens."[35] Justice is discovered only contextually, and carries with it rights and responsibilities that all moral agents have the obligation to respect. Rights, including human rights and legal rights, give content to justice, and can be overridden only for compelling moral reasons. Nash argues that, at least in Western cultures, rights are important, for having no rights suggests having no moral consideration. Increasingly people and societies are recognizing human environmental rights in the form of guarantees to environmental sustainability and integrity as an essential part of human rights. Nash sees this as wholly consistent with Christian faith and ethics.

More controversial is whether rights can be extended outside the human community to biotic rights. After discussing some of the conceptual problems of such a move, Nash argues that biotic rights do in fact exist—not as mere human constructs but in recognition of the moral claims that inhere in living beings. Given the intrinsic value of all creatures, advocacy for biotic rights is not only an expression of benevolence, but also an obligation of justice.

Finally, examining the debate between animal rights activists and ecological holism, Nash maintains that a Christian ecological ethics must preserve the creative tension between moral respect for individual organisms and holistic respect for communities and species. Against biotic egalitarianism, which places all species on the same moral plane, Nash argues for a graded model where all species have intrinsic value, but not equal intrinsic value:

> Yet, it also seems important to avoid biotic egalitarianism and to insist on a gradation of value among rights-bearers. . . . Here the same criteria are extended to all species, creating an ascending/descending scale of intensity for rights and responsibilities. Among species, the moral significance of rights is proportionate to the value-experiencing and value-creating capacities of their members, and the corresponding responsibilities of moral agents are proportionate to this significance. Other things being equal,

> sentient creatures, for instance, are to be preferred over nonsentient ones in conflict situations.[36]

To answer charges that this is merely hierarchicalism and "speciesism," Nash argues that there is an important moral distinction between *intra*-species rights, human rights that apply to all humans as moral equals, and *inter*-species rights, "which are not equal and which do not prevent, though they clearly restrict, humans from destroying members of other species."[37] Nash contends that a graded model is a more accurate reflection of ecological insights into the biophysical world as well as faithful to biblical understandings.

Nash concludes his discussion of rights with a proposed Bill of Biotic Rights.[38] Nash considers these to be prima facie rights—rights that we have strong reasons for honoring unless there are compelling moral reasons not to do so. Biotic rights may be overridden only for just causes and then only within the limitations of proportionality and discrimination.

Two further observations conclude Nash's discussion of biotic rights: First, while a biocentric/ecocentric ethics that extends rights to nature may be seen by some as excessively sentimental, what is key is its ethical sensitivity in moving beyond a solely anthropocentric framework. Second, moral entitlements that are not also social entitlements have relatively little functional value in Western industrialized societies—they must be structured into the ethos, policies, and laws of human communities.

The final chapters of Nash's book indicate the focus of his praxis in constructing an ecological ethic. He pays more attention to participation in politics than to changing values and lifestyles (although he acknowledges all three are important), and focuses Christian praxis on public participation in policy formation on ecological issues. To "the strict constructionists and sectarians who yearn for ideological and verbal purity in the environmental movement," Nash responds that they would do well to pay more attention to actual lived values and commitments and participation in the risks that come from specific actions and policies than the meaning of particular words such as "steward."[39] Grounding his praxis in policy-oriented, Western society, he focuses on legal rights and social entitlements as the best strategy to protect ecological integrity. Central to this are six ecological middle axioms that can guide Christian participation in policy debate, particularly resolving the economics/ecology dilemma and approaching all problems holistically and relationally.[40]

Nash ends this final section and the book by dismissing both optimism and pessimism as inappropriate responses. Rather, "the best we can do is hustle and hope . . . [and] struggle in the confidence that with each step forward, God the Politician and the Lover of life is ever creating new possibilities to realize the integrity of God's—and our—beloved habitat."[41]

How does Nash's ethics of ecological integrity and Christian responsibility compare to the goals and assumptions of a liberationist ecological ethics? To answer this, as in the case of Hall, it is helpful here to look more closely at the methodological approach of Christian liberalism on which Nash draws and the resulting views of humanity, nature, God and their interrelation, and then assess these in light of the criteria for an ethic of Gay and Gaia outlined in the previous chapters.

Nash's approach to ethics is perhaps best described as belonging to the progressive end of Christian liberalism and realism. In contrast to the God of liberation theologies who actively intervenes in history on behalf of and with the oppressed to create a more just society, Nash places more emphasis on the eschatological dimension of God's future justice that vindicates the moral dimension of God while calling Christians to responsible ethical behavior in the present. Infusing Nash's analysis is the realist emphasis on the sinfulness of humanity that conditions and limits all human efforts toward justice and guards against utopian or romanticized visions of social harmony. Hence his is an ethic of significant reform within the pragmatic realm of what is possible rather than a fundamental revisioning and restructuring of current social realities:

> We cannot, however, infer reality from hope. . . . Cosmic redemption seems essential as a matter of internal consistency in Christianity, but it provides no argument for external correspondence with reality. . . . In the midst of the moral ambiguities of creation, we can experience only promising signs—not the full harmony—of the New Creation, the Peaceable Kingdom. The very fact that eschatological visions are necessary precludes romantic illusions about historical possibilities. Nevertheless, the vision represents the ultimate goal to which God is beckoning us. Our moral responsibility, then, is to approximate the harmony of the New Creation to the fullest extent possible under the constricted conditions of the creation. The present task of Christian communities . . . is to anticipate and contribute to the promise of ultimate liberation and reconciliation in human communities and with the rest of nature.[42]

It is clear from this comment that Nash distances himself from neoliberal followers of Christian realism who align themselves with societal power to provide an apologetics for the status quo as the best available compromise given humanity's sinful condition.[43] He is aware that "realism can degenerate into an apathetic acceptance of the status quo when it does not allow and press for the extension of the parameters of the politically possible."[44] Instead, citing recent rapid social change in the former Soviet bloc as an example, Nash contends that Christian ecological ethics should be strategic in assessing what is realistically feasible given current societal constraints, while remaining open to potentialities previously declared unrealistic.[45]

Nash's realist, Neo-Orthodox hermeneutics shapes his use of the Bible and Christian tradition. In exploring the ecological implications of the classical doctrines, he begins each discussion with explicit discussion of biblical verses that ground the doctrine. Realist themes infuse his exegesis: the story of the flood in Genesis 6–9 is "morally ambiguous"; divine promise in the Hebrew scriptures always entail "human obligation" and "responsibility"; the creation is a world of both "interim goodness" and "systemic alienation"; some "domination *of* nature by humans is necessary to prevent the domination of humans *by* nature," and so forth.[46] In addition to the Bible, Nash draws heavily on the themes and emphases of the classical Christian doctrines and creeds that emerged as authoritative in the Protestant Reformation. His presupposition and contention is that the vast majority of authentic doctrine is compatible and consistent with the findings of ecology and the other natural sciences when reinterpreted, revised, and extended in light of contemporary ethical insights and ecological data.

A major strength in Nash's liberal ethics of responsibility is his explicit use of the physical sciences, particularly ecology and biology, to engage the Christian scriptures and tradition. Because of his theological and ethical claim that the ordered world as created by God has moral meaning, science can help us understand the world in order to better understand its moral implications. He maintains that Christian teaching should be reinterpreted, revised, and extended in light of the findings of ecology, and that when this occurs, the vast majority of authentic Christian doctrine will be found to be ecologically compatible with the physical sciences. Thus, in contrast to Hall, Nash draws explicitly on the insights of science—particularly ecology and evolution—to inform and complement his biblically based understandings of nature, human nature, and their relationship. There is a much closer interweaving of insights from the physical sciences with those from the Bible and tradition in formulating norms and principles to guide his ethics than in the stewardship model. It results in ecologically sound views, such as humanity existing fully within nature and the limits of dominion as a model for understanding the historic evolution of humanity in its relation to the rest of nature.

The primary problem with Nash's model from a liberationist perspective lies in the justice side of the ecojustice dialectic. Like Hall, Nash does not incorporate a deconstructive, critical phase in his approach to his sources, particularly to classical Christian teaching rooted in the Bible and tradition. While he argues for reinterpreting and revisioning Christian doctrine in light of contemporary ecological data, there is no parallel commitment to a similar revision or reinterpretation in light of the voices of socially marginalized perspectives—particularly those where the Christian tradition has contributed to their marginalization. Hence, while distributive justice is an important norm in his ecological ethics and the situation of the poor and impoverished is always part of his social and ecological analysis, there is no epistemological privileging (or, in some cases, even acknowledgment) of these perspectives in his theological and ethical analysis.

Thus, while Nash draws heavily on the ecological sciences to indicate the variety of ways humanity's relation to the natural world has become distorted, he makes very little use of the social sciences or cultural criticism. This has two important effects methodologically. First, it ignores the perspectives of women and racial, ethnic, and sexual minorities that highlight relational inequities within the human community, critical to any liberationist ethics of ecojustice. Second, it hides the historical, anthropological, and sociological presuppositions that inform Nash's social and theological analyses. Because Nash draws so little on the social sciences, he also lacks an adequately critical view of the physical sciences. He has a good sense of how ecological insights must engage and inform theology and ethics, but shows little critical awareness of the socially constructed nature of science and the role it has played and continues to play in the human justice side of ecojustice in such debates as those on race, gender and sexuality.

Another strength of Nash's approach is the responsibility he takes for examining, reclaiming, and owning the Christian environmental legacy in the face of complaints such as those voiced by Lynn White Jr. He begins this recovery with an explicit acknowledgment and confession of "guilt" for Christianity's responsibility for part of the current ecological crisis, but then moves on to set forth a more nuanced and complex understanding of Christianity's role and stance, particularly by drawing on minority strands within the tradition that are more earth-centered and friendly.

This, however, illustrates another problem with Nash's approach that stems from ignoring the epistemological claims of the socially marginalized: conspicuously absent from the minority traditions Nash recovers are the voices of women and racial, ethnic, and sexual minorities. The result is an uncritical recovery of strands of Christian faith that have "coexisted comfortably and coherently with ecological values."[47] For example, in response to the ecological complaint against Christianity, Nash draws on the work of Paul Santmire to highlight the "ecological motif" in the thought of Christian thinkers such as Augustine, Luther, and Calvin. Yet how ecological can this motif be, when the persecution of earth-centered folk—typically women, ethnic, and sexual minorities—all too often was justified by the theological thought of these same Christian leaders?[48] I am not arguing against examining the thought of writers such as Augustine, Luther, and Calvin for recoverable resources that can aid us today, but rather against an uncritical recovery that ignores the voices and experiences of the socially excluded that can give us a more complete and complex understanding of the many dimensions of the Christian tradition—and one that includes all sectors of humanity in the ecology we seek to protect.

When Nash does on occasion draw on the work of women and minorities, he pays little attention to the central themes of their work. Thus, for example, he cites Carolyn Merchant's work in *The Death of Nature* to refute much of the ecological complaint against Christianity as overly simplistic, yet only in a footnote does he mention that, "Merchant rightly stresses the historical and ideological connections between the subjugation of nature and the subjugation of women."[49] Merchant's gender analysis of the scientific revolution is not peripheral, but rather the central thesis of her work. Yet Nash's own ethics shows very little impact or influence by feminist ecological critiques of gender.

Finally, a telling result of Nash paying little attention to liberationist voices of critique from the margins is his decision to develop his ethic primarily within the traditional confines of Christian doctrine and biblical witness. This follows both from his conviction that the Christian tradition as it stands can provide an adequate moral grounding for an ecological ethics, and his decision to chart a middle ground that may appeal to practitioners of mainstream Christianity. Yet this is highly problematic for a liberationist ecological ethic, as the insistence of the many lesbian and gay voices cited previously on the need to open the canon attest. More convincing is Rosemary Ruether's argument that *all* ancient religious traditions are inadequate as they stand to support an ethics of ecojustice because they were formed under patriarchy and under conditions vastly different from the global ecocrisis and awareness we face today. Hence they all, including Christianity, need significant reworking to form an adequate grounding for ecological ethics. This means learning from and paying attention to voices from outside our own communities with whom we stand in multiply situated power relations. The liberationist attention to power and difference is a critical part of this task, but is missing from most of Nash's ethics.

Hence Nash's lack of awareness or acknowledgment of the socially constructed nature of either theology or science is a significant shortcoming in using his work to ground a liberationist ecological ethic. He is most interested in finding an ethic that can support needed public policy changes, but his approach precludes developing a radical perspective and critique of the political, economic, and cultural structures that have produced the ecocrisis. By not acknowledging constructionist insights on power differences, Nash implicitly accepts the epistemological limits of the Christian tradition and Western science and works for a reformed perspective within these.[50] This approach is understandable for persons coming from social locations of power and privilege, but it is inadequate for the socially marginalized. Despite the many valuable contributions and insights of Nash's ethics, in this area of method it lacks a sufficiently critical perspective to ground a liberationist ecological ethic.

The ambiguous character of Nash's methodological approach for contributing to a liberationist ecological ethics is reflected in the views he develops of humanity, nature, and God. To spell out his understanding of human nature and humanity's place in nature, Nash turns to the Genesis creation accounts and the theological doctrine of *imago Dei*: Human beings are both created by God and created in the image of God and therefore have sacred, inherent worth. Nash draws on ecological and theological insights to argue that human beings are not a special creation, however. In contrast to Hall, Nash argues human beings must be understood as fully *in* nature: "Humans are totally immersed in and totally dependent on the biophysical world for our being. We cannot talk about humans *and* nature, but only humans *in* nature. We have evolved with all other creatures through adaptive interactions from shared ancestors. We are biologically (and theologically) relatives—albeit remote—of caterpillars, strawberries, the dinosaurs, the oaks, the protozoa, and all other forms of being."[51]

Human beings are unique as are other creatures, and while human beings undoubtedly have qualities and powers superior to other creatures, other creatures have some qualities and powers superior to those of human beings. Human superiority over other creatures is therefore restricted and not rigidly demarcated. Nevertheless, human rational and moral powers so radically exceed the powers of any other creatures that they must be considered significantly different qualitatively. There are important moral and ethical implications that derive from this basic biological fact. Even so, Nash argues that human beings do not "transcend" nature: "We can never transcend nature, contrary to that mainstream theological tradition which contrasted nature and spirit. Human psychic/spiritual capacities are not additives to nature, but derivatives from nature. In history, we are inextricably immersed in nature. We can, however, transcend some instinctive necessities and realize some of the rational, moral, and spiritual *potentialities in* nature, far beyond the capacities of any other creature."[52]

What is important from this reflection ethically is that human beings are the only creatures with moral agency—the relative freedom and rational capacity to transcend instinct sufficiently in order to define and choose good or evil, right or wrong. Human nature becomes distorted through sin when we exercise our agency for evil, alienating ourselves from God (and otherkind) by appropriating autonomy and sovereignty for ourselves. Yet a theologically and biologically realistic human self-understanding rejects both the self-aggrandizement of humanity over nature and the self-deprecation of human beings as less than or just another part of nature, ontologically and intrinsically equivalent to a rock or a mosquito.

Here Nash's ethics addresses a number of the deficiencies found in Hall's ethics of stewardship. Key among these is Nash's reliance on insights from evolution, ecology, history, and anthropology to inform his theological analysis of human nature. Hence, in contrast to Hall, Nash sees human beings as fully part of nature. Apparently unique human qualities, such as our rational, moral, and spiritual capacities, are in fact expressions of potentialities that exist *in* nature, not apart from it. Nash argues this is a basic biological fact, with important moral and ethical implications that derive from it. This use of scientific insights to engage critically the theological tradition is a necessary part of a liberationist ecological ethics, and Nash provides several constructive examples of how it can help to rethink assumptions undergirding ethics.

As with Hall (although for different reasons), it is in Nash's theological analysis of humanity that liberationist insights from social constructionism reveal weaknesses in his model. Somewhat surprisingly, given his attention to the importance of social location in how we understand the ecology/economics dilemma, Nash seems blind to the importance of social location in his own theological analysis of humanity. This stands in sharp contrast with his analysis of the ecological crisis. There Nash integrates the perspectives and concerns of socially marginalized persons and pays primary attention to issues of distributive justice in resolving the ecocrisis. These perspectives are conspicuously absent, however, in his theological analysis.

A clear example of this is Nash's understanding of human sin, which forms the theological core of his understanding of humanity's relation to the ecocrisis. From Christian tradition he sees the root of sin as egoism and pride. Nash follows Christian realism in seeing the question of power as central to understanding sin, but too often it is a view of power divorced from awareness of the importance of social location in understanding it. Hence sin is primarily pride of power, "a declaration of autonomy from the sovereign source of our being."[53] It comes from taking ourselves too seriously, making ourselves the central frame of reference. Ecologically, sin is our refusal to act in God's image as responsible representatives who value and love all God's creatures.

The problem with these understandings of sin is that they assume a universalized human experience of autonomy and agency that is in fact not universal. They may be appropriate understandings of sin for those whose social location reflects a position of power or access to power, but sin may look very different for those at the bottom of the social hierarchy. For people in power, understanding the root of sin as egoism and pride—taking one's autonomous self too seriously—is appropriate because it alienates them from God by violating creaturely relationality and causing injustice. For socially marginalized or powerless persons, however, sin is often rooted in the *inability* to take ourselves seriously, in not having enough "pride." This is the point expressed in slogans such as "Black Power" and "Gay Pride." Often dismissed as sinful expressions of prideful self-assertion, they instead are recognitions that without a minimum level of moral agency and autonomy there can be no participation in full humanity. Rather than sin being a declaration of autonomy from God and others, sin for marginalized people may be remaining in destructive patterns of dependency and victimization.

Nash is insightful at one level when he names ecological sin as "grasping more than our due,"[54] whether at the level of individuals, corporate bodies, nations, or species. Again, however, this claim is problematic when it assumes a universalized human experience. From a liberationist perspective, sin for the oppressed is at least as much a willingness to settle for *less* than our due, for less than full participation in humanity, than overstepping our bounds.[55] Here "refusing to act in God's image" means not claiming our full humanity, rather than overstepping its limits. For lesbians and gay men, this has meant not seeing our experience as an integral and valid part of nature and human nature, giving in instead to internalized homophobia and believing ourselves to be inferior to heterosexual persons. The social constructionist insights of scholars such as Brooten and Weeks form a critical step in exposing hidden power assumptions in our understandings of humanity and nature, and must be fully incorporated into theological and ethical analysis.

In contrast to Hall, Nash gives extensive attention to how we understand nature. He defines nature most simply as the biophysical world, which includes human beings. He grounds his understanding of nature in his mixed approach drawing on ecocentric, biocentric, and anthropocentric perspectives. He thus argues that nature and its creatures must be seen to have both intrinsic and instrumental value: "the biosphere cannot be reduced to an instrument for human welfare; it has its own integrity which, in the final analysis, is also the integrity of the human species as part of nature."[56]

Nash maintains that while all nature has intrinsic moral value, an important characteristic of the biosphere is that all species necessarily also have instrumental value to other species. Ecologically, plants and animals provide essential food for other creatures, including human beings. Theologically, parts of nature with instrumental value to human beings can be understood as gifts to be used without guilt, but with frugality and awareness that their necessary instrumental value does not remove or reduce their intrinsic worth as created by God.

For Nash the inherent value of nature is best understood as grounded in Christian understandings of love: "The value of all beings is objectively and ultimately grounded in Love, and all deserve, therefore, to be treated not merely as means to human ends, but as ends in themselves."[57] Divine love manifests itself through grace, which too often has been understood in Christian circles to operate only within the human realm, redeeming human beings *out of* nature. Drawing on the work of Joseph Sittler, Nash calls for the elimination of the traditional theological dichotomy opposing nature and grace in favor of seeing the whole of nature as an expression *and* beneficiary of grace: "The elimination of a nature/grace dichotomy, and its replacement with an understanding of nature as a manifestation and beneficiary of grace, endows all of nature with an intrinsic moral significance."[58]

Grounding the intrinsic worth of nature in Christian understandings of divine love is buttressed for Nash by the doctrine of creation, which affirms that the whole of creation is seen by God as good, apart from its instrumental worth to humanity. Divine valuation of nature hence is cosmocentric and biocentric, not primarily anthropocentric, as distorted understandings of human dominion often presuppose. Yet there is an ambiguous reality to nature: evil is an inherent part of the system. Ecologically, nature is built on predation, which presupposes suffering and destruction of other life-forms. Theologically, nature is part of a fallen world of systemic alienation. Hence Nash argues from the Christian doctrines of creation and redemption that creation has an interim goodness that still awaits the fullness of liberation and redemption.

Theologically, therefore, Christian ecological ethics must maintain an eschatological perspective. There is a moral ordering to the world, and sin manifests itself as disdain for the limits and relationships inherent to that order. Yet relationships in nature are also morally ambiguous, both harmful and helpful, destructive and creative. Ironically, Nash notes, the ambiguous relationships among the individual parts, various creatures, are essential for the wellbeng of the wholes, ecosystems and the biosphere. A Christian ecological perspective understands this as part of the interim goodness of creation awaiting the eschatological fullness of redemption.

With respect to its views of nature, Nash's ethics of responsibility is more nuanced and developed than Hall's stewardship ethics, and avoids some of its pitfalls. He more consistently critiques the vestiges of anthropocentrism in Western views of nature. Also helpful is Nash's insistence that humanity be understood to exist fully *in* nature, recognizing that our rational, moral, and spiritual abilities are potentialities that exist in rather than transcend nature. Here Nash moves Christian ethics a long way toward overcoming the rigid culture/nature dichotomy that has typified Western thinking.

Yet it is just in this area that Nash's work, while a substantial improvement over Hall, still shows significant shortcomings. Despite his strong emphasis that humanity must be seen as fully part of nature, and any limited human transcendence takes place within nature, a subtle nature/culture epistemological dualism continues to inform and shape his work.

Consider the images in his discussion of the nature/culture relation involved in dominion. After arguing for a helpful narrow interpretation of dominion—the protection of the earth and its creatures by humans against human exploitation[59]—Nash goes on to maintain: "None of this denies, however, that humans must 'subdue'—yes *trample, conquer,* and the other strong connotations of the Hebrew word *kabash* . . . the earth's resistance in order to survive and maintain cultures."[60] Given Haraway's insights on how metaphors and images both constrain and enable human possibilities, Nash's choice of such coercive and violent verbs to describe culture's relation to nature is telling. Nash's protests against dominion including an exploitative component are contradicted here by the value-laden emotional content of actions such as trample and conquer. As ecofeminists and Native Americans have argued, language that acknowledges humanity's need to alter its environment, but evokes care, respect, cooperation, and gratitude is more appropriate in an ecological attitude and ethics.

In maintaining this view, Nash argues that some degree of domination *of* nature by humans is necessary to prevent the domination of humans *by* nature. Ecofeminist Janis Birkeland critiques this logic as dualistic morality based on a concept of power as "power over." She argues that ecological ethics must instead be grounded in a morality based on reciprocity and responsibility, which she calls "power to."[61] In addition, note the way Nash's claim collapses differences within the human community in its relation to nature. Yet this same logic of "dominate or be dominated" is used to justify First World white heteromale oppression of women, people of color, sexual minorities, and non–First World peoples and nations, especially and increasingly whenever these groups threaten First World [male] access to nature's "resources." It presupposes fixed relations where some "one" must be "on top," so that mutuality—the continuous movement toward real reciprocity—becomes unthinkable.

Greta Gaard argues that such logic is rooted in the Western epistemological dualism of self/other.[62] It shows up in a number of places in Nash's theology and ethics where he premises claims on strong and rigid distinctions between categories. For example, Nash sees humanity's relation to the rest of nature to reside in the *imago Dei* and to be expressed through a dominion of love. As in the case of Hall's stewardship ethics, Nash sees dominion as humanity reflecting and protecting the interests of God by acting benevolently and justly toward nature. Again, what is missing is a sense of nature's agency in this relation that blurs the self/other boundary in favor of a web of interwoven relations, dependencies, and responses. This is grounded theologically by Nash's strong rejection of anything that smacks of pantheism: there must be a clear separation of God from God's creation. The danger with this is that the dualism of self/other in the creator/creation relation gets passed through the *imago Dei* to separate humanity from the web of nature.

A final problem with Nash's view of nature and the humanity/nature relation is his understanding of death and finitude as part of natural evil to be overcome in the resurrection. Haraway's critique of ending stories through which we derive meaning for ourselves and nature is applicable here. While Nash's position is far from the apocalyptic desire for escape from body, earth, and death that Haraway (and Rosemary Ruether) critique, it shares the assumption that "conclusive death is inimical to the ultimate good of creatures."[63] Creation therefore has only an interim goodness, and nature is part of a fallen world of predation and systemic alienation. Against this I would argue with John Cobb Jr. and Ruether that while death may be tragic, it is not to be associated with evil. Premature and painful death, such as the gay community is encountering with AIDS, raises profound questions about hope amidst suffering and finitude, yet the answer for an ecological ethics is not to equate death itself with evil to be overcome.[64]

In fact, viewed ecologically, we are connected to other lifeforms through time as much by death as by living. Nash himself indirectly acknowledges this in maintaining that we are biologically relatives of dinosaurs and other ancient as well as contemporary creatures. Even more directly, however, our bodies are literally made up of the contributions of thousands of different organisms that have lived in different epochs in the past, and we may in turn contribute our bodies to the bodies and well-being of future generations of life. Death is what connects us through time to the living. Death, especially premature or painful death, may have a tragic dimension, but this should not obscure the blessings of death nor lead to an equation of finitude with natural evil.

Our final consideration of Nash's ethics has to do with his views of God and how these are reflected in the humanity/nature relation. In articulating his understanding of God and its implications for ecological ethics, Nash begins with classical Christian images grounded in Scripture and the creeds, especially the Apostles' and Nicene creeds. Drawing on the radical monotheism of H. R. Niebuhr, he emphasizes the distinctiveness of the Creator over and against the creation: "The Creator alone is worthy of worship."[65] Nash opposes any move that threatens the dialectical tension in the creator/creature relation, either by collapsing the relationship into pantheism that identifies God and creation, or by dualistically severing the relationship as in deism that sees God as distant and unrelated to creation.

Rather, while "only the Creator is worthy of worship, all God's creatures are worthy of moral consideration, as a sign of the worthiness imparted by God and, in fact, as an expression of the worship of God."[66] Radical monotheism does not desacralize nature; nature is of sacred worth not because it is identical with God but because it has been created good by God and placed under divine sovereignty. Hence nature is sacred, with inherent, intrinsic worth, but it is not divine. Only God is divine. Thus Nash affirms panentheism rather than pantheism. Nature is infused by God's spirit: "God is *in all* and all is somehow *in God* without being part of God."[67] Understood from a trinitarian perspective, God exists *in* the creation as the Holy Spirit, and is *incarnate* in creation as the Christ as cosmic representative.[68]

It is this intrinsic relationality within divine love that underlies the fundamental interrelatedness of all creation. Nash thus focuses much of his discussion of the humanity/nature relation on the doctrine of *imago Dei* and human dominion: what does it mean for the human/nature relationship to understand humanity to be created in the image of God as divine love and to relate to the rest of nature through the divinely mandated task of dominion?

Key to an ecologically and theologically sound interpretation of dominion is to observe from the beginning what dominion is not: it is not a license for human domination and exploitation of the biosphere. Rather dominion as an expression of being created in the image of God means that humans are to act toward nature as God acts toward the world: through love. The human practice of dominion means being God's representations of nurturing and serving love: "Humans practice dominion properly when they care for God's creation benevolently and justly in accord with the will of the ultimate owner."[69] Similar to Hall's understanding of stewardship, Nash argues that humans act in the image of God when we act as responsible representatives, reflecting and protecting the interests of the Creator.

Theologically, Nash thus strongly critiques any understanding of dominion as domination as a distortion of biblical views of dominion and humanity's appropriate relation to the rest of nature. Ecologically, such views are a violation of biological relationality and ecological limits. Ethically, this distortion of dominion denies human moral obligations to nonhuman creatures.

Given these realities and the enormous expansion of human power and technology in recent years, Nash argues for a narrow interpretation of dominion:

> Any contemporary, empirically sensitive reinterpretation of human dominion must be a narrowly defined one. It is important to note, for example, that dominion was neither possible nor

> necessary until those late-coming moral agents, *Homo sapiens*, with their creative and destructive capacities, entered the evolutionary scene. Until then, the planet thrived biologically without human assistance—and its greatest threats have come recently only as a consequence of human exploitation. God displayed cosmocentric and biocentric values and involvements long before humans arrived, and continues to do so today.... In this context, the idea that the earth, let alone the whole creation, was made for "man" is not only ludicrous but sinfully arrogant. It is a cultural addendum to the Christian faith, and a violation of the integrity of that faith.
>
> These realities suggest that dominion has narrow implications: it is primarily the protection of the planet and its inhabitants *by* humans *against* human exploitation. Furthermore, these realities suggest that the primary goal of dominion may be to preserve and restore as much as possible, compatible with human physical and cultural needs, the natural systems and dynamics that would prevail without the presence of modern humanity.[70]

Hence as humans are moral representatives created in God's image, dominion in an ecological context implies human duties and responsibilities on behalf of nature's welfare at least as much as rights to nature for human benefit. Rather than dominion justifying exploitation, authentic dominion judges and condemns all exploitative relations with nature and otherkind. Genuine dominion grounded in respect for the intrinsic worth of nature and God's biocentric valuation of all creation in turn is best understood in the role of guardian for biodiversity.

With this understanding of the humanity/nature relationship rooted in the *imago Dei* and dominion, Nash explores other Christian doctrines for insights they provide. Taking the incarnation out of a narrow, solely anthropocentric perspective, he explores the ecological implications of those parts of Christian tradition that understand Christ within a cosmic framework:

> Jesus is not only the Representative of God, in being the decisive reflection of divine love, but also the Representative of Humanity, in being the decisive expression of basic humanness and the fullness of humankind's historical potential for love. In identifying with this Representative of Humanity, however, God entered into solidarity not only with all humanity, but also with the whole biophysical world that humans embody and on which their existence depends. The Representative of Humanity, therefore, is also the Representative of the biosphere, even the ecosphere, indeed the universe. The Universal Representative is the cosmic Christ, not only as the one who illuminates the love of the Creator of the cosmos, but also as the one who unites all things and holds all things together before God. (Eph. 1:10; Col. 1:15–20)[71]

One implication of this is that it is the nature of being human to exist not only as the *imago Dei*, but also as the *imago mundi*, "as embodiment of the biophysical world." Hence humans exist in nature as part of nature, and we derive our dignity *from* nature's dignity. This is in marked contrast to theological perspectives that presuppose strong human/nature and spirit/matter dualisms, assigning human dignity and self-worth to those parts of humanity that are distinctive from nature. In these schema nonhuman nature can only have instrumental value to humans.

Hence anthropocentric oppression of nature is both a denial of the link of our self-worth with the worth of nature, as well as sinful usurpation of divine sovereignty. It reverses the *imago Dei* to make God into the image of arrogant humans. For this reason Nash reacts strongly to anthropocentric images meant to convey the dominion relationship such as caretaker, gardener, and manager because of the instrumental evaluation of nature they presuppose. He is much more cautious than Hall about the possibility and advisability of recovering the metaphor of stewardship because of its historic associations with anthropocentric and instrumental management of the biosphere as humanly owned property and resources. The realist in Nash notes the tendency toward anthropocentric abuse that each of these terms may facilitate or mask, while at the same time he is more concerned with evaluating the concrete ethical and ecological implications of human practices than engaging in debates on "ideological and verbal purity."[72]

In exploring the ecological implications of ecclesiology, Nash extends Carl Braaten's understanding of the church as an agent of "eschatopraxis"—doing the final future now—into the ecological realm to argue for including nature in our understanding of neighbor: "Since God's ultimate goal is the perfection of just and harmonious relationships (shalom) among all creatures, the church's historical mandate includes the pursuit of justice, peace, *and ecological integrity*. . . . A commitment to ecological integrity on the part of the church must be understood in the context of the church's eschatological orientation. In this context, ecological responsibility is a sign of the church's apostolicity and catholicity."[73] While most of the explicit New Testament statements about love apply only to divine/human and interhuman relationships, Nash argues it is reasonable to extend these to include all God's beloved creatures as our neighbor: "The 'love of nature' is simply the 'love of neighbor' universalized in recognition of our common origins, mutual dependencies, and shared destiny with the whole creation of the God who is all-embracing love."[74]

Because Nash's ethics of Christian responsibility, like Hall's stewardship ethics, is rooted primarily in biblical exegesis and classical doctrinal affirmations, it suffers from some of the same limitations. Yet Nash develops his theological insights on the divine in dialogue with extensive awareness of the current ecological crisis which gives his ethics a richer, more nuanced theological grounding.

Like Hall, Nash grounds his ethics in Christian understandings of divine love. His emphasis on the relational dimension of love that seeks to establish connections and build caring relationships complements well lesbian and gay emphases on love in friendship and the erotic as love. Within the six ecological dimensions of love he explores, he explicitly includes eros as the grounding of other-esteem and receptivity.

From a lesbian and gay liberationist perspective, what is missing in Nash's understanding of love is not just that love contains elements of eros, but that love itself—the power that enables and embodies caring connection—is rooted in and flows out of the erotic. The forms of divine love that Nash lifts up—reconciliation, communion, community, harmony, and shalom—take shape as the erotic as love in mutual relation flows from within and between persons and other beings. Love is experienced through, not apart from, our sensual embodiment of relationality.

Thus the erotic is central to experiencing divine love and fostering moral agency and ways of knowing. Beverly Harrison echoes Audre Lorde's emphasis on the erotic by arguing that all knowledge is rooted in our sensuality. It is embodied, mediated by our bodies, made "sense-ible" by the erotic experienced through feeling: "Feeling is the basic bodily ingredient that mediates our connectedness to the world. When we cannot feel, literally, we lose our connection to the world. . . . Our power to value the world gives way as well. If we are not perceptive in discerning our feelings, or if we do not know what we feel, we cannot be effective moral agents. . . . Failure to live deeply in 'our bodies, ourselves' destroys the possibility of moral relations between us." Love as "the power to act-each-other-into-well-being" is rooted in the erotic.[75] Our feelings help us discern when love as mutuality in right relation is fostered or distorted in our actions with others.

With this grounding in the erotic, liberationist thinking goes beyond Nash's conception of love in explicitly incorporating anger as a critical and constructive component of love as justice (note the absence of anger or conflict in any of the dimensions of love Nash lifts up). When incorporated into the praxis of right relation, anger is that part of erotic love that signals violation of right relation. Hence Harrison argues: "Anger is not the opposite of love. It is better understood as a feeling-signal that all is not well in our relation to other persons or groups or to the world around us. Anger is a mode of connectedness to others and it is always a vivid form of caring. To put the point another way: anger is—and it is always—a sign of some resistance in ourselves to the moral quality of the social relations in which we are immersed."[76]

The praxis of erotic love as justice generates conflict as well as harmony and reconciliation. It has been critical to liberationist thinking and praxis as a way of recognizing that love is never spiritualized, but always a human praxis that involves differences in power in our relationships and interactions. Anger provides the starting point and energy to begin to transform relations of alienation toward reciprocity. In ecological ethics, incorporating anger as a dimension of love is critical both in empowering agency for the praxis of ecojustice as well as an ecological dimension of the divine that signals human wrong relation with other parts of nature.

Another problematic dimension of Nash's understanding of God that informs his ethics is his rigid separation of God and creation. Nash is concerned that God and creation be neither collapsed into pantheism nor separated nonrelationally. He argues for a form of panentheism where God's Spirit indwells creation, giving it a sacramental and sacred dimension by association with the divine. Hence he argues against metaphors of the divine/creation relation that identify the two sides more intimately, such as Sallie McFague's image of the earth as God's body.[77] Among the concerns he lists is that it compromises divine independence, may contribute to "animism and idolatry," and seems to promote narcissism because God would be seen to be loving God's own body. As an alternative, he suggests that creation be seen as "God's beloved habitat."[78]

It is instructive to evaluate Nash's conception of the God/creation relation in light of Hunt's attributes rooted in friendship. Hunt understands love in friendship to emphasize that we are more united than separated in our relations with each other and the divine. The attribute of embodiment affirms with Harrison that everything, including the divine, is mediated bodily and through our connection to the earth. Hence, differing from Nash, Hunt finds that experiences of the divine rooted in friendship emphasize divine interdependence over independence and do not equate loving one's own body with narcissism but rather as a necessary precondition of loving others. In addition, Nash's quick dismissal of the metaphor of the earth as God's body for fear that it may contribute to animism and idolatry reveal an exclusivist (nonecological) Christian bias closed to authentic encounter with other spiritualities such as Native American and pagan religions within the biblical world.[79]

Behind Nash's concern for keeping the divine/creation boundaries clear and rigid are remnants of two Western dualisms reflective of masculinist experience seen as universal human experience. The first is the pervasive spirit/matter dualism that rigidly separates the divine from any direct participation in the material realm. The second is the self/other dualism discussed earlier with respect to Nash. It is interesting that Nash picks instead a metaphor—habitat—that typically has reflected an anthropocentric and utilitarian environmental image. It suggests that creation exists passively for God's use and pleasure. A more helpful panentheism for ecological ethics that avoids these dualisms and maintains the embodied relationality of God and creation are Heyward's divine-as-relational matrix or Clark's integration of spiritual and physical energy to infuse the cosmos.

A final concern with Nash's view of God is his determination to protect the moral character of God in the face of finitude and evil by rooting ecological ethics in an eschatological perspective. Nash finds ecological value in this. He argues that only an eschatological perspective can protect the intrinsic value of all creation in the face of human needs and instrumental valuing of nature. Yet it also leads Nash to the problematic claim that creation has only an interim goodness. More helpful for ecological ethics are Ruether's and Clark's critique of any views of divinity premised on fear or rejection of death and finitude. It places ethical emphasis firmly on this life and world as the only locus of our reality and living. The eschatological dimension may be affirmed in other ways than the hope for overcoming death such as Ruether's affirmation of trust in the future of the matrix of life: "Mother, into your hands I commend my spirit. Use me as you will in your infinite creativity."[80]

To summarize this assessment of models of Christian liberalism for ecological ethics, while both Douglas John Hall and James Nash come from within the broad framework of Neo-Orthodox and liberal biblical theology, Nash has moved much farther than Hall in engaging this theological grounding with the reality of and debates around the ecological crisis. Nash has broadened his perspective from his original theo-ethical anchoring because he regularly confronts the ecocrisis in a variety of circles, and he has to make a Christian case for ecological policies and rights to *both* Christian communities and public policymakers. He has taken the best of the heritage of liberal realism and developed an ethic of ecojustice that gives concrete guidelines for mainline churches to have input and impact on public policy around ecological issues while remaining grounded in the mainstream of Christian doctrine.

In this, Nash's ethic of integrity and responsibility demonstrates both the strengths and the weaknesses of liberal Christian ethics that continue to work within Western epistemological, political, and theological frames of reference. For those who traditionally have been excluded or oppressed by this set of relations and values, this is the primary problem with Nash's approach. As John Cobb Jr. noted in his own appreciative appropriation of Christian realism, the limits of this approach are that "its maximum achievement will be ameliorative" because "it accepts the existing structures of power."[81] This is the importance of explicitly incorporating a critical, deconstructive moment in an ecological ethics—it exposes the historically constructed nature of current power arrangements and allows the possibility of imagining and working toward alternative configurations. Yet this is precisely what Nash refuses to do.[82] While recognizing some of the short-term strategic value of Nash's approach in the current social arrangement of power, a liberationist ecological ethic must move beyond it to a radical critique of the current assumptions and practices that shape our relations to each other and all the earth.

Particularly important in this is recognizing the different understanding and analyses of power between liberal and liberationist approaches. Christian realism and models that share its theological and social presuppositions pay close attention to societal power dynamics, but typically take one set of experiences and perspectives as normative—that of Western white heterosexually identified men—and uncritically universalize this in their analysis of power. When this universalized view of power is combined with realist emphases on humanity's fallenness and the inevitable corruption of power, the effect often is to undercut the agency of oppressed groups seeking power for social change. Liberationist approaches insist on paying attention to multiple power differences along lines of social difference within the human community and show how these are reproduced through cultural values and social relations of production and reproduction. While recognizing the corrupting aspects of power, liberationists focus in addition on empowering moral agency to changed oppressive social relations.

For similar reasons, Nash's explicit location within and adherence to the framework of traditional Christian doctrine divorced from constructionist critique results in other problematic positions. These include Nash's view that death is part of "natural evil" and therefore a problem to be overcome, and the need to protect the moral character of God by appealing to eschatological views of redemption that give the creation only an "interim goodness." His concern that the God/creation relation not be corrupted or reduced to pantheism results in a too-rigid separation between the two that unconsciously perpetuates a spirit/matter dualism. More promising from a liberationist perspective are Cobb's and Ruether's views that emphasize the interweaving of divine, human, and otherkind agency in a matrix of relations. Nash's acceptance of the Genesis emphasis on humanity's right to subdue the earth—to either dominate or be dominated by nature—reproduces a logic of domination and self/other all too familiar to members of the human community who have been on the dominated end of such relations. A liberationist ecological ethic needs to find ways to overcome these dualisms and logics rather than to justify them or unconsciously pass them on.

Thus while Nash's ethic contains numerous resources and insights for ecological ethics, the primary approach and logic informing it needs to be critiqued and transformed to meet the goal of an ethic of Gay and Gaia: constructing an ecological ethic grounded in and responsive to liberationist concerns, particularly those of lesbian and gay perspectives. For this the models of John Cobb Jr. and Rosemary Ruether, which seek to critically engage and transform the mainstream Christian tradition, are more adequate than either of the liberal models provided by Hall and Nash.

8 Process Theology: An Ethic of Liberating Life and Sustainability

We call for the liberation of life. . . . There is the liberation of the conception of life from its objectifying character right through from cell to human community, for the concept of life itself is in a bondage fashioned by interpreters of life ever since biology and allied sciences began. Secondly, there is the liberation of social structures and human behaviour such as will involve a shift from manipulation and management of living creatures, human and nonhuman alike, to respect for life in its fullness.

—Charles Birch and John B. Cobb Jr., *The Liberation of Life*

With its roots in the philosophical thought of Alfred North Whitehead and Charles Hartshorne, process theology shifts attention from individual substances to interrelated events as the fundamental basis of reality. In contrast to classical theism with its stress on God as immutable and impossible (and thus external to and separate from creation), process theology sees God as *internally* related to all that is. Hence, like the ecological sciences, process theology stresses the influence of context and the interrelatedness of all things. Thus, in recent years process theology has proved especially conducive to Christians seeking a theological grounding for environmental activism and ecological ethics.

John Cobb Jr., for many years professor of theology at the School of Theology at Claremont, California, is perhaps the best-known proponent of process theology.[1] Since 1969 he has been active in environmental issues and has worked within a process framework to provide a Christian grounding for ecological ethics. Convinced that "changes are needed in all areas of our thinking if we are to shift away from attitudes and practices that continue to lead us toward self-destruction,"[2] together with David Griffin he organized the Center for Process Studies in Claremont to promote process thinking as an alternative to dominant Western dualistically informed paradigms.

Cobb has written widely on process thought and ecological ethics, particularly in two books on which I focus here, in addition to references to his other writings. In 1981 Cobb and Australian biologist Charles Birch wrote *The Liberation of Life: From the Cell to the Community*, in which they develop an ecological model of life consistent with the current state of scientific knowledge and set within a process framework. From this model they explore a number of "transformative implications" for how we understand the relation of human beings to the rest of nature, for theoretical ethics, and for human social practices.[3] In a 1992 publication, *Sustainability: Economics, Ecology, and Justice*, Cobb gathers together papers and talks he has given over the past twenty-three years to develop a Christian ecological ethic rooted in sustainability and justice, key to which is examining the relationship between economics and ecology.[4] Together, these two books illustrate Cobb's development of an ethic of sustainability and interdependent community grounded in process thinking.

While the ethics developed in *The Liberation of Life* and *Sustainability* are related, the emphases are somewhat different. In *The Liberation of Life*, Cobb and Birch give primary attention to developing an ethic from an ecological model of life that emerges from the biological and ecological sciences. They pay very little explicit attention to Christian theology and ethics, although a process theological perspective pervades the discussion. However, in *Sustainability*, Cobb develops this life ethic within an explicitly Christian framework.

Key to the argument in *The Liberation of Life* is examining different paradigms for understanding life in order to critique the dominant, anthropocentric paradigm. Cobb and Birch call for a "liberation of life" in two senses, as described in the epigraph to this chapter: both in the way we conceive of life under traditional metaphysics, and in the way we live. Their premise is that to liberate life itself we must first liberate the concept of life. Thus the primary focus of the book is developing a liberating and ecologically sound understanding of life that can inform our efforts at social change.

Woven through both books are three themes Cobb makes explicit in *Sustainability*. First, Cobb sees the basic problem in environmental writing and activism as the absence of a positive scenario or vision that can motivate people to make the difficult lifestyle, economic, and political changes needed to respond to the ecocrisis. He argues that Christian tradition and practice have an important contribution to make in developing this vision.

Second, like Nash, Cobb maintains that the question of ecology is closely tied to the question of economics: Without a change in how we conceive economic progress, no other changes can have much effect on public policy to protect and preserve ecological integrity. Hence an important consideration is whether an economically and ecologically sustainable society can also be a just society. Here again Cobb argues that Christian tradition, witness, and lifestyle changes can make a positive contribution.

Third, Cobb argues that fundamental to both of these is changing our way of thinking in order to shift away from attitudes and practices that continue to lead us to self-destruction. He proposes process thinking as the best grounding for this shift to a new Christianity, because it rejects the history/nature and mind/thought dualisms that underpin traditional thought and have led to ecologically destructive practices. Process thinking also provides an ecological perspective that sees each event in reality as constituted largely by its relations with other events.

Cobb's primary thesis is that an ecological model of life is needed to give a better view of life and ground an ethics of life that can lead to liberation. Such an understanding must be built on the insights of modern biology. Yet it is the way we think about life, rather than the products of science and technology, that is critical: "In the end, our conceptual environment, not science and technology, will determine the future."[5] Dominant formulations of Christianity built on mechanistic and managerial assumptions about life encourage basic attitudes that have supported unsustainable exploitation of the natural world. What is needed is a new Christianity, informed by a different life metaphysic. Cobb argues that process metaphysics and theology are best suited for this task. Christian ethics in turn must adapt to the reality of life in a world of ecological limits.

Cobb and Birch see *The Liberation of Life* as part of a tradition of writing about human problems from perspectives informed by the natural sciences, especially biology and ecology. The basic structure of their work starts with the insights of biology about the nature of life in order to derive what they term the ecological model of life. Ethical guidelines are then derived from the model and aspects of Christian tradition and teaching. It is a praxis-based approach in the sense that it flows from a lived commitment to an ecological ethics from which Christian insights are explored and developed to give further ethical guidance.

Cobb is clear about the role his own social location and experience as a member of an affluent developed nation has played in his thinking. He recalls the experience he and his wife had trying to live communally with others in order to model a more ecological lifestyle, and the discouragement that resulted from their failure. This experience has deepened his conviction that what is most needed in environmental activism is a positive scenario that helps economically privileged people in particular to envision a society more enjoyable and livable than the current one. Cobb is less clear about the influence of gender, race, and sexuality on his social location and thinking, although his writings commonly incorporate insights from various liberation theologies, particularly feminist theology.

Cobb shares many of the presuppositions that inform Nash's work, although his are interpreted through process and scientific frameworks. First, like Nash, he argues that ecological perspectives and actions assume values that can be evaluated as morally right or wrong. In deriving ethical guidelines from ecological and biological insights, Cobb locates himself in the ancient tradition dating to Aristotle that understands ethics to be appropriately derived from science.

Second, in a postmodern era when most attempts to do metaphysics are seen to have failed epistemologically, Cobb argues that metaphysical assumptions matter. All views of life are grounded in assumptions that can be understood in terms of metaphysics; what is needed is to move away from the Enlightenment's dualistic and mechanistic metaphysics to one that is ecologically informed and sound.

Third, as mentioned previously, Cobb, like Nash, sees the intersection of economics and ecology as the crux of the ecological crisis and central problem of ecological ethics. Neither economic issues nor the ecocrisis can be addressed without regard for and input from the other.

Finally, Cobb is more thorough than Nash in rejecting anthropocentrism in favor of ecocentrism or "life-centrism" as the appropriate framework for ethics. This flows from his commitment to process metaphysics, which sees all life as inseparably interconnected and relational. An anthropocentric perspective limits subjectivity to human beings, which Cobb argues is a distortion of what we now know about life.

Since first discovering the urgency of the ecological crisis in 1969, Cobb has shifted the primary focus of his praxis from directly discussing environmental issues to a more indirect approach. Working from his conviction that the way we think is at root in the crisis, he has focused on integrating a process perspective with disciplines such as biology and economics to begin to explore ways to shift our thinking away from dualistic and ecologically destructive attitudes and practices. His academic research and writing has focused primarily on two books and sets of issues. *The Liberation of Life* combines ecological insights from science and process philosophy to provide a different model of living things that can form the basis of the needed new way of thinking. *For the Common Good,* written with economist Herman Daly, pays explicit attention to the crucial intersection of economic issues and ecological concerns. In *Sustainability,* Cobb draws on the themes of both these earlier works to set his ethic within an explicitly Christian framework.

Cobb continues to be preoccupied with economic issues as the key to resolving ecological and justice issues: "As long as we collectively suppose that meeting economic needs and having full employment require a growing economy, we will collectively support policies that put greater and greater pressure on an already overstressed environment. We will also continue to support policies whose results are greater and greater injustice, with the rich getting richer and the poor getting poorer both within each country and among the world's nations."[6] Within the Christian community he has worked to weave together theological insights that can help provide a positive vision to ground needed lifestyle and policy changes. He sees the attention to public policy by Christian realists like Nash as one important contribution by Christians toward what is needed to reverse the ecological crisis, but believes an even more important contribution is changing our way of thinking and providing leadership through modeling needed lifestyle changes while articulating a compelling vision of an ecologically sustainable society. Reflecting an awareness of his own social location, Cobb addresses his efforts primarily towards affluent Christians, while trying to be attentive to the perspectives of the poor and socially marginalized.[7]

With this overview of Cobb's approach, let us turn now to the structure and content of his argument. *The Liberation of Life* begins with an examination of biological insights about life in order to construct an ecological model of life. Cobb and Birch move from this to consideration of the relationship of humanity to the rest of the natural order in order to develop an ethics of life. In the final chapters they explore some of the ethical implications of this model by looking at issues related to the biological manipulation of life, justice and sustainability, and economics and ecology.

Cobb and Birch begin their exploration of the understanding of life by noting how difficult it is to draw any sharp line between the living and the nonliving. Yet there are differences, and they have important implications. Living and nonliving things may be made up of the same atoms and molecules, but what distinguishes them is how their basic components relate to and are shaped by their environment—that is, their ecology. Cobb and Birch explore this relationship between component and environment at three different levels of ecology: at the level of the molecule, the individual organism, and whole populations.

In each of these levels the authors emphasize the importance of the ecological context for understanding life at that level. For example, at the molecular level, "what the DNA does is dependent upon the environment in which it occurs, in other words it depends upon the ecological context."[8] At the level of the organism, "there is no such thing as an isolated individual. The individual is what it is in great part through its relationship with the external world around it. Its world is one of physical and chemical and psychological pressures, to which it must adjust appropriately if it is to survive and flourish."[9]

From these studies they derive the principle that maintains that life is interdependent with other forms of life and its environment at all levels. Ecological systems are self-sustaining when this interdependence allows the resources and minerals used by living forms to recycle through the system through natural processes of living and dying. Human activities that recycle little of our by-products often convert sustainable systems into unsustainable ones. Cobb and Birch note that human societies become unsustainable in three ways: when they deplete nonrenewable resources, when there is greater demand for renewable resources than the environment can sustain, and when pollution exceeds the absorptive capacity of the environment. They conclude that a critical part of any human ethics is that society be so organized that it is sustainable indefinitely into the future.[10]

After concluding that ecology is the most appropriate category for considering each level of life, Cobb and Birch turn to the other great find of modern biology: evolution. Evolutionary changes in life-forms occurs through natural selection, "the differential survival and reproduction of individuals in a population."[11] Throughout most of the earth's history, life has evolved through a combination of chance and necessity, taking advantage of random mutations to adapt to changing environmental conditions. With the evolution of higher animal forms, many scientists have argued that purposive behavior may also become a significant factor.

Cobb and Birch argue that human beings are fully a product of natural selection, evolving through chance, necessity, and purposive behavior over the last several million years. Two evolved features are particularly important in Homo sapiens: the development of a brain with the capacity for abstract thought, and the capacity to communicate by means of language: "Human beings are the only biological species with a highly developed capacity for symbolic thought and the use of language."[12] As a result, in the human species there has been a shift to the predominance of cultural evolution over genetic selection in human development in the last fifty thousand years. Yet, like Nash, Cobb and Birch emphasize that human beings are fully a part of nature, even though a unique part.

Arguing that the insights and assumptions of ecology and evolution belong together in human vision and imagination, Cobb and Birch acknowledge that the two schools of thought have tended to be separate, often resulting in quite opposite social and spiritual effects. Evolution, with its emphasis on natural selection and survival of the fittest, became the foundation for nineteenth-century doctrines of progress and the alleged superiority of the human species over "lesser" species in a competitive world. Human indifference to the fate of others—both otherkind and weaker human beings—was one result. Ecology, on the other hand, has downplayed the separate development of the species and encouraged human beings to see themselves simply as one part among many in the whole web of life rather than as unique agents of purposive change.

Cobb and Birch reject both of these images as inadequate. They argue instead: "Evolution is not a process of ruthless competition directed to some goal of ever-increasing power or complexity. . . . A species co-evolves with its environment. But equally, there is no stable, harmonious nature to whose wisdom humanity should simply submit. Intelligent purpose plays a role in adaptive behavior, and as environments change its role is increased. Human culture is an immensely important factor in ecology and one which necessarily introduces profoundly new elements into the web of life."[13] This conclusion about the unique positioning of humanity within nature as an agent of purposive behavior grounds the rest of the discussion about deriving a life ethics.

With this grounding Cobb and Birch assess and critique different models that have been used to account for living organisms and behavior. In proposing an ecological model as a better alternative, they are concerned with two issues: (1) finding a model that allows us to better understand life, and (2) shifting our pattern of thinking from what they term "substance" thinking to "event" thinking. They argue that implicit in any model is a metaphysical framework. It is therefore not enough to change the model without changing the assumptions that inform it; therefore, we need new metaphysical assumptions.

The framework implicit in most biological models since the Enlightenment has been mechanism. Organisms and their constituent parts have been understood as machines or studied as if they were machines. Mechanistic biological models developed historically within an Aristotelian-Newtonian framework that conceptualized ultimate constituents such as atoms and molecules to be particles, best understood as bits of matter: "The principles which govern material particles are the laws of mechanics, and the behavior of larger entities composed exclusively of material particles must be machine-like. Further, such machine-like behavior is fully deterministic. Hence mechanism, materialism and determinism belong together."[14] Philosopher René Descartes made the doctrine of mechanism in physics and biology explicit. Because human thought did not fit Descartes' mechanical explanation, he developed what became known as the Cartesian doctrine of the bifurcation of nature into mind and matter, thereby reinforcing the mind/matter dualism in Western thinking.

Cobb and Birch argue that a problem inherent in mechanistic models of science emerges when human beings are considered. On the one hand, human beings are seen as fully natural and hence fully determined by nature. On the other hand, human beings through cultural evolution are seen to be in control of their destiny. A self-contradiction thus emerges between determinism and responsibility for the future in discourse about the human: "Science appears here in two roles, as a body of conclusions about the nature of things, including human beings, and as a means through which human beings can exercise control over the world, including themselves. In the first of these roles it is presented as teaching strict determinism. In the second, it is offered as providing human beings with a radical control over human destiny that implies radical freedom. In this discussion the two doctrines are unreconciled."[15] Mechanistic models predicated on scientific determinism must see human beings either as totally determined by genetic coding or else posit a supernatural element in human beings. In light of scientific evidence and human experience, a more plausible belief is that human beings are partly determined and yet partly able to determine their own destiny. A different model of life is needed that allows consideration of both machinelike and non-machinelike traits of organisms.

For this the authors propose an ecological model that emphasizes the interrelatedness of all things. A fallacy of the image of the machine for modeling biological life is that, in theory at least, a machine can be abstracted from its environment without changing its nature. That is, it is seen as an independent entity having an internal structure whose behavior flows from it, independent of the environment. Yet scientists are finding thc opposite to be true of organisms, and put increasing emphasis on the importance of environment. Hence an ecological model of living things pictures organisms inseparably interconnected with their environment:

> The ecological model holds that living things behave as they do only in interaction with the other things which constitute their environments. It does not deny, of course, that many features of their behavior are determined by structures of the organism itself or that much has been learned when this structure has been

> examined by scientists with the mechanistic model in mind. But the ecological model proposes that on closer examination the constituent elements of the structure at each level operate in patterns of interconnectedness which are not mechanical. Each element behaves as it does because of the relations it has to other elements in the whole, and these relations are not well understood in terms of the laws of mechanics.[16]

Yet finding a new model is only half the battle. The shift from a mechanistic to an ecological paradigm can only succeed when accompanied by a shift from substance thinking to event thinking. Here Cobb and Birch draw on the process metaphysics of Whitehead to critique the dominant paradigms of Western thinking which, since Aristotle, have conceived of objective reality primarily in terms of substantial and enduring objects. These objects are seen to be acted on by external forces and interact much as billiard balls do, through external movements that do not affect the internal structure of the object. Substance metaphysics underwent radical criticism by David Hume. His critique was taken into account in subsequent human philosophies stemming from Kant and Hegel, but the assumptions of substance metaphysics were left in place subconsciously in inherited thought patterns among practitioners of the natural sciences. Most scientific models thus implicitly presuppose a substance metaphysics.

Cobb and Birch argue that what is needed is a shift from substance to event as the primary way of looking at and understanding the world:

> Instead of taking for granted that the world is composed of the substantial objects of sense experience or of the substances which underlie them, one may think instead of a world composed of events and the smaller events which in turn constitute them. What is to be explained, then, is why things happen as they do. And the explanation will consist in analysis of the causal relations among events and the component occurrences which make up larger ones. . . . Substance thinking recognizes

> the occurrence of events and undertook to explain them in terms of substances. Event thinking must recognize the existence of relatively enduring "substantial objects" and undertake to explain them in terms of patterns of interconnectedness among events.[17]

Thus, events are seen to be primary, and substantial objects are viewed as enduring patterns among changing events.

An example may help clarify the difference between the two perspectives. From an events perspective, an atom is no longer seen as an entity or substance, but as a multiplicity of events that are interconnected with each other and with other events (atoms and molecules) in a describable pattern. Cobb and Birch stress that mechanistic models can be valid and illuminating under limited conditions where the structures studied are relatively independent of environment. The problem occurs when the adequacy of mechanistic and substance thinking is extrapolated beyond a limited context.

One important implication of this model is that it does away with the dualisms of mind/matter and living/nonliving that inform mechanistic and substance models. The lines between living and nonliving are blurred.[18] The term *ecological* highlights the belief of this model in the internal relation of *every* entity to its environment, living and nonliving alike. By internal relation, Cobb and Birch mean the relations that are constitutive of the character and existence of some things. While substance thinking sees relations as external to substances, event thinking sees relations as internal to events. Because all things are related internally to their environment, both living and nonliving things can be studied and understood within an ecological model.

Cobb and Birch next turn to rethinking the relationship between humanity and the rest of the natural order in light of the ecological model. They are particularly concerned with "testing the adequacy of the ecological model to human experience against the deep-seated objections in Western philosophy and religion against considering human beings as essentially part of a continuous natural process."[19] Central to these objections has been the assertion that only human beings have the experience of transcending our biologic selves. Yet from an ecological perspective, all living things share in the experience of transcending. Life itself is a transcendence of physical causality: "Living things inherit from their physical environment, but their response introduces something new." Instead of separating us from otherkind, human acts of transcendence instead confirm our connectedness: "Our intelligence, our language, our culture enable us to transcend far more radically than can any other living things of which we know. It is very unlikely that the sense of the responsibility for society as a whole, possessed by some human beings, could be even remotely approximated in other creatures. But the act of transcending which expresses itself in these rich and distinctive ways in us establishes, not our discontinuity with other living things, but our continuity."[20]

If human beings are fully part of the human world, then what about the problem of human evil? Surveying the debates in sociobiology and paleoanthropology about the source of human nature, Cobb and Birch note that some identify the source of evil in humanity's genetic nature and seek a remedy through strengthening human culture, while others reverse this, seeing the problem as lying in human culture's distortion of nature. Against both of these views, Cobb and Birch follow the insights of Reinhold Niebuhr, who, in *The Nature and Destiny of Man*, argues that while the Bible recognizes the tension between nature and culture, it refuses to identify either side with evil. Rather, they contend, "neither nature nor culture is bad, and that we participate in both is our glory. But this glory is at the same time the condition that leads us continually to sin."[21] One important ethical ramification of this is that human evil is not located outside ourselves in nature or culture where we can avoid taking responsibility for it by assigning it to fate.

Cobb and Birch agree with Niebuhr's description of the fall as a phenomenon of human existence that results from a new level of freedom. But they correlate the symbol of the fall with particular points in the evolutionary historical process as the occurrence of a new level of order and freedom bought at the price of suffering.

This they term a "fall upward." For example, until the coming of animal life there was little or no suffering, but there was also little value of experience. Each evolutionary increase in the capacity for richness of experience has been paid for with equally great increases in the capacity and experience of suffering. This insight has critical implications for ethics, because each new liberation in areas such as technology, politics, education, or sex produces new forms of enslavement and suffering. Human progress is therefore fundamentally ambiguous.

The authors next turn to the question of whether or not animals experience, and how answers to this question shape our relationships to them. They note that critical to the debate is how one understands experience. Against those who narrow their understanding of experience to fully self-conscious experience, they include unconscious or nonconscious experience: "For us to say that something experiences is to say that it is not merely an object in our world of experience but also a subject of relations in its own right. It is acted upon and it acts. The ecological model is a model of living things which are acted upon and which respond by acting in their turn. They are patients and agents. In short, they are subjects."[22] If this is so, then a better way of asking the question is not whether living organisms experience, but do they participate in subjectivity? Cobb and Birch conclude yes. Recognizing the subjectivity of all living organisms must therefore fundamentally shape our interactions with them.

Extending this recognition of experience in all things to nonliving entities, Cobb and Birch reject all radical discontinuities, either between the living and nonliving, or between different levels of living organisms. They disagree with the view common among many indigenous groups that all things, living and nonliving, have psyches, souls, or minds, but nevertheless maintain that experience includes "the non-conscious taking account of the environment which characterizes molecular, atomic, and quantum events."[23] Because in a process ecological model, all things are characterized by both internal and external relations, this understanding of experience provides for continuity in internality among all things.[24]

One of the important implications of this proposal is that the subjective side of evolution has yet to be fully understood or described, although it is critical to understanding the place and role of humanity in the natural world. Cobb and Birch argue that when unified human experience began to make its own enjoyment its primary end, and to instruct the body accordingly, the human being as we understand it had arrived. The human psyche is the result of this unified human experience, its continuity through time, and its dominance over the body. The emergence of the human psyche is a case of the "fall upward." It introduces both the possibility of far richer experience in the natural world, but also the capacity for fundamental dis-ease and dis-harmony. Yet while we are unique in evolutionary terms, human beings are still fully natural creatures and exist in continuity with the rest of the natural world, animate and inanimate.

Having used the ecological model to examine the humanity/nature relationship, Cobb and Birch turn to developing an ethic grounded in an ecological understanding of life. Their premise is that every view of life has ethical implications. One critical implication of their model is that all things have some intrinsic value, either in themselves or in their constituent parts. This flows from understanding all reality to be composed of events rather than substance. Since subjectivity or experience is attributed to all such events, and experience is always valuable, all events have intrinsic value.

Cobb and Birch develop two important ethical ramifications of the ecological model: (1) Our understanding of rights and duties must be rethought once we view animals as true subjects; (2) our whole understanding of ethics and the traditional categories we have inherited must be reconceived in light of a deep sense of the interconnectedness of all things. The goal of ethics must be the enhancement of the "richness of experience," for "to have richer experience is to be more alive."[25]

In exploring ethics beyond the framework of anthropocentrism, Cobb and Birch follow the lead of the early land ethicist, Aldo Leopold, in seeing the extension of ethics to the whole natural world as a process of ecological evolution. Leopold described this in both biological and philosophical terms: "An ethic, biologically, is a limitation on freedom of action in the struggle for existence. An ethic, philosophically, is a differentiation of social from anti-social conduct."[26] Leopold illustrates this using the example of property. In antiquity, it was seen as quite appropriate that other human beings be held as property. Eventually the extension of ethical criteria to all human beings made this unethical, and human slavery and human property were abolished under the new criteria. In our relation to the rest of nature, we now need to abolish the property-mentality that sees the land and its inhabitants only for their instrumental value.

Cobb and Birch argue that the problem with the anthropocentrism that informs traditional ethical theories, whether based on Kantian or utilitarian principle, is that it distorts the nature of reality by seeing *nonhuman* entities only as means to human ends rather than also as subjects with their own interests:

> If subjectivity were confined to human beings, if all else existed only as objects of human knowledge and use, then anthropocentric ethics and economics would conform to reality. But such a view simply does not fit with what we know about life and about ourselves as one form of life. We are subjects in a wider community of subjects as well as objects in a wider community of objects. If we, as subjects, are of value, and we are, there is every reason to think that other subjects are of value too. If our value is not only our usefulness to others, but also our immediate enjoyment of our existence, this is true for other creatures as well.[27]

A new ethic is needed, Cobb and Birch contend, one that recognizes in every animal, including humans, both end and means. From this they derive an ecological ethical principle: "We should respect every entity for its intrinsic value as well as for its instrumental value to others, including ourselves. Its intrinsic value is the richness of its experience or of the experiences of its constituent parts. We deal with entities appropriately when we rightly balance their intrinsic and their instrumental worth."[28]

The criteria for evaluating intrinsic worth for matters of ethics is the capacity for subjectivity or richness of experience. Hence there is a graded scale of intrinsic value from rocks and other abiotic entities on one end, whose capacity for subjectivity is so slight that it can be ignored for practical, and therefore ethical, purposes, to human beings whose capacity for subjectivity is so great that in ethical debate they can never be considered solely as means apart from their intrinsic value as ends in themselves. While plants can be treated primarily as means—although extremely important means that we abuse at our peril—in animal life a new level of experience arises that requires that we respect them as ends as well.

Hence Cobb and Birch support the development and application of animal rights within the ethical and legal spheres. Yet while sensitive to the charges of speciesism, they, like Nash, nevertheless oppose an absolute egalitarianism between species that ignores differences in intrinsic value. They argue that it has not been wrong to view animals as means to our ends—this is what it means to participate in the natural order where predation and symbiotic existence are built-in components. What *has* been wrong has been the tendency to view animals as only means to human ends. As an alternative, Cobb and Birch develop the standard of richness of experience to give practical guidance in such areas as factory farming and research involving animals. Advocating a case-by-case method rather than absolutist principles, they nevertheless maintain that "it requires a great human advantage to compensate for animal suffering and loss."[29]

Following their conviction that human beings are fully a part of finite nature, Cobb and Birch maintain their opposition to absolutist approaches in their view of human rights. Reversing the emphasis of their views on animals rights, they argue that it has not been wrong to emphasize that all human beings should be treated as ends. Rather, what has been wrong has been to ignore that human beings are also always means to others' ends—not only for the welfare of other human beings, but for the welfare of other creatures. This entails duties and obligations, as well as rights.

Because Cobb and Birch locate intrinsic worth in richness of experience and capacity for richness of experience, they argue that some human experiences are of more intrinsic value than others. They thus oppose absolutist ethics that grant infinite intrinsic value to all human life without distinction, because no finite thing is of infinite worth. This becomes important in ethical issues such as abortion where they argue that the intrinsic worth of the woman outweighs that of the fetus, and in cases of terminal illness where bodies can be kept alive at great financial and emotional cost to relatives and society, but where the capacity for richness of experience is negligible.

At the same time, Cobb and Birch's awareness of the interrelatedness of all things leads them to argue against individualist ethics alone in favor of biospheric ethics. Here the rights of species to survival is more fundamental than the right of individual members to life. Yet at the level of the biosphere a rights approach to ethics itself becomes inadequate. Rather, "a sense of the web of life, the value of diversity, and the way people are constituted by internal relations to others" will guide our imaginations to give ethical guidance.[30] In the long run, there is congruence between meeting human needs and animal needs, yet in the short term it may be impossible to resolve conflicts in favor of both parties. In such cases Cobb and Birch argue that human beings must compromise human interests for the sake of the interests of other species. Here the role of legal rights and representation for nonhuman species has an important role to play.

Where does religious faith come into this? Cobb and Birch acknowledge that ethical claims are more persuasive for most people when they move beyond the rational realm to find grounding in religion, or the ways people most deeply experience meaning in the world. The ecological model suggests a new cosmology, a new grounding for ancient religious truths. What is needed now is a religion of life. Here Cobb and Birch follow Whitehead's process metaphysics closely in articulating a view of God as Life, a cosmic power working everywhere that conditions permit to bring enrichment of experience. The appropriate response to God as Life is trust, and openness to transformation through trust in Life. Trusting Life is not a passive stance, but rather a renunciation of control, a shift from control to support in our attitudes toward other living things. For Cobb and Birch trusting Life is the essence of faith: "We find ourselves choosing to trust Life. That means not only that we try to open ourselves to what Life can bring us and do with and through us. It means also that we live in the expectation that death's word is not the last. What form Life's victory will take we do not know. . . . For us, faith in Life means more the confidence that Life is not finally defeated than the belief that we can state accurately the form of its victory."[31] This understanding of God as Life raises a number of important issues for Christian theology and ethics that will be considered later in this chapter.

From this "ethics of life" come several key components for ethics. In terms of *values,* all life has value, but not all life is of equal value. There is a hierarchy of *intrinsic value* from the simplest forms of life through human beings, measured by the richness of experience and capacity for richness of experience of each form of life. All living organisms including humans also have *instrumental value,* their value to others either of the same or different species. The guiding *ethical requirement* is to provide circumstances that promote richness of feeling. *Ethical dilemmas* occur when there are conflicts between the intrinsic and instrumental values of organisms and species.

Rights come from the intrinsic value of all experience. Individuals with the capacity for experience have the right to enjoy richness of experience as they can. Because experience is not infinite, rights are not absolute. The only absolute is *respect for life itself,* which forms our ethical grounding. This view of ethics implies agency: "If we are called, ethically, to promote richness of experience, we are at the same time called to promote the ability to respond freely to situations rather than being passively shaped by them. Based on the ecological model, ethics requires that we respect and encourage the free decisions of those whose richness of experience we would promote."[32]

Yet ethics must recognize the riddle of inequality: through different genetic and cultural endowments we are born unequally. Ethical action thus is committed to *equality in opportunity* for all human beings—the maximum opportunity to develop fully one's talents and promote the richest possible experience for all. Because all are interrelated, if there is to be any sense to the unity of humankind, *justice* requires that we share one another's fate. Human moral agency is rooted in our capacity for transcendence; this implies responsibility for our actions, including being responsible to generations yet unborn. *Sustainability* is an ethical principle that implies community through time with future people and otherkind.[33]

Ethics also involves *responsible risk-taking*: life is an adventure that requires risks, but to be ethical they must be assessed in terms of cost and benefit. Ecological ethics emphasizes addressing *causes over symptoms*. It looks at organisms in their wholeness and interrelatedness with their environment. Finally, ecological ethics recognizes the *relative nature of ethical guidelines*. Guidelines are "created goods" rather than absolute values. They should be kept open for modification and transformation in light of fresh insights, perspectives, and analyses.

With this overview, Cobb and Birch direct their analysis to such critical ethical issues as the need to create a just and sustainable world and the shape of economic development within an ecological perspective. Here I sketch out some of their responses in order to indicate how their ecological model might function concretely on some of issues central to ecojustice.

Noting that problems of social justice are equally problems of ecological sustainability, Cobb and Birch ask what would a just and sustainable society look like from the perspective of an ethic of life. Their starting premise is that the enhancement of the richness of experience of individual human beings and other animals is of prime concern. For human beings this means primarily the realization of existing potentialities. Moving beyond the local context to thinking about this at the global level requires additional ethical guidelines and principles. In this discussion they focus on the guidelines suggested by the World Council of Churches: justice, participation, sustainability. The discussion is made more urgent by increasing awareness of how remote our current situation is from these norms.

In discussing justice, they note that in today's global reality, justice and sustainability require each other. By justice, Cobb and Birch mean the idea of the good society that requires at least the three elements of equality, participation, and personal freedom. Justice in this sense does not require absolute equality, but it does require that we share one another's fate and provide equal opportunity for each person to develop her or his talents. It entails that people be able to participate in the decisions that affect their destiny, which requires that certain basic personal freedoms be respected.

Sustainability for Cobb and Birch means to be capable of indefinite existence. They use it in its widest connotation by subsuming economic and political sustainability within ecological sustainability. Seen from this angle, for example, neither capitalism nor socialism as practiced in their twentieth-century forms are sustainable political-economic systems. For a society to be sustainable it must respect the finite limits of the earth in three areas: the earth's limited capacity to produce renewable resources, such as timber, food, and water, its limited amount of nonrenewable resources, such as minerals and fossil fuels, and its limited capacity to absorb human waste products.

For economic systems to be sustainable they must shift their logic from limitless growth to long-term sustainability. In our present global state of highly inequitable divisions of wealth and poverty, Cobb and Birch recognize that many poorer nations will need to grow to reach economic maturity and be able to sustain their populations. But this growth must be aimed at meeting real needs rather than being a built-in requirement of an economy built on rampant consumerism. Affluent nations must shift immediately from overconsumptive, growth-based economies that are fundamentally nonsustainable and rapidly are depleting the earth's resources and absorptive capacities. The only way poor nations can grow is for wealthy nations to curb their own growth.

From these considerations Cobb and Birch develop several characteristics a sustainable society will have. They observe that in a sustainable world with a more just distribution of wealth, nations will be considered overdeveloped when their citizens consume resources and pollute at a rate greater than would be possible for all the people of the world. By this criterion, virtually all of the nations of the world currently considered developed are actually overdeveloped. Vast levels of military spending only exacerbate the situation. In the context of national and international economies premised on ideologies of unlimited growth in a world of increasingly apparent ecological limits, international militarism in the form of war and preparations for war is the greatest ongoing threat and obstacle to sustainability.

Cobb and Birch note that two very different ways of viewing the world are at stake in defining how we approach ethical issues of justice and sustainability:

> To those who are sensitive to fundamental features of living things the fact that there are limits to economic growth is so evident that it is difficult to appreciate the sincerity and conviction with which others deny that there are such limits. Two structures of experience, two faiths, confront each other here. The first sees humanity as a part of the natural world, a very remarkable part in which the element of transcending, present in all life, is extremely prominent. For them human beings transcend most successfully as they respect most sensitively the conditions of all life.... By attentiveness to the strategies of life, they can learn what true growth is, they can develop appropriate goals and find deep satisfaction in their attainment. This is the religion of life....
>
> The other sees humanity as outside of nature, having nature as its object, its possession, to be used for human ends. Its adherents recognize little difference between living things other

> than human beings and inanimate materials. Both are resources for human management, manipulation and consumption. Human desires are, and rightly so, unlimited. And human ingenuity is similarly unlimited. To place confidence in human intelligence, and in the science and technology in which it is expressed, will enable growth in numbers and in consumption for as long as these are desired.[34]

From within the perspective of the ethics of life, the problems encountered within modern economic paradigms are not so much challenges to be overcome by further science and technology, but limits to be understood and respected. The question ultimately is not whether we will accommodate ourselves to the earth's capacity to support life, but when and how.

In turning to the issue of economic development, Cobb and Birch emphasize six features of the ecological model that contrast with the nonsustainability of dominant economic models:

1. *The intrinsic value of every person*, measured in terms of richness of experience. Intrinsic value is found only in individuals; any policy which is not directed toward the enrichment of experience of individuals is misdirected.
2. *Relationality:* Individuals are constituted by their relations. Our richness of experience is richness of relations and therefore depends on the community and the richness of relationships that make up that community.
3. *Transcending:* Although individuals are constituted by relations with others, no one is merely the product of these relations. Each creates a creative synthesis from relationships and experience.
4. *The limits of transcending:* Human beings and thought are conditioned and constrained by their historical situation and interests. Acknowledging the limits to transcending is as important as acknowledging its reality.
5. *Continuity between human beings and the rest of the natural world:* Intrinsic value is not limited to human beings, and

we are constituted by our relations with human and non-human beings. "Human beings are impoverished by the decay of their non-human environment, and furthermore, what happens to the non-human world is important in itself."[35]

6. *The possible symbiosis of desirable goals:* It is possible to have both sustainability and justice, life for humanity and life for the rest of the natural world. Goals that appear competitive in the light of the dominant model may be mutually supportive in the ecological model.

Implicit in the shift to an ecologically informed socioeconomic model are new definitions of common terms. Community now is understood to be a truly global community of ecologically interdependent communities, "that group of people and other creatures who most deeply affect one another, whose lives are most richly intertwined."[36] Good technology is evaluated ecologically as that which aids the production of truly needed goods while using the least amount of resources and avoiding disturbing the life-sustaining character of the natural environment and human communities. Human economic freedom is no longer primarily the freedom to choose between existing goods, but the ability to envision new ecologically sustainable goods and to shape life in order to attain them. Attention must be paid to the connections between ways certain human groups and the rest of the natural world have been exploited, particularly the connections between the objectifications of women and nature. And extending subjectivity to include all fellow creatures has important ramifications for how community is understood and the role it plays in economic analysis.

Cobb and Birch stress that key to resolving development issues is finding a just and sustainable role for women. In virtually every contemporary ecological issue in the developing world, from population growth to agricultural practices, improving the situation and agency of women is one of, if not *the,* critical factor. Other areas that need to be addressed include developing just and sustainable energy usage through "soft energy technologies" (reliance on renewable energy sources that are diverse, flexible, and require relatively low technology) and exploring just and sustainable forms of transportation and urban habitat.

While the authors acknowledge that no one can be an expert in all these areas, they contend that it is a right and duty for nonexperts to challenge the experts to think in new ways. Critical to this is getting experts to examine the worldview and models implicit in their disciplines for their implications for justice and sustainability. They conclude with a threefold call: for needed lifestyle changes to be able to meet future crises constructively; to liberate our thought about life in order to liberate life from current deadly policies; and to orient ourselves to and trust in the "silent working of Life in the hearts of men and women" in making these changes.[37]

Because John Cobb starts from an explicitly postmodernist stance in developing his process ethic of sustainability and community, his methodological approach parallels a liberationist ethical methodology at several important points.[38] Cobb grounds his work in an explicit critique of Enlightenment values and epistemology that presuppose anthropocentrism, dualism, and individualism, while wishing to retain such gains of modernity as the political meaning of the dignity of human life and human rights. He explicitly incorporates a constructionist view of all knowledge and truth claims to argue for a "constructive postmodern vision" that avoids falling into a "'most modern' nihilism" of total relativism.[39]

One sees this in Cobb's understanding and use of the sciences. Of the authors of the four models presented here, Cobb probably has the most sophisticated and historically critical perspective on the physical sciences, both in terms of methodology and the content of their claims. He takes seriously the claims of science, understanding them within a scientific framework, yet also engages the sciences in critique involving both facts and values. Like Haraway, he avoids seeing the sciences as either inevitable progress or a passive reflection of "the way things are" (modern perspectives), or as merely a discourse of power (a radical constructionist postmodern view). Science is one (important) way of knowing, but needs to be engaged by other ways of knowing.

Similarly, Cobb's approach involves a serious engagement with the Bible and Christian tradition, with attention to liberationist and postmodernist insights. From a conviction expressed twenty-five years ago that Christians must "explicitly criticize our own Scriptures" and "rethink the whole matter of biblical authority" to remove barriers that prevent development of an ecological ethics, Cobb has shifted radically as he discovered "that much of the [ecologically] destructive effect of the Bible's influence has been the result of misunderstanding rather than accurate interpretation."[40] Hence for ecological ethics, Cobb's efforts have shifted from critique and dismissal to recovering and reinterpreting biblical insights. Unlike Hall and Nash, however, he recognizes the importance of a critical, deconstructive engagement with biblical texts as part of the task of ecojustice. While recovery of biblical insights not distorted by Enlightenment assumptions is necessary for a postmodernist Christianity, Cobb argues that "we also need a postpatriarchal Christianity, and that will have to be a new Christianity . . . [involving] a far more radical task, one that must work against the Bible, however much inspiration it can still find there."[41]

Cobb can argue for such deconstructive engagement with the Bible because he takes a praxis-based approach to Scripture. The primary goal of this praxis is enhancement of life, that is, ecojustice, and the Bible and tradition provide resources for the task. For Cobb this approach enhances rather than diminishes biblical authority, because it recognizes that authentic authority is relational. He maintains that biblical thinking is neither anthropocentric, individualistic, nor dualistic, nor is it literalist, objectifying, or foundationalist—all modes of thinking characteristic of modernity. Rather, "for postmodern theology, the authority of the Bible is the authority of its felt truth, its power to illumine our lives and our history. The Bible can never be accepted as heteronomously imposed."[42] Cobb understands the unity of the Bible to be "more the unity of a sociohistorical movement than of a coherent system of teachings." Hence its authority is experienced relationally in our praxis, when "different parts of the biblical record take on relevance and importance as the church faces different situations."[43] To be faithful to the biblical witness, we must take it seriously, engage it, and push it further in light of God's ongoing revelation.

Cobb's process model is also similar to liberationist approaches in its explicit openness and commitment to learning from other perspectives and disciplines.[44] From a process, ecological perspective of interdependence and interrelationality, this necessarily implies openness to being transformed through engagement with other voices. He argues that particular attention must be paid to the voices of those most marginalized and excluded from the discussion, particularly poor humans and nonhuman animals. Cobb also maintains that theology and ethics must not only learn from, but must engage critically and contribute to secular thought and disciplines such as "economics, political theory, sociology, psychology, and even the natural sciences," breaking down the artificial Enlightenment compartmentalization of disciplines and rooting out "modern assumptions that no longer carry conviction."[45] All of these are consistent with the liberationist insights discussed in earlier chapters.

Cobb's process ecological ethic offers many rich insights for developing a liberationist ecological ethic. The primary difference from and obstacle for a liberationist perspective is one of location. While being very open to learning from marginalized voices, Cobb continues to maintain a sense of being located at a normative center to which these other voices may relate, reshaping and reforming the enterprise. Consider this quote where Cobb argues for a Christocentric theology informed by diverse "theologies of . . . ," each reflecting a particular location and experience:

> Such a theology would affirm precisely the need of each to hear the other and to be transformed through what one hears. That process of transformation would affect the center as well. The meaning of "Christ" cannot remain the same after the impact of black theology or of the recognition of how often and how easily Christocentrism has been used to evoke and justify the persecution of the Jews. But such a center would provide a way of hearing the many voices in which their tendency to exclude one

> another would be overcome and they could be more fully incorporated into the ongoing creation and transformation of theology as such.[46]

From a liberationist perspective the center/margin model of theology and ethics is itself a primary part of the problem. Rather than invite previously excluded voices to speak to and through an "unmarked" center, the challenge is to "pivot the center" to allow all voices to be heard, reshaping the understanding of center by "'decentering' the dominant group."[47]

The continuation of a center/periphery assumption in Cobb's thinking is not necessarily inherent to a process ecological ethics, but it needs to be identified and critiqued for use in a liberationist approach. It appears in Cobb's appreciation for and use of the voices of the socially marginalized. Cobb has a high degree of sensitivity to and respect for the epistemological privilege of particularity in specific locations, situations, and experiences, but in his ethic they serve primarily to inform rather than reorient the project to the margins.[48] This may be a result of Cobb's decision, like that of Nash, to appeal to the mainline churches as his primary audience. Unlike Nash, however, Cobb is far more attuned to the perspectives and insights of women and minority voices. Given his audience and his own location, it may not be possible for him to privilege "subjugated knowledges" and shift his ethics to the margins, but it is important for a liberationist ecological ethics to do so.

With respect to the humanity/nature dichotomy that has characterized Western ethics, Cobb pays close attention to many facets of humanity's relationship to the rest of nature. These insights can be organized under three rubrics: human beings in nature; perspectives beyond anthropocentrism; and the question of value and worth.

According to Cobb's view on *humans in nature*, any understanding of humanity's relationship to nature must begin by affirming that human beings are fully part of nature. This means human culture is also fully part of nature. Human culture with its element of intelligent purpose is an immensely important factor in ecology which introduces new elements into the web of life. Dualisms that see humanity as wholly distinct from nature, or human culture distinct from and superior to "natural" human instincts fundamentally distort reality and need to be overturned.

Cobb is especially concerned with Christian insights that uncritically absorb these dualistic presuppositions. He critiques theological interpretation of the *imago Dei* that accommodates itself to Cartesian dualisms and mechanical views of the world, resulting in a view of nature as merely the stage for the drama of human history and redemption. Dominant Christian tradition has too often interpreted the *imago Dei* to mean that since humanity alone participates in the image of God, the destiny of human beings is of a wholly different order than that of all other creatures. Fellowship with God has been understood as made possible through spiritual transcendence of ahistorical nature.

Like Nash, Cobb is concerned to heal the history/nature dualism that informs these views. All of nature is evolving, thus all of nature has a historical dimension. From a process perspective there is no radical discontinuity between humanity and other creatures. Because all living creatures have capacity for some degree of experience, Cobb argues that the human acts of transcendence through intelligence, language, and culture are a rich form of experience that establishes our continuity with other living things. For Christians, the *imago Dei* can help us to recognize the unique role we have in the web of life because of the added element of intelligent purpose found in humanity, but it cannot be used to radically sever our continuity with and relationship to the rest of the natural order. Humanity is an inseparable part of nature.

In order to move *beyond anthropocentrism*, Cobb, in light of the above, critiques all forms of anthropocentric thinking in favor of an ecocentric framework. Like Nash, Cobb is able to argue his case from anthropocentric and instrumentalist angles when the context requires. Yet unlike Nash, who sees a mix of anthropocentric, ecocentric, and biocentric perspectives as the best fit for reality, Cobb argues for the need to shift our thinking completely beyond anthropocentrism. His main critique of anthropocentric thinking is that it does not fit reality but rather distorts our understanding of reality. If all creatures have intrinsic worth, then anthropocentric perspectives are inadequate.

Cobb is aware that all statements about reality are actually statements about our perceptions of reality. Yet he argues that seeing intrinsic worth in all creatures is not a question of human perception: "I am convinced that [other creatures] have value whether I perceive them to have value or not."[49] To support this claim, Cobb draws on biblical perspectives that oppose anthropocentrism and grant value to all of creation apart from their instrumental value to humans. Examining the Genesis creation accounts, he notes that "the same creation story that includes the doctrine of the *imago Dei* also states that God saw that the other creatures were good quite apart from human beings."[50] Significant in this story is that it does not say God *declared* creatures to be good. Rather, God *saw* that they were good. Their inherent worth does not depend on a declaration from God. Instead, in the Genesis accounts, creatures have value in themselves which is recognized by God. "We *are* of value. That is why God treats us as having value."[51]

Cobb argues further that *questions and conflicts having to do with value and worth* between humanity and nature must be grounded in a recognition of the intrinsic value of all creatures and a deep sense of the interconnectedness of all things. Both are important. Recognizing the intrinsic value of all creatures means that we can never treat other creatures solely as means for human use. Recognizing our interconnectedness means that in addition to intrinsic value, all creatures, including humans, have instrumental value for others. We are interdependent within and between species for our survival and capacity to thrive.

Other creatures do exist as means as well as ends with respect to humans (as we do with respect to them, Cobb adds). Cobb recognizes a hierarchy of intrinsic value in the natural world based on richness of experience. Ecological ethics must give guidelines for balancing considerations of intrinsic and instrumental worth based on what happens to richness of experience and capacity for richness of experience of the parties involved. With respect to animal/human conflict, Cobb argues that because of the significant intrinsic worth of animals, one such guideline is that it requires great human advantage to compensate for animal suffering and loss. Such human activity as killing individual animals or destroying species habitat for recreational purposes are proscribed by this guideline.

Yet Cobb is opposed to absolutist ethics based on equal or infinite intrinsic worth. Like Nash, he opposes animal rights positions that presuppose absolute egalitarianism in intrinsic worth of animals and other creatures. In response to charges that hierarchical views of intrinsic worth are just a form of speciesism, Cobb acknowledges that the logic and form of the arguments of animal rightists seem quite valid. There is no logical reason why an absolute respect for human life should not be extended to animal life. His disagreement with the argument is with its starting premise that there should be absolute respect for human life.

Cobb argues against absolute respect because he believes that differences in intrinsic worth are found not only between species, but also within species, including within the human species. Again, the criteria are richness of experience and capacity for richness of experience. Since the human capacity for richness of experience so far outdistances our actual experience, for the vast majority of human beings our intrinsic worth is, practically speaking, equivalent. It is in the extremes of life, such as humans existing in permanent vegetative state following an accident, that the distinction becomes morally relevant. Here the lack of any further capacity for richness of experience makes it justifiable—and in some cases morally imperative—that the person be allowed to die.

John Cobb's process ecological ethics makes at least two significant contributions that address shortcomings in Hall and Nash from a liberationist perspective. First, Cobb emphasizes the need for a new way of thinking that overcomes dualistic Enlightenment presuppositions. He argues that this is best done through a grounding in process metaphysics that shifts our epistemological assumptions from substance thinking to event thinking. Process thought may provide a promising framework for the needed shift in a liberationist ethics from essentialist forms of thinking—grounding our beliefs in assumptions of timeless, changeless "nature" or substance—to constructionist approaches that emphasize change and evolution of new patterns through the interaction of individuals, communities, and social values and practices.[52]

In addition, like Nash, Cobb emphasizes the ecology/economics interrelation as the crux of justice and ecological problems. He moves further than Nash, however, in drawing attention to the way economic forces, structures, and interactions have shaped and distorted our socially constructed understanding of humanity and the human/nature relation as inherently competitive, inherently individualistic, and inherently conflictual.[53] Cobb examines the modern economic assumptions and practices that underlie these views, and how they limit the possibilities of creating and maintaining human communities that relate to the land and surrounding biotic communities in ecologically sustainable ways.

These insights need to be expanded and reshaped, however, for an adequate liberationist ethics. With respect to the first point, Cobb is certainly right that we need fundamental changes in our ways of thinking. While Cobb is always sensitive to the consequences of any ethics for the poor and socially marginalized, what is often missing from his writing is explicit input and insights of nonprivileged perspectives.[54] Cobb is strongest in drawing on ecology and process thinking, yet both of these disciplines have emerged from and are located primarily within the dominant social locations of Western society. The insights they provide are important and should not be overlooked. The difficulty lies rather in the patterns Cobb's ethics reproduces. Cobb's approach subtly maintains a center/periphery framework where primary agency and epistemological perspective remain in the center with the affluent, rather than relocating the vision and reshaping the whole enterprise from the periphery among the marginalized of the human and wider earth community.

What is needed is more attention to the "insurrection of subjugated knowledges" of socially marginalized persons and communities that also is challenging modernity's Enlightenment assumptions and is moving us into and through the postmodern period.[55] As Brooten's feminist perspective demonstrates, these "subjugated knowledges" reveal different aspects about the socially constructed nature of our views of humanity and nature and provide important insights that incorporate the historical knowledge of struggle. This is what lies at the heart of the environmental justice movement where communities of color are coining terms such as environmental racism to connect their struggles against the environmental degradation of their communities with the forces and practices of economic and racial oppression.[56]

A similar critique needs to be made of Cobb's economic analysis and the insights it gives for understanding socially constructed views of humanity. His economic analysis needs to be broadened to include analysis of all social relations of domination along lines of social difference and how they intersect—namely through class, race, gender, and sexual orientation. We need richer, more complex, and multilayered perspectives on our economic practices, both of the constructed views of human nature they reflect, as well as how these views and the social praxes that inform them affect human relations with the rest of nature. Cobb makes room for this with his emphasis on justice and the perspective of the poor, but it does not seem to enter his analysis in his ecological ethics in any integral way.[57] Yet feminists and liberationists have insisted that these issues must be addressed simultaneously if we are to return to an economics for community that is more earth friendly; only by addressing social relations of domination built on gender, race, class, and sexuality can we have a true economics for community.

With respect to nature and the humanity/nature relation, Cobb offers a more promising grounding for a liberationist ethics than either Hall or Nash because his thoroughly ecological framework calls into question any fixed boundaries as natural. Because process ethics views all creation participating to some degree in experience and subjectivity, it is congruent with Haraway's insight that nature is made up of a matrix of human and nonhuman actors who all experience some sense of agency.[58]

Cobb's approach offers a helpful perspective on the shifting nature/culture boundary consistent with many of the points raised by Haraway. He argues that culture must be seen as fully a part of nature and consistent with natural evolution. Within nature, humanity has a unique positioning as an agent of purposive behavior, yet even this must be seen in continuity with rather than in distinction from other creatures. Human culture has introduced new elements into ecology that affect nature's agency. Yet a process perspective rejects radical discontinuities and sharp dualisms and boundaries, between human and nonhuman, between living and nonliving, between individual and community.

Cobb distances himself from the metaphors of competition that inform Nash's work to argue that within an ecological framework, symbiosis and cooperation are more important. Cobb reclaims theological positions such as the *imago Dei* and dominion to draw attention to humanity's distinctive positioning as an agent of intelligent purpose within the web of life, without, however, denying the uniqueness of other creatures, nonhuman agency, or our continuity with other creatures. For Cobb, dominion expresses a reality statement about humanity's positioning with respect to the rest of nature; denying dominion does not change that positioning or humanity's need to demonstrate ecological responsibility within our place in the web of life. Here Cobb is in agreement with Haraway's claims that ecology means taking seriously "social nature" that includes human interaction, rather than "saving nature" as a pure entity divorced from human activity.

Finally, a caution for any ethics such as Cobb's that draws heavily from the sciences for normative ethical criteria is Haraway's attention to the interrelation of science, cultural values, and moral criteria in naturalizing ethics. Cobb's process grounding with its attention to constantly evolving and changing experience and insights is a built-in guard against this, but it needs to be made explicit in making the connections between science and ethics. One needs constantly to examine how scientific representations interact with cultural constructions of race, gender, sexuality, and class, and how these then influence ethical criteria.

Because process thought calls for a fundamental rethinking of traditional Western influences on theological assumptions, Cobb devotes a good deal of attention to reworking understandings of God and the God/nature/humanity relationship. Working first within the ecological model, Cobb argues that God is best understood as Life itself. Life brings order out of chaos, and is the only power that creates value and freedom. Yet the power of God as Life is not limited to living things: "Indeed, there is no definite boundary between living and nonliving things, and we may think of Life as exerting its gentle pressure everywhere, encouraging each thing to become something more than it is." Life is a cosmic principle that works for higher order in the midst of entropy—the universewide process of movement toward ever-greater disorder. "Wherever it works it can be trusted to transform the inanimate into living matter and to quicken the spirit."[59] Trusting God thus means trusting Life, which can help us change from attitudes that lead us to distrust and try to control nature to supporting other living things and recognizing how God as Life connects all living things. One ramification of this in Western cultures would be to overcome anti-body dualisms, helping us to overcome estrangement from our bodies and the Life that works in them.

How does this understanding of God as Life relate to traditional Christian understandings of God and God's attributes? Here Cobb's process theological categories become more explicit. For the traditional claim that God is omnipotent, for example, if omnipotence means the ability to bring about new things without relevance to the context and surrounding conditions, Life is not omnipotent. Yet Life can be understood as omnipotent through process understandings of God: it is the cosmic power that works in gentle and persuasive ways wherever conditions are possible to expand life, order, and freedom. Similarly, Life is Creator in that it is the only power that creates value, freedom, and order, working everywhere and always to this end. "Every power other than Life erodes order. Only Life creates it. If by the world we mean the encompassing meaningful order, then Life is the creator of the world. Life creates by bringing order out of chaos."[60]

Cobb's process view of God as Life calls into question some traditional Christian doctrinal views such as eschatology and teleology. Rather than seeing the process of the world aimed inevitably at some remote omega point, Cobb argues that its teleology is simply the creation of values moment by moment. This means that Life does not aim specifically at the creation of human beings, nor does it have any one particular goal for the course of evolution on our planet. In fact, there will come a day when earth will no longer provide the conditions that make living things possible, and Life will not prevent that eventuality. Yet Life as God can be trusted to continue to have the last word over death. We cannot know what form Life's victory will take, but faith in Life is the confidence that Life is not finally defeated.

Cobb explicitly opposes traditional Western theism that since Aristotle's "Unmoved Mover" has seen God as unaffected by other things. God has been defined in terms of divine immutability, impassibility, and aseity—possessing being in and for itself. There is no true relationality between God and the world: everything is caused by God but there is no reciprocal effect of the world on God. In contrast to this, Cobb sums up the contribution of Whitehead's philosophy to understanding God as Life as follows:

> We agree with Whitehead that this [Western theistic] synthesis is profoundly perverse. It pictures the living God of the Bible as an object having only external relations. True perfection consists not in excluding everything but in including everything. The Primordial Nature of God is the inclusion within God of the entire sphere of possibility, realized and unrealized in the universe. God so orders possibilities that they may be realized in the world, each in its due season. . . . But like all living things, God not only acts on others, but also takes account of others in the divine self-constitution. The Consequent Nature is God's perfect responsiveness to the joys and sufferings of the world. God is not the world, and the world is not God. But God includes the world, and the world includes God. God perfects the

world and the world perfects God. There is no world apart from God, and there is no God apart from some world.[61]

Cobb's theology is not pantheistic. He does not equate God with the world. He argues that whereas no world can exist without God, God can exist without *this* world. "But since God, like all living things, only perfectly, embodies the principle of internal relations, God's life depends on there being some world to include."[62] The fundamental relationality of divine reality requires that there is some world to which God relates. In this sense God is best described as panentheistic. God "is not to be found somewhere outside the organisms in which it is at work, but it is not to be identified with them either."[63]

While Cobb draws on process metaphysics to develop the ecological view of God as Life, he continues to work within an explicitly Christian framework. In addressing how the view of God as Life relates to the Bible, he notes that the Bible witnesses throughout to a changing understanding of God, and subsequent diverse Christian theologies have developed quite different doctrines of God. Hence, no one today can simply believe in "the" biblical God: "Merely to repeat biblical ideals in the context of a quite different world view would not be faithful to the spiritual dynamic that is manifest in the Bible. We are more concerned that faith in our time give us the same freedom which, according to Paul, faith brought to believers in his day."[64]

Yet Cobb is concerned that witness to God as Life be an appropriate continuation of the biblical witness. With Birch he notes that of the many images the Bible employs of the divine, no image is more central than Life. It is closely bound to both the Spirit and the Word. Much like the biblical image that the Spirit breathes life and is itself the Life of God, Cobb argues that it is the divine Life that works within us and makes us alive.

Cobb draws on this understanding of God working within us through persuasion to explore the relation of human moral agency to grace. Moral agency is made possible through human transcendence over genetic endowment and cultural conditioning. Yet we recognize in ourselves profound resistance to change, so that human freedom is not simply choosing between good and evil. Human self-centeredness distorts our use of freedom, "but we discover that there is a power at work in us that can transform even our distorted wills. This transformation is not subject to our control but comes as a gift. We call it grace, and we can place no limits on the extent to which grace can make us into new men and new women."[65] It is the transformative power of grace, of Life working within us, that provides grounds for hope.

In similar fashion, Cobb connects this process view of God as Life working in the world with the Christian tradition of discernment of Christ. Cobb maintains that instead of aligning themselves with an omnipotent God seen as forcing divine decisions on the world, Christians need to be about discerning where Christ as the Incarnate Logos is at work in the world in "quiet works of creative love." Christ works through the persuasive powers of Life:

> If our eyes are opened by faith, we see Christ wherever we look. We see Christ in the aspirations for justice and freedom on the part of the oppressed and in the glimmering desire of the oppressor to grant justice and freedom. Christ appears most strikingly in the miracle of conversion when something radically new enters a person's life and all that was there before takes on changed meaning. . . . Wherever human beings are reaching out from themselves, wherever there is growth toward spirit, wherever there is hunger for God, wherever through the interaction of people a new intimacy comes into being, we discern the work and presence of Christ. . . . In quietness and in unexpected places Christ is bringing something new to birth, something we cannot foresee and build our plans upon.[66]

Christians need to maintain an attitude of expectancy to discern Christ's work and be open to being transformed by it. This is given further content in the cross, which testifies that hope stands in closest proximity to sacrifice and solidarity with the disadvantaged.

To support moving beyond anthropocentrism, Cobb looks to the Bible to develop a theocentric perspective that can support an ecocentric framework. Like Nash, he notes that the Bible's focus on human beings is derivative from its understanding of God: "It is because God made us in God's image, because God cares for us, because God sent Jesus to suffer for us, that we are to appreciate our own worth and care for one another."[67] Cobb warns against two common misunderstandings that frequently result from distortions of the biblical theocentric perspective. One is that human beings and other creatures have worth only because God declares us to have value. Instead, Cobb argues, creation really does have value, and in Genesis God is said to see this, as mentioned previously. A second distortion results from so separating God from the world that service to God can be separated from service to fellow creatures. Rather, Cobb argues, it is precisely in and through service to other creatures that God is served.

In these discussions Cobb affirms and draws on the biblical doctrines of *imago Dei* and dominion. Against interpretations of the *imago Dei* and dominion that assign intrinsic worth to human beings and only instrumental worth to other creatures, he argues that the very passage in Genesis that assigns the *imago Dei* only to human beings also asserts clearly the intrinsic value of other creatures. God *saw* that they were good. He notes also that dominion in the biblical context is exercised for the sake of the ruled, not for the sake of the ruler alone. It is assumed that what is ruled has value, and not simply instrumental value to the ruler.

Hence Cobb invokes the *imago Dei* to argue against the deep ecology assertion that human beings are merely one species among many, no more and no less privileged than the others. "Even before the fall human beings, although certainly one species among others, are also differentiated. We were not *simply* one species among others. We were created in the image of God. We were assigned a particular privilege and a particular responsibility."[68] Too often in Christianity people have misread this specialness and abused human privilege and responsibility. Yet the idea that human specialness is itself the problem—that we do not in fact have both privilege and responsibility with respect to other species—is foreign to the Christian scriptures and belief.

Related to this, Cobb argues that there can be no turning back to a pre-fall state of natural innocence in our relations with other creatures. The way forward is through a deepening of knowledge rather than through trying to return to a purely natural attitude that may have preceded the "fall upward" portrayed in the eating of the tree of knowledge of good and evil. Cobb contrasts these two positions: "The difference here could be put in theological terms as follows. For the particular deep ecology position I have described, the fall is an unmitigated disaster. The only possible form of health is the one that was lost in that fall. Our only hope is to return as far as that is possible to the earlier condition. For Christianity the fall is ambiguous. Something of great value was lost. Life since the fall has been beset by great evils. But the salvation that is mediated to us by Christ exceeds in value the innocence that preceded the fall."[69] The difference for Christians is that in Christ they find a deepening wisdom that involves a heightened self-transcendence rather than a return to innocence.

In looking at the implications of this for humanity's relation to other creatures, Cobb explicitly affirms the biblical doctrine of dominion. He acknowledges the deep ecology critique of Western perversions of dominion, but insists on the validity of its basic insight. For Cobb, human beings have dominion whether we want it or not. The question is not whether to maintain it or relinquish it, as some would argue, but *how* human beings should exercise it. Human beings are no longer simply one species among others in a mutual interspecies rivalry, but are uniquely positioned because of the enormous advantages we hold in that rivalry. The doctrine of dominion helps to focus on the important issues at stake: what are the appropriate limits of humanity in its dominion over other creatures, for the sake of their well-being as well as ours, as responsible representatives created in God's image?

In making these claims about the intrinsic value of all creatures and humanity's relation to otherkind, Cobb draws on the Christian belief that there is a privileged source or perspective, that of God.

Some interpretations have argued the privilege of God's perspective based on belief in God's power to force others to conform. Cobb repudiates this view. Rather, combining biblical and process insights, he argues that God's perspective is privileged because it includes all other perspectives. One ethical ramification of this is Cobb's belief that the more inclusive the view, the more privileged it is. In addition, having a divine, inclusive perspective can help move ethical dilemmas past the innumerable, fragmented, and partial perspectives that conflict by providing a broader framework. One can ask, what is the contribution to the whole—to God—in a particular dilemma. An example is whether or not to protect biodiversity for its own sake. Cobb's intuition that biodiversity contributes to the value of the whole beyond the mere sum of its parts is supported by a holistic, inclusive view of God. Hence, a privileged inclusive perspective can better protect nonhuman nature from a solely anthropocentric, human instrumental assessment and practice.

In light of lesbian and gay insights, the process understanding of God has numerous helpful elements for a liberationist ecological ethic that move us past some of the problems noted in the ethics of Nash and Hall. The process rejection of radical discontinuities and rigid boundaries gives a metaphysical grounding for rejection of hierarchical dualisms. While God and creation are understood to be distinct from each other, emphasis is placed on the interconnections between them. A process understanding of panentheism does not equate God with the world, but likewise maintains that God depends on there being a world with which to relate. It thus affirms the liberationist ecological emphasis on full relationality and interdependence.

Cobb's process perspective provides a helpful interpretation of the relation between divine omnipotence and limits. Power as the ability to empower and cause things is never divorced from context. Hence God as the cosmic power works wherever conditions are possible to expand life, order, and freedom. God works with our agency and through the subjectivity of all creatures and created things to bring life, order, and relationship out of chaos. But where conditions do not exist or are thwarted, God does not impose a will that violates the context. This view provides a helpful grounding for Clark's emphasis on the divine as both co-suffering companion and source of empowerment. It is consistent with the ecological and justice emphases of a liberationist ethics.

Finally, a process view of God critiques traditional understandings of teleological eschatology such as Nash's that ground their hope in a final overcoming of finitude and death. There is no omega point towards which creation is moving inexorably to encounter redemption. Rather God as the cosmic principle of life "aims" at increasing the creation of values moment by moment. Such a view stresses both the ecological interdependency and the importance of human agency and subjectivity within the larger fabric of life. Nash's critique that "a God who saves only by memorializing 'has beens' in a flawless memory is not the Suffering Servant, but rather the Supreme Ego who makes all creatures into suffering servants, sacrificed for the sake of God's greater glory,"[70] is premised on maintaining rigid boundaries between God and individuals and between individual creatures, rather than preserving the tension between the interconnectedness and personal subjectivity of all that process thought advocates. Faith in the future in a process view holds that while we cannot know what form it will take, God as Life will continue in the face of chaos and death and we will participate in some form in this ongoing development of God.

While Cobb's process view of God is suggestive in many areas for a liberationist ecological ethic, a few caveats remain that must be further developed. Cobb's emphasis on the central metaphor of God as Life has a powerful resonance in ecological circles and among "two-thirds world" liberationist communities who long have termed their theologies of liberation as theologies of life.[71] The image of Life has different connotations in the United States, however, in the wake of two decades of acrimonious debate on abortion between pro-life and pro-choice forces. Clearer attention must be paid to the social consequences of placing at the center the metaphor of Life for God in a liberationist ecological ethic.[72]

Cobb has a constructive interpretation of the traditional Christian belief in the privilege of God's perspective. Rather than basing this in the classical attributes of divine omnipotence or omniscience, Cobb develops the criterion of inclusiveness: from a process view, God's perspective is privileged because it includes all other perspectives. While inclusiveness is a helpful criterion, it still begs the questions of relative power in inclusion. In a limited world where total inclusion is not possible in trying to discern God's perspective, whose voices are included, and what relative weight is given to them?

Two additional criteria from feminist and liberationist thought suggest a way to complement Cobb's view. Liberation theologies have drawn attention to the epistemological privilege of the poor and socially marginalized to address the question of inclusion and weightedness. In an ecological context this needs to be expanded to include the voices of other creatures in nature.[73] Similarly, Haraway's work on situated knowledges and partial perspectives as a surer guide toward accountable perspectives offers a caution on moving too quickly to a presumed universal and inclusive (divine) perspective that actually functions to mask and obscure its partial and limited nature.[74] Her insights on articulation in coalitions rather than "voiceless representation" offer suggestive possibilities for developing the criterion of inclusiveness as a way of discerning divine perspectives.

Finally, a danger with process views of God from a liberationist perspective is the disappearance of the prophetic voice. God as Life works through "gentle and persuasive ways" and Christ as the Incarnate Logos is found "in quiet works of creative love."[75] One wonders if there is a reflection of liberal, middle-class values and constructs in the predominance of these images of God in Cobb's work.[76] More adequate for a liberationist ecological ethic is Rosemary Ruether's emphasis (discussed in the next chapter) that we need both the prophetic and ecological voices in our images and understanding of the divine, both the thunderous, prophetic voice of God and the beckoning, persuasive voice of Gaia.

To summarize, like Nash and Hall, John Cobb Jr. develops his ethics of sustainability and interdependent community primarily for a mainline North American church audience. Yet because his methodological approach radically critiques all hierarchical dualisms that inform human thinking about humanity, nature, and God, and because of his sensitivity to hearing marginalized voices both within and outside the North American context, Cobb's ethic has a stronger appeal to a liberationist approach than either an ethics of stewardship or realist responsibility.

Cobb's starting point for his ethic is a theme critical to any ethics of ecojustice: the need to change our way of thinking by challenging inherited values and assumptions—particularly those of the Enlightenment that have supported modern ecologically damaging practices. He demonstrates the importance of paying attention to metaphysical assumptions and groundings that (often unconsciously) shape our views of important theological and ecological categories such as humanity, nature, God, and the relations between them. Cobb explicitly embraces a postmodernist stance that incorporates constructionist thinking (without falling into a radical constructionist relativism) and argues persuasively that this stance is a more appropriate and promising one from which to engage the Christian scriptures and tradition than modernist perspectives.

Cobb's combined attention to scientific and biblical claims grounded in a process hermeneutics leads him to helpful theological reinterpretations compatible with ecological insights. Because Cobb's model sees God as internally related to the world, his panentheism avoids reproducing self/other and spirit/matter dualisms that appear in other models, without collapsing the relationship into pantheism. This highlights the important liberationist ecological emphasis on full relationality and interdependence.

The primary inadequacy with Cobb's process ethics of sustainability and interdependent community lies in its continuation of a center/periphery dynamic where marginalized voices at the periphery must still relate to the more dominant (usually unmarked) voices at the center. A liberationist ecological ethic must make an epistemological shift away from a center/periphery model. Even a supportive model such as Cobb's where peripheral voices are welcomed and heard is inadequate without decentering the model of relation and shifting to an interconnected web of relations where some "nodes" are epistemologically privileged because of their ability to see more critically and accurately from the vantage point of their location in the web.

Cobb is correct that fundamental to any ecological ethics must be a new way of seeing, but this new vision will be inadequate without giving epistemological priority to those subjugated knowledges too often missing from the dominant discourse. The epistemological dialectic of ecojustice must be maintained in an ethic of ecojustice. Cobb's process-ecological model provides an important contribution to the ecological side of this dialectic, but without the explicit inclusion of socially dominated voices, our views of *both* ecology and justice will be distorted and inadequate.

None of these problems with Cobb's process ethic of sustainability and interdependent community are necessarily inherent to the model itself; with the explicit incorporation of a liberationist hermeneutics and praxis, this model offers many possibilities as a grounding for a liberationist ecological ethic. The most adequate of the ecological models, however, is Ruether's ecofeminist ethic, to which we now turn.

9 Ecofeminist Theology: An Ethic of Justice, Interrelatedness, and Earth Healing

[In God and Gaia] we hear two voices of divinity from nature. One speaks from the mountaintops in the thunderous masculine tones of "thou shalt" and "thou shalt not." It is the voice of power and law, but speaking (at its most authentic) on behalf of the weak, as a mandate to protect the powerless and to restrain the power of the mighty. There is another voice, one that speaks from the intimate heart of matter. It has long been silenced by the masculine voice, but today is finding again her own voice. This is the voice of Gaia. Her voice does not translate into laws or intellectual knowledge, but beckons us into communion. . . . We need both of these holy voices.

—Rosemary Radford Ruether, *Gaia and God*

Making the connections between the historical manipulation and devaluing of women and nature, ecofeminism combines ecology and feminism to explore the interconnections of male domination of women and domination of nature in cultural ideology and social structures.[1] *Ecofeminism* as a term has been used to describe both the range of women's efforts to save the earth and the transformation of Western forms of feminism that result from new views of women and nature.[2] Ecofeminist theology has emerged in recent years as the effort to ground theological reflection in the insights and praxis of ecofeminism.

Rosemary Radford Ruether, professor of theology at Garrett-Evangelical Theological Seminary, has a long history of connecting ecological and feminist issues in her writings and speaking that predates even the coining of the term ecofeminism. In 1975 she published *New Woman, New Earth: Sexist Ideologies and Human Liberation*. Its final chapter, "New Woman and New Earth: Women, Ecology, and Social Revolution," was one of the earliest theological explorations of the connections between the oppression of women and domination of nature. Her 1983 book, *Sexism and God-talk: Toward a Feminist Theology*, continued this examination in a chapter titled "Women, Body, and Nature: Sexism and the Theology of Creation."

In her 1992 book, *Gaia and God: An Ecofeminist Theology of Earth Healing*, Ruether engages in extensive critique and assessment of the Western Judeo-Christian legacy for both its contribution to the current ecological crisis and for constructive resources it might provide for reversing the ecocrisis. Drawing on her training as a historian, she explores the creation and destruction myths of the Ancient Near Eastern, Hebraic, and Christian traditions and their teachings on evil and sin in order to recover usable portions to propose an ethic of earth healing. Because of its explicitly liberationist stance and commitment to deconstructive as well as constructive engagement of the Western and Christian traditions, Ruether's ecofeminist approach offers the most promise of the current models of ecological ethics for an ethic of Gay and Gaia.

Are Gaia and God on speaking terms with each other? With this question—do the living and sacred earth and the monotheistic deity of the biblical traditions have anything to say to each other?—Ruether begins her evaluation of the heritage of Western Christian culture. Her goal is a theology and ethic of earth healing, possible only through recognizing and transforming the ways Western Christian culture has justified and perpetrated relations of domination between men and women, classes and nations, and humans and the earth. Grounded in an ecofeminist perspective and praxis, Ruether describes her purpose in *Gaia and God* to be "to assess the cultural and social roots that have promoted destructive relations between men and women, between ruling and subjugated human groups, and the destruction of the rest of the biotic community, of which humans are an interdependent part."[3] Out of this assessment may come usable ideas from the legacy of the Western and Christian cultural heritage that can help nourish healing relations with one another and with the earth.

Ruether's starting premise is that destructive and dominating relations to the earth are interrelated with gender, class, and racial domination in the human community. Addressing this interweaving of social relations of domination demands ecojustice: simultaneously addressing domination of the earth and social domination and understanding the ways they mutually reinforce each other. Ruether maintains that those of us in the West inherit from our traditions both "cultures of domination and deceit and cultures of critique and compassion."[4] Our task is to build community on the second culture in order to unmask and overturn the power of the first. Such healed relationships, however, must be accompanied by a new consciousness and transformed inner psyches. Ecological healing thus is a theological and psychic-spiritual process that requires new ways of symbolizing relationships between women and men, humanity and the earth, humans and the divine, and the divine and the earth.[5]

To evaluate the heritage of Western Christian culture, Ruether combines ecology and feminism in the unified perspective of ecofeminism. Ruether understands *ecology* in its broadest sense, as a combined socioeconomic and biological science that examines how natural communities function to sustain their health and how they become disrupted, particularly through human misuse of "nature." *Deep ecology* takes ecology to another level to examine the symbolic, psychological, and ethical patterns of destructive human relations with nature. *Feminism* seeks the well-being and full equality of women. It has many different forms, including *socialist feminism* that argues such equality is not possible apart from transformation of the social relations of ownership of the means of production and reproduction, and *radical feminism* that digs more deeply to examine the patterns of culture and consciousness that maintain male domination over women. Ruether's use of *ecofeminism* draws on ecology and feminism in their deep forms to examine how male domination over women and domination of nature are interconnected in cultural ideology and social structures.[6]

This is a particularly important task theologically, because the Judeo-Christian understanding of a male monotheistic creator god has served symbolically to reinforce relations of social domination, including men over women, masters over slaves, and humans (mostly ruling class males) over animals and the earth. Here an ecofeminist perspective is critical to understand the linkage of sexism with exploitation of the earth: "Domination of women has provided a key link, both socially and symbolically, to the domination of the earth, hence the tendency in patriarchal culture to link women with earth, matter, and nature, while identifying males with sky, intellect, and transcendent spirit."[7]

Ruether notes that sacralizing relations of domination in classical traditions is only part of the story. In addition these traditions struggled with what they perceived to be sin, evil, and injustice, and sought to create what they understood to be just and loving relations between people in their relation to the earth and the divine. Noting that "it would be surprising indeed if there were no positive insights that could be reclaimed from three thousand years of collective human struggle about the meaning of life and the way to live justly and well," Ruether looks for glimpses of a "precious legacy" of transformed relations that can be separated as recoverable resources "from the toxic waste of sacralized domination."[8]

Like Nash and Cobb, Ruether is explicit about the role her social location plays in shaping her theology and ethics. She recalls the "deep ecumenism" of growing up in a combined Protestant and Roman Catholic family with a variety of religious identities in her extended family that made a multiplicity of perspectives living together seem normal. Critical reflection on her experience as a woman inside and outside the church has led her to affirm an explicitly feminist stance, which has broadened to include sensitivity to issues of race and ethnicity, class, sexuality, and the earth. She holds herself accountable to Western Christian culture and society while being open to learning from others—both those outside her Western context and minority or subjugated perspectives in the West. Within Western society and the church she has been most active in communities of women, and it is to them that she holds her primary accountability.[9]

In her focus on critiquing and assessing classical religious traditions, Ruether is explicit about a number of presuppositions that shape her work. Each of these in turn form an important grounding for an ethics of Gay and Gaia.

First, because the ecological crisis is new to human experience, there is no ready-made ecological ethic or spirituality in past traditions. All previous traditions are inadequate in the face of current vast ecological devastation. Any elements drawn from older traditions must be reinterpreted to be made useful today.

Second, Ruether maintains that each tradition is best explored by persons who claim community in that tradition; hence she begins with Western Christian culture. Critique and reinterpretation of our traditions must begin in the context of communities of accountability in order to be able to then draw on the spiritual resources of "a global community of interrelated spiritual traditions."[10]

Third, the Christian tradition does contain important and valuable themes for ecological spirituality and ethics. It also has severe problems and defects, so critical assessment is necessary. Yet because it is the religious idiom of over one billion persons worldwide, any attempt to develop a global ecological consciousness must find ways to be intelligible in a Christian paradigm.

Fourth, healing is possible only through recognition and transformation of the way Western culture has enshrined social relations of domination. In this process, spirituality and social change are inseparably intertwined. Transformed psyches are needed to reshape both ideas and social relations.[11]

Fifth, the most helpful way of doing this is by bringing together the deepest forms of ecology and feminism to explore the connections of male domination of women and nature. In focusing on the Western and Christian tradition, Ruether makes no a priori claims for truth or privilege of this culture, but argues instead for a plurality of responses and ecofeminisms as what is needed for healing at this time.[12]

Finally, Ruether brings a positive evaluation and openness to nonbiblical traditions and religious insights. Against the ahistorical ideology of Christian doctrinal purity, Ruether argues that Christianity's historical syncretistic ability to synthesize many elements from diverse sources should be recognized as a strength to be built on to help provide the creativity needed to address today's ecospiritual crisis.[13]

With this overview of Ruether's methodological approach, let us turn now to the structure and content of her argument. Ruether structures *Gaia and God* in four sections under the headings "Creation," "Destruction," "Domination and Deceit," and "Healing." She notes that these themes correspond to traditional Christian theological categories of creation, judgment, sin and fallenness, and redemption, but her broader headings allow her to include nonbiblical traditions in her evaluation of the Western heritage. In each section she pairs analysis of historical myths and narratives with contemporary alternatives in order to examine the social values inherent in each. Her goal is to critique and assess the contributions of each tradition in order to suggest a more holistic meaning of these concepts for a healing ecological ethics.

In the first section, Ruether looks at Western creation myths as the societal blueprints for cultures that give a combined scientific, social/ethical, and theological/spiritual worldview. Creation stories serve both to reflect the worldview of a culture as well as to frame the parameters the worldview passes on to its heirs: "They reflect the assumptions about how the divine and the mortal, the mental and the physical, humans and other humans, male and female, humans, plants, animals, land, waters, and stars are related to each other."[14]

Ruether begins by examining the Babylonian, Hebrew, and Greek Platonic creation stories as the primary narratives that have shaped the biblical and Christian tradition. Her purpose is to show how these stories functioned to codify and sacralize patriarchal relations of domination. While the Hebrew creation story in Genesis provides the normative creation account for Christianity, behind this story is the more ancient Babylonian account found in the *Enuma Elish*. The composition of the Genesis account by the Hebrew priestly authors reflects their selective appropriation and correction of the earlier Babylonian account. Similarly, while Christians drew explicitly on Genesis accounts in formulating the Christian doctrine of creation, the Greek creation myth found in Plato's *Timaeus* provided the cosmology considered scientifically normative in the West through the Middle Ages. It therefore furnished the cosmological filter through which Christians read the Genesis creation account for fifteen hundred years.[15] A review of Ruether's study here is instructive for how this kind of constructionist engagement of the tradition and what lies behind the tradition reveals the obstacles and resources available either to facilitate or obstruct a liberating social and ecological praxis.

Ruether begins her discussion of the *Enuma Elish* by noting that this Babylonian creation account was itself derived from earlier stories in the Sumerian world that began with a primal mother, seen as the origin of both the cosmos and the gods. The primal mother gives birth to successive generations representing the generation of the cosmos: first, the primal parents, Heaven and Earth; second, the primal cosmic forces, water, air, and vegetation; and finally, the anthropomorphic gods and goddesses who represent the ruling classes of the new city states. The *Enuma Elish* reworks these stories to portray the ascendancy of its male god, Marduk, over the surrounding cities and land. Ruether notes how these creation accounts reflect actual struggle at the time with other humans and with the land: "[Like the Sumerian stories] the Babylonian story also assumes an intergenerational struggle between the older and younger deities, a struggle that represents both the political conquests of younger states over older states and villages, and also the struggle to harness and organize the machinery of political control and control of land and water against the 'chaotic' social and natural forces that erupted periodically against this order."[16] A significant change from the earlier stories is that the mother goddess, Tiamat, and her consorts now represent the forces of "chaos" that threaten Marduk, who represents the new Babylonian dynasty. Marduk responds by capturing and killing Tiamat and her consorts and using their dead bodies to fashion the world. To free the gods for leisure, he forms humans from blood and clay and imposes servitude on them.

Ruether finds several powerful social messages in this account. First, the early Babylonian rulers knew that they and their god, Marduk, did not preexist the world. Rather they arose from a recent generation of power out of earlier stages of development. Second, the earlier world from which they arose was seen as matriarchal, with a dominant female goddess and subordinate male consorts. This is reversed so the male god Marduk is now dominant with his subordinate female consorts. Third, earlier models of generation of the cosmos were seen as a result of a combined male and female effort: "Apsu, the primordial begetter of all things, commingles in a single body with Tiamat, who bears all things. The gods and goddesses gestate within this commingled male-female union."[17]

In contrast to this, Marduk's male power is military and architectonic: he fashions the world from the dead primordial matter of Tiamat's and her consorts' bodies: "This transition from reproductive to artisan metaphors for cosmogenesis indicates a deeper confidence in the appropriation of 'matter' by the new ruling class. Life begotten and gestated has its own autonomous principle of life. Dead matter, fashioned into artifacts, makes the cosmos the private possession of its 'creators.' Even though the new lords remember that they once were gestated out of the living body of the mother, they now stand astride her dead body and take possession of it as an object of their ownership and control."[18] Finally, the Babylonian story also mandates class hierarchy of rulers and slaves. Leisure, based on expropriated labor, becomes the mark that sets the aristocracy apart from workers and slaves, and identifies the aristocracy with the gods.

What continuities and differences does the Hebrew story share with the *Enuma Elish*? The Genesis account begins with the strife between the creator and the primal mother already eliminated: the mother has been reduced to formless matter that the creator shapes into creatures. Yet unlike the Babylonian story, the creator here both works and rests, and commands humans to do the same. The seventh day of the week becomes a day of rest for all. Like the rest of creation, humans are formed from the primordial stuff; unlike the other creatures, however, humans are created in the creator's image and given dominion over the other animals, plants, and the land.

Among the social messages in the Hebrew story, Ruether notes that God is now modeled after the intellectual power of the priestly class, calling things into being through ritual naming. Second, the leisure/labor distinction between God and humans and between rulers and workers is abolished. Human beings are seen as servants of God, but not slaves: it is a servanthood of royalty that manifests itself as representatives of God's rule on earth through dominion over the rest of creation.

What about messages of domination in the story? There is no explicit hierarchy of one class of humans over another as in the Babylonian story that mandates male over female and aristocrat over slave. Yet in terms of gender relations, while both male and female are seen as created in the image of God, Ruether argues that the male pronouns used for God and for Adam suggest that males are the more appropriate representatives of God. Females share the benefits of corporate human sovereignty, but also fall under the rule of the male patriarch who heads the family. The appropriation of an earlier folk story into the second Genesis creation account makes this message explicit. Ruether argues that the specific intent of this older story is to mandate the patriarchal relationship of husband and wife.[19]

In response to the environmentalist critique of the biblical portrayal of humanity through Adam being given dominion over the rest of creation, Ruether first concedes that the story is explicitly anthropocentric. The natural world culminates in the creation of humanity, which is given sovereignty over the rest of creation. Yet to be fair to the story, Ruether argues that neither exploitative nor destructive rule over the earth is intended. Yahweh—not humanity—remains the owner or possessor of the earth and its creatures. As Hall noted, human dominion is given in the form of royal stewardship, which implies careful tending on behalf of the true owner. Ruether notes that restriction of human use of animals in the Levitical laws, such as the inclusion of animals in the sabbatical rest, makes explicit the limit of human rights to the lives of other creatures. The Hebrew word itself for human, *adam*, derived from *adamah*, the earth, reinforces this sense of kinship with the earth and otherkind.

In turning to Plato's creation story, the *Timaeus*, Ruether notes its very different style: it is more abstract and philosophical in contrast to the mythic personification of the Babylonian and Hebrew stories. Plato's account begins by defining primal reality as dualistic, divided into the eternal and invisible realm of thought and the temporal and visible realm of matter. Like the gods in the Babylonian and Hebrew stories, the Demiurgos creates by "making": "The metaphor for cosmogenesis is taken from the work of the artisan, who shapes things from dead stuff, not from the reproductive process of begetting and gestating."[20] Ruether notes that this ontological shift from reproduction to making has important ecological implications in that it gives the cosmos the status of a possessed object, distinct from the self-subsistent life of the divine.

In shaping the primal elements of fire, air, water, and earth into the spherical body of the cosmos with earth at the center, the Demiurgos fashions a universe that is geocentric and hierarchic. Next the Creator shapes a world soul which becomes the principle of life and motion in the cosmic body. Beings appropriate to each realm are formed: gods in the planets and stars, birds in the air, fish in the sea, animals on earth, and so forth. The creation of human beings follows:

> The Demiurgos then proceeds to shape human souls from the same elements from which he had mixed the world soul, but in

> a more diluted form. This soul mixture he divides up into parts and places in the stars, where they receive a celestial education in the eternal nature of reality. Creating the bodies of these beings is too low a task for the Demiurgos and is assigned to the planetary gods. Once the souls have received their celestial infusion of truth, they are incarnated into the male bodies. There their task is to control the chaotic sensations that arise from the body.
>
> If the souls succeed in this task, they will shed the body and return at death to their "native" star, there to have a "blessed and congenial existence" (like the gods). But if one fails to attain this control over the body and its sensations, the soul will be reincarnated and pass in a second birth into a woman. If in that state he [*sic*] doesn't desist from evil, he will be reincarnated as a "brute" who resembles the "evil" nature into which he has fallen. This round of incarnation will continue until the soul masters the body and returns to his "first and better state," that is, as a (ruling class) male human, winning finally his return to his original disincarnate state in his star above.[21]

At least two important social messages come out of this story. First, reality is seen as fundamentally dualistic, divided between mind and body. Only the first, mind or consciousness, is primal, eternal, and good. The body, visible corporeality, is secondary and derivative; it is the source of evil in the form of physical sensations that must be mastered by the mind. The true home of the soul or mind is the world of the stars away from earth and body; incarnation in the body is merely a preparatory "testing place." Second, the primal dualism of mind over body is reproduced in the hierarchy of male over female, human over animal, and rulers over workers. Hence Plato sees male domination, class hierarchy, and human superiority over animals as manifestations of the primal division of reality into consciousness over body. They are to be accepted as a given part of the natural order. Pervading this mandate for social domination is the additional cultural attitude of alienation from body and earth. Salvation consists of the male mind freeing itself from bodily mortality in order to secure the immortal life of the soul.

In reading the Genesis creation account through the eyes of the accepted Greek science and philosophy of the day, early Christianity created a cosmological synthesis of ancient Near Eastern, Hebrew, Greek, and Christian ideas. One result is an unresolved ambiguity in Christian views of the divine/cosmos relation. To protect God's absolute sovereignty, Christian philosophical theology argued that God must be the source of all matter, which God creates out of nothing.[22] Hence matter is seen both as an emanation from divine being and grounded in divine being, but also as a kind of created being outside of God, neither divine nor immortal. The result is an ambiguity in the Christian view of God: "The Christian view seems to want to span two concepts of the divine-cosmos relation, seeing God as a totally distinct, eternal, and self-subsisting Being, over against the non-divine, dependent status of created being; and yet also, in some sense, seeing creation as welling up out of and being sustained in existence through the being of God."[23] This unresolved ambiguity has been played out historically in Christianity's ambiguous attitude toward and relation with the rest of the natural world.[24]

Christianity attempted various syntheses of the Platonic belief in the immortality of the human soul with the Hebrew belief that the soul as the life principle of the body will be resurrected in the body on a renovated earth. Christianity retained the Platonic symbolism of the rational soul as masculine and the body and its passions as feminine, but rejected reincarnation as an explanation for social hierarchy. Rather each soul is seen as a unique creation of God with equal capacity for holiness and redemption. One result of this is an unexplained division in Christianity between equality of souls in their relation to God and yet inequality of persons and status in society due to gender and class hierarchy. Female subordination in this world is a "natural" result of the inferiority of the female body and personality and as punishment for causing original sin, but paradoxically women are seen as equal before God with regard to salvation: "Thus God seems to will different principles in creation and redemption. In creation women are inferior and under male subordination, yet this inferiority is abolished in their equal capacity to transcend creation and be transformed by grace."[25]

Christianity retained and strengthened some of the dualistic conceptions of these traditions. The understanding of the soul to be created but still capable of immortal life apart from the body reinforced in Christianity a sharp distinction between human beings and other life-forms. Only humans have a soul that is rational, and hence immortal. This human/animal dualism is in turn reinforced by the Platonic soul/body split. The essence of being human is "a transcendent, disembodied, immortal 'soul' that can kick aside the physical world of bodily life."[26] Despite official Christian doctrine that retains Hebrew concepts of the resurrected body and renewed earth, Christian eschatology in practice posits an immortal soul that attains salvation by escaping from mortality. It is neither connected to nor limited by the fate of the rest of earth's creatures.

Despite incorporating antinature and anti-body Platonic assumptions, Christianity still affirmed the Hebrew teaching of the original earth as paradise. Nature has "fallen" through the consequences of human sin, which brings death into the entire created world. Christ rescues humans from this state of sin and death, but the fate of the rest of the world is more ambiguous. This results in Christianity maintaining two seemingly contradictory stances toward the rest of creation. On the one hand, humanity is responsible for the inadequacies of nature—namely, its mortality—as a consequence of human sin. On the other hand, humanity bears no ultimate responsibility for the rest of creation. Since animal and plant life lack a rational soul and hence have no personhood of their own, and since we share no common fate with them, humans can treat them as possessions and exploit them at will.

These ambiguities in the Christian worldview frame Ruether's examination in the rest of *Gaia and God*. Before proceeding she articulates several underlying assumptions that flow from this section and inform her exploration. First, "nature," in the sense of all cosmic life, was *not* originally paradisiacal. It is therefore not capable of completely fulfilling human hopes for the good—that is, benign regard for individual and communal life, which Ruether sees as the human ideal. Like Cobb, she argues there is no going back to an original innocence in paradise. Second, humans are the "evolutionary growing edge" of an impulse in nature toward consciousness and kindness.[27] The possibilities of human consciousness and extending loving care toward others allow humans to stand out from our mortal limits, yet also demands greater kindness than exists in most interanimal and interhuman relations.[28] Echoing process theological insights, she sees these experiences as pointing to a source of life that is also an impulse to increased kindness and consciousness still imperfectly realized.

Ruether argues that an ecological ethic must be grounded in acceptance of both sides of this human dilemma: that we represent the growing edge of the source of greater awareness and benignity in the cosmos, and yet with plants and animals we share the common mortality of all organisms. She maintains that human achievements are not lost at death, but neither are they retained individually through transcending death: "We pass on our ideals to the future not by escaping personal death, but partly by reshaping 'nature' to reflect these human ideals. But this reshaping is finally governed by the finite limits of the interdependence of all life in the living system that is Gaia. Ecological ethics is an uneasy synthesis of both these 'laws': the law of consciousness and kindness, which causes us to strain beyond what 'is,' and the laws of Gaia, which regulate what kinds of changes in 'nature' are sustainable in the life system of which we are an inextricable part."[29]

Recognizing the centrality of the sciences in helping us to discern the "laws of Gaia," Ruether addresses the question of whether contemporary science can provide us with a more adequate creation story. She argues that one important implication of the Enlightenment breakdown of the viability of the Christian creation story is that it destroyed the basis for a whole moral and spiritual system predicated on an earth-centered view. For many Christians, the most threatening part of this unraveling came with Darwinian evolution, which included humans in the earth's evolutionary history. The absolute distinctiveness of humans from other animals could no longer be maintained. Another critical ramification was the separation of science and religion, and hence questions of fact versus values, along the dualistic lines presumed by Enlightenment mechanistic and reductionistic science stemming from Newton and Bacon.

Yet as Cobb and Birch chronicle, the unchallenged authority of "objective" science itself has broken down in recent years under increased awareness and criticism of its presumed neutral values and its methodology of reductionism. Following twentieth-century findings in subatomic physics, the paradigm for scientific knowledge has undergone change. One can no longer make hard and fast distinctions between observer and observed, for the observer is an integral part of the reality observed and inevitably affects that reality. Classical Greek dualistic oppositions between spirit and matter, objective and subjective, and fact and value seem more like endpoints on a spectrum or a matter of context and perspective rather than ontological opposites.

One important consequence of these changes is that one can no longer abstract fact from value, or science from ethics:

> What is constituted as the web of relationship will also be shaped by how we relate to it. The knower must take responsibility for shaping the reality that is known in ways that can be benign or destructive. Deciding what is benign or destructive is itself a matter of definition. Ultimately all reality continues to be interrelated, yet the effects of one kind of relationship rather than another can appear on the biotic level as flourishing life or as deadly poisons. Thus the knower cannot avoid ethical responsibility, not only in terms of how the knowledge may be subsequently used, but also in terms of what kind of relationship is implied by the stance of the knowing itself.[30]

Even so, Ruether acknowledges that this understanding of science remains a minority position. Most scientists continue to function with mechanical, "value-free" models, while most religious and ethical thinkers have long since abdicated themselves from needing to know how science functions.

Can we build on these insights for a new understanding of the cosmos? Here Ruether turns to the scientific story of cosmogony that continues to emerge from the sciences, particularly astral and subatomic physics. According to this view, the universe began some eighteen billion years ago in a cataclysmic beginning referred to as "the Big Bang." Some thirteen billion years later, the earth and the rest of the solar system began to take shape out of the remnants of exploding stars. One ramification of this claim is that all atomic elements that make up the earth, from rocks to plants to the human body, began as "stardust." Ruether traces the evolution of life on earth, from the appearance of the first organic molecules a little over three billion years ago to the arrival of Homo sapiens only four hundred thousand years ago, and concludes: "Clearly the anthropocentric claims to have been given 'dominion' over the earth, and over all its plants and animals, appears [*sic*] absurd in the light of 4,599,600,000 years in which earth got along without humans at all!"[31] Yet human beings seem to be making up for lost time in terms of influence on the planet. The rate of ecologically damaging change of the earth under human influence has accelerated to the point that humanity faces the possibility of causing its own extinction and severely damaging the whole biosphere.

What are the implications of the scientific story and perspective for ecological ethics? Because ecology as the science of biotic communities demonstrates how nature generates and sustains life, it can give guidelines for how humans can participate in our biotic communities as sustaining, rather than destructive, members. Hence Ruether suggests that unlike other modern sciences that claim to be only descriptive, "ecology suggests some restoration of the classical role of science as normative or as ethically prescriptive."[32] While human capacities such as consciousness and the impulse to care represent the "growing edge" of nature, humanity is still bound by the laws and limits of the natural world. From the cosmic story Western consciousness must relearn the lesson of ecological limits. Yet Western consciousness also must heal the splits it presumes that have separated knowledge from wonder, reverence, and love if we are to be able to tell the cosmic story in a way to foster a healing relation with the earth.

Ethics and spirituality can learn two important lessons from ecology: all things are interrelated, and the basic elements of life on the planet—plants, air, water, soil, and animals—have evolved together. Hence humanity's profound kinship with the rest of the planet extends through space and time. Interdependency and sustainability through time must form the heart of an ecological ethic.

Against those who see competition as the law of the natural order, Ruether argues that cooperation and interdependency are the primary principles of ecosystems. Competition between species and populations is a subcategory that helps to maintain the interdependency by sustaining a balanced relation of each population in relation to the whole. The human cultural construct of competition is false when applied to the natural world because it is mutually exclusive. It sees the other side as an enemy to be eliminated, rather than an essential part of an interrelationship upon which it itself is dependent. Ruether argues that human ethics should be a more refined and conscious version of natural interdependency, always looking for ways to make the interrelation cooperative and mutually life-enhancing for both sides. All too often human cultures have done the opposite, imagining the "other," whether human or animal, as a threat to be contained or destroyed.

Not only do creation stories contain implicit social messages, so do stories of the end-times, as Donna Haraway observed.[33] Hence in the second section of *Gaia and God,* Ruether turns her attention to Western religious narratives of world destruction. She asks why these narratives have played such a large part in religious consciousness, and what role apocalyptic narrative plays today in the face of the escalating ecocrisis.

Ruether notes that the roots of ancient Near Eastern myths of world destruction lie in actual experience of real destruction, both from nature and from other human groups. One particularly great flood that caused near total destruction was long remembered and passed down by many groups. The Hebrews took these destruction prototypes and revised them to reflect their own monotheistic theological and cosmological views. For them, God transcended both the cycles of nature and changes in history. The Hebrews moralized many of these stories and placed them in the context of Israel's history. World destruction and its threat came to be seen as punishment by God for failure to obey God's covenantal laws.

Over time, then, the view of mortality and death in these stories changed. In the early Hebrew religion, mortality was a natural part of the cycle of life rather than a problem to be overcome. Blessedness consisted of a long, healthy, and prosperous life rather than an escape from mortality. This began to change with the introduction of apocalyptic forms of thinking. Early Hebrew apocalyptic narratives introduced the theme of the resurrection of the body not to overcome human mortality, but as a way to settle the accounts of unrequited injustice and unrewarded unrighteousness. It was only as Hebrew religion entered the Hellenistic period that the focus began to shift to immortality as a way to overcome the problem of mortality.

Development of apocalyptic narratives reaches its fullest form in the Christian Book of Revelation, written during a period of persecution near the end of the first century C.E. In this narrative, trials and tribulations of the Christian saints under successive evil emperors are seen to give way to battles between the Messiah-warrior from heaven and Satan and his armies. The final battle and great judgment culminates in a New Jerusalem replacing the first heaven and earth, and all creation is made new. Evil and mortality are banished, and God dwells in direct, unmediated form in the midst of the saints.

In analyzing the function of apocalyptic narrative, Ruether notes the importance it has held to socially marginalized groups as a means of social protest. Apocalyptic narratives often have provided oppressed groups with a way of coping amidst incomprehensible and seemingly intractable social chaos. Apocalyptic thought thus is an offshoot of prophetic thought. The problem with apocalyptic theory, however, is its dependence on a dualistic worldview. Rather than seeing good and evil to be located in different kinds of relationships, as did earlier understandings of justice and righteousness, the apocalyptic world is polarized into different kinds of beings, "us versus them," absolute good against absolute evil. Good and evil become reified, and too often evil is projected onto those who are seen as physically or socially other. Furthermore, based on a fantasy of escape from mortality, death now is seen as the last enemy to be overcome.

Ruether locates the foundation of this desire to escape from the body, earth, and evil in a certain model of God. The God of apocalypticism is an entirely transcendent, unrelational God, unrelated to earth, body or mortality. Followers of this God imagine themselves to be safe from world destruction, which will only affect those who are evil. Thus, paradoxically, apocalypticists often are greatly encouraged by signs of world destruction because they see it as the means by which they will escape death. Ruether likens this style of thinking to some militant forms of environmentalism that envision "Mother Earth" rising up against arrogant humanity to return the earth to a precivilized state, thus restoring paradise. Yet any ecological disaster that would destroy most of human civilization is unlikely to restore a paradisaical earth. Instead, Ruether argues, we must acknowledge the deep fears and desperate hopes that inform apocalyptic thinking and shape an alternative ethics and spirituality to channel these fears and hopes more realistically and more lovingly.

Understanding the structure and characteristics of apocalyptic thought provides a useful way to examine the current ecological crisis as a new narrative of world destruction. Contemporary narratives of "ecological apocalypse" would have seemed inconceivable to earlier generations, and most people today still have difficulty comprehending the enormity and acuteness of the problem that confronts us. Yet unlike their religious forbears, contemporary narratives of world destruction coming from ecologists and other scientists carry no assurance of salvation and subsequent renewal. Ruether argues that since we no longer can conjure up a God outside nature who will rescue us from our miscues, what is required instead is an immediate ecological metanoia: "conversion of our spirit and culture, of our technology and social relations, so that the human species exists within nature in a life-sustaining way." Time is short, so this metanoia must be effected quickly. We must recognize that we have no home outside the earth, so that destruction of the earth becomes destruction of ourselves as well. "The capacity to be the agents of destruction of the earth also means that we must learn how to be its cocreators *before* such destruction becomes terminal."[34]

Thus, an ecological ethic must confront the "four horsemen of ecological destruction" that profoundly jeopardize the biosphere today: (1) the explosion of human population at the expense of the plants and animals of the earth; (2) ecological damage to air, water, and soil; (3) the misery of the growing masses of poor people; and (4) global militarization to retain inequitable resource distribution and consumption for a wealthy elite. Our task in response must not be to indulge in apocalyptic despair, "but to continue the struggle to reconcile justice in human relations with a sustainable life community on earth."[35]

In order to reconcile justice and ecological sustainability, ethics must confront the reality and meaning of evil in relation to both human nature and the biophysical world. Given the history of scapegoating certain groups of persons as sources of evil, it is here that a liberationist stance is especially important to explore how evil gets named, by whom, and to what ends. Ruether looks at the classical heritage of Jewish, Greek, and Christian understandings of evil and asks to what extent these constructions of the relation of evil to nature have misnamed evil in ways that reinforce dominating relations of women and subjugated peoples, as well as negation of the earth.

Ruether begins her examination of classical narratives of sin and evil by observing that awareness of good and evil arises from human consciousness that things are not as they should be. She argues that categories of good and evil, like destruction narratives, arise from "concrete experiences of negativity and preferred alternatives."[36] The danger of this historical process lies not in the ways these narratives have allowed communities to imagine better or worse alternatives to their present reality, but in the tendency to translate these historically conditioned capacities into universal absolutes and then imagine that these absolutes actually exist. When the evil side of this polarity is identified with other people and things, the problem is only compounded. Paradoxically this false naming of evil as physical and social otherness and the efforts of dominant males to protect and separate themselves from this perceived otherness ends up creating ideologies that justify doing evil against these others in the name of overcoming evil!

The Hebrew, Platonic-Gnostic, and Pauline-Augustinian traditions of naming evil have greatly influenced the overall Western cultural tradition. All three "have been constructed from the vantage point of dominant males in a way that has functioned to justify their violent power over women, other subjugated people, and the earth."[37] Yet within each tradition are recoverable elements that can contribute to a liberationist ecojustice ethics.

Ruether argues that Jewish thought in the Hebrew scriptures and later in the Talmud does not see the fall from original goodness as rendering humans incapable of doing good or pleasing God, as Christianity would later assert. Rather, after the fall, people retain their freedom to choose good or evil and accordingly to please or displease God. Where Jewish concepts of evil become problematic is when they mix ethical and ethnocentric-cultic judgments. Cultic concepts of evil, for example, are deeply entwined with notions of purity and pollution in Hebraic thought. This led to carefully delineated separations in Jewish life between holy and unholy according to physical and social differences along lines of gender, ethnicity, and physical ability or "normality." This culminates in its most extreme form in the holy wars and slaughters of the book of Joshua, whose example has served ever since as a mandate for later groups to conquer, exterminate, or enslave other "impure" peoples encountered through conquest and colonialization.[38]

From Platonic and Gnostic traditions, Western culture gained its view that evil resides in the physical body and the material world. Evil results when the soul loses its contemplative union with the intellectual essences and, instead of governing the body, becomes governed by the body. Evil is thus connected to mortality: "the dissolution side of the life cycle was, for the Platonist, the manifestation of evil, expressed in everything that was 'out of control' in body, mind, and emotions."[39] One important ramification of Platonic and Gnostic views is that evil is not seen relationally, as evil done to others, but as something residing in the soul itself. The soul can ascend to heaven only as it purifies itself of bodily and material influences, there to be united with the godhead in immutable and eternal form.

Ruether sees in the Pauline-Augustinian tradition of Christianity a fusion of Jewish ethical and Greek metaphysical views of evil. Evil comes to be seen as located both in humanity's free choice to rebel against God, and in a flawed ontology of mortal being. A result of this fusion is the Christian view of the human dilemma: humans are culpable for sin and evil, yet incapable of escaping it through their own capacities. Paul understood this to be the result of a profound dualism of two modes of existence: existence according to the flesh, that is, slavery to sin and death, and existence in the Spirit, which frees the Christian through rebirth in Christ.

Problematic for Ruether in this view is the extent to which Paul identifies evil with natural or created life. This makes redeemed life something fundamentally different from and transcendent to humanity's original created state. Ruether thus sees in Paul a quasi-Gnostic cosmology where human beings anticipate shedding the mortal body, subject to decay, in order to be transformed into an immortal, spiritual body (1 Cor. 15:35–50). Such a view results in identifying sin with death and goodness with immortal life.

Yet Paul also puts a strong emphasis on the morality of loving relation, grounded in his understanding of the church as the body of Christ. In this context, sin is understood as that which violates relationship with other people. In addition Paul disconnects sin from many of the purity regulations of Jewish laws such as dietary restrictions, holy day observances, and circumcision. He encourages an ethics of mutual tolerance and forbearance between "strong" and "weak" members of the community on these matters. Central to this is Paul's conviction that salvation comes through

Christ as a gift from God given in faith, not from keeping the commandments.

Ruether concludes that the classical Christian understanding of sin as both ethical and metaphysical, as both disobedience and finitude, results in a mixed heritage to the Western world. Most problematic is the notion that humanity is culpable for its own finitude. In response Ruether writes, "Although we may evaluate our mortality as tragic, or seek to embrace it as natural, what mortality is not is sin, or the fruit of sin."[40] Ruether sees the preapocalyptic Hebrew view of mortality as a natural condition that we share with other creatures to be a more authentic ethics for ecological living. Here redemption is understood as the fullness of life within the limits of the earth. Ruether notes that in the Western tradition, despising nature as finite but renewable life has been closely related to despising women as birth givers. Women have been scapegoated for both sin and death, as the source of both impurity and finitude. Religious mandates to separate holy from unholy, spiritual from carnal, and immortal from mortal life created phobic patterns toward death and decay that get extended toward classes of people viewed as other: "These phobic patterns have been used to structure social apartheid along gender and ethnic lines."[41]

Nevertheless, Ruether also finds several important usable resources in the Western heritage. Among the constructive recoverable elements she includes (1) an understanding of sin and evil as unjust relations; (2) an understanding of repentance as special advocacy of those victimized by systems of oppressive power; (3) the Pauline-Augustinian existentialist recognition of the divided self which acts against its own interests and desires; and (4) the concept of inherited sin that recognizes that we inherit alienated structures of culture and social organization.

In light of the above analysis, Ruether argues that reconstruction of the ethical tradition must begin by clearly separating questions of finitude and mortality from those of sin. Rather than being located in finitude, sin must be seen as human misuse of freedom that violates the basic relations that sustain life through exploitation of other humans and the earth. Sin as evil lies in the distortion of relationship and the absolutizing of the rights to life and power of one side of a relation to the detriment of the other parts with which it is interdependent. Sin becomes structural when it leads to systems of social control and cultures of deceit to maintain and justify unjust power relations.

If sin is historical, rooted in social relations rather than natural finitude, how do contemporary feminist views account for it? Here Ruether examines what she terms "the new fall story" about the origins of patriarchy that has emerged in contemporary ecofeminism. She sees several presumptions as characteristic of this account, including: both humans and nonhuman nature have an inherent capacity for biophilic mutuality; women are better than men at mutuality, either through social development or by inherent nature; and predating patriarchy was an original social order defined by female modes of relationality, mutuality between the genders and between humans and nature, and active agency and leadership of women. This order disappeared with the emergence of early urban civilizations and the development of patriarchy.[42]

While Ruether agrees with many of the underlying values of this new fall story, she warns against the dangers of reversed forms of scapegoating and the implications of maintaining untenable assumptions about an originally paradisaical nature: "It too easily allows somewhat marginalized Euro-American women and men to identify themselves with a lost innocence and to fail to take responsibility for their own complicity in the evils they excoriate.... Without more careful evaluation, this 'story' can mislead us about how we understand ourselves as Western people and our capacities for both good and evil."[43] To sort out the useful from misleading elements, she examines the roots and major forms of the "lost paradise" hypothesis in light of findings of contemporary paleoanthropology to see what parts can be verified historically.

The roots of lost paradise stories in the West lie in the biblical story of Eden and the Greek story of the Golden Age. Ruether notes that both narratives were shaped by men to blame women for the troubles of physical labor and illness: "The stories seem to be compounded of two elements, an idealized memory of preagricultural societies and idealized (male) childhood. The adult male resents the wife, whom he must support by his labor, and idealizes his lost nurture by an all-giving mother. Woman-blaming for the lost paradise may have psycho-familial roots, roots that go back to primal human social patterns."[44] In contrast to classical male versions of the narrative, many ecofeminist theories of lost paradise posit the idea of an original matriarchy.

While acknowledging the undeniably powerful mythic appeal for our times of such reconstructions, Ruether nevertheless asks whether they are best understood as history or as symbolic myth. One problem in particular she notes is their "post-Christian Romantic construction of the 'feminine' as essentially and unchangingly nurturant, benign, and peaceful."[45] In addition to the stereotypic linking of qualities of women and nature, Ruether finds the projection of contemporary values onto ancient peoples to be methodologically problematic.

To test the historicity of these ecofeminist reconstructions, Ruether examines several studies of paleoanthropology and gender to explore the role of gender in human cultural evolution. She finds a more complex pattern in the shift from female-centered Neolithic farming villages to early patriarchal urban civilization than portrayed in the "fall into patriarchy" stories. Social roles originally dominated by women, such as agriculture, shift to men under a variety of historical changes, leading eventually to the new male aristocracies of patriarchal cultures that define legal social systems based on male control over land, labor, and females.

Ruether also seeks to understand whether there are other factors that may have contributed to undermining early matricentric societies, paving the way to patriarchy. She examines comparative studies in anthropology and gender to conclude that matricentric societies, such as those postulated to have predated patriarchy, may have an inherent psychosocial weakness in their difficulty in constructing meaningful, nonexploitative roles for adult males. This she sees as a result of different female and male roles tied to the processes of biological and social reproduction: "While the female role is built into the process of life-reproduction and food-gathering, the male role has to be constructed socially." Ruether thus sees the inherent matricentric pattern of societies as the breeding ground of male resentment and violence that lead to and undergird patriarchy. "Societies that fail to develop an adequately affirmative role for men, one that gives men prestige parallel to that of women but prevents their assuming aggressive dominance over women, risk developing the resentful male, who defines his masculinity in hostile negation of women. The symbolic negation of women in conflictual societies provides the myths through which actual dominance over women is promoted and justified."[46]

Ruether ends this section with several conclusions. First, there can be no literal return to the culture of the Neolithic matricentric village as a basis of gender equality in society today. Reclaiming the memory of earlier cultures and learning from contemporary indigenous cultures and spiritualities with closer gender parity may provide valuable resources for seeking gender parity, but are insufficient by themselves. Second, the inherent weaknesses of the matricentric core of human society and its consequent vulnerability to patriarchy need to be addressed in constructing an ecofeminist ethics. Critical to this will be creating new male and female psyches and social roles that provide holistic alternatives to male socialization into violence through mother negation.

Ruether then turns to examining the ways systems of domination historically have been constructed and justified. In ancient Mesopotamia the rise of the institution of slavery made it the dominant metaphor for relationship, including the divine/human relation. This division also took place along gender lines. In each of the subordinated categories, women predominated; even male household slaves were female-identified. In the peasant pastoral society of ancient Israel, the rigid class hierarchy of urban civilization was modified or suppressed, but women continued to be rigorously marginalized. The symbolic lines between divine and human became more rigidly drawn, and gender hierarchy demarcated the division of the divine and holy from the realm of human and unclean, with the female representing the creaturely and the unclean.

In ancient Athens, understandings of human identity were shaped by the conflict between the Greek male citizen identity and its "excluded opposites": the female, the barbarian, and the animal. Greek male identity was established by excluding the non-Greek, the nonmale, and the nonhuman from citizenship in Greek society. Plato and Aristotle later reworked this relationship through the metaphor of hierarchy. "Rather than powerful but excluded others, over against the Greek male, the new metaphor incorporates the female, the alien, and the animal as 'natural' inferiors in a hierarchical 'chain of being' that stretches from immaterial divine Logos at the upper end of the hierarchy, to unformed matter at the lower end."[47] Under Aristotle's influence, slavery became the central metaphor of hierarchy, and procreative power was appropriated as a solely male capacity.

This new metaphor of ontological hierarchy as social hierarchy with descending levels of rationality was in turn appropriated by Christianity and shaped Western culture into the modern period. Out of it developed the protest traditions of asceticism and apocalypticism with their emphasis on flight from women, the body, and the earth into ascetic negation. Augustine was the great synthesizer of this for the Christian West with his fusion of patriarchy, asceticism, and apocalypse. Asceticism has left an ambivalent heritage. In its social protest against imperialism and excessive consumption, it provides a positive legacy. Yet tightly fused with this is an attitude of hostility toward women, sexuality, and the body, and contempt for the natural, material world. Medieval Christian asceticism was deeply ambivalent toward women and physical nature, often seeing nature as the "enemy without" and women and sexuality as the "enemy within."

In Calvinism the split between nature and grace becomes radicalized, and nature since the fall is seen as totally depraved, unable to give knowledge of God or sustain relation to God. Calvin's antinature epistemology meant that saving knowledge could come only from beyond the earth, descending in God's revealed Word in scripture. Francis Bacon and others in the scientific Enlightenment reversed this antinature epistemology, seeing nature as the reliable source of knowledge. Yet this did not result in greater appreciation for the natural world. Rather, scientific knowledge developed as a tool to control the world, and nature is seen as female, to be subjugated and ruled over, forced to yield "her" secrets.

This view of nature as other to be subjugated becomes strengthened with Descartes' radical dualism between mind and nature. Employing the metaphor of "machine" for nature, Descartes saw the natural world as mindless and soulless, made up of individual mechanical parts that obey the laws of causality. The human mind and soul no longer had any connection with the natural world: "All innate spiritual elements having been eliminated from nature, human spirit need no longer interact with nature as a fellow being, but could see itself, like the clock-maker God, as transcendent to it, knowing it and ruling it from the outside."[48] With colonialism and industrialism, the new science of the Enlightenment became bound to the technology of war, resulting in the destruction and decisive reshaping of the human, plant, and animal ecologies of colonized and industrialized regions.

Ruether concludes this survey by noting that the cultures of deceit that served to justify and maintain these systems of domination have passed through three mythic patterns:

> In the first mythic pattern, the male acknowledges his dependency on the mother (as woman and as earth), but he also coopts her as the means to his power and the lap upon which he is enthroned as king of the universe. This ancient myth of the Goddess with enthroned male king on her lap continues in the Catholic iconography of Mary and the infant Christ.
>
> The second mythic phase negates this dependency by seeking to separate the immortal soul or spirit from the mortal body. The positive dependency on mother and nature is denied, but their potency is recognized negatively as demonic power. In the third mythic phase, there is an effort to sterilize the power of nature altogether, imagining it as dead stuff totally malleable in the hands of men in power. Although this is the myth of dominance that has triumphed in modern science, it was al-

> ready suggested in the Babylonian myth of the universe constructed out of the dead body of Tiamat.[49]

Modern science and technology have added an enormous increase in both productive and destructive power to these earlier systems of hierarchic domination and exploitation, and have combined with multinational political and economic power to construct a global system of domination. To counter this, an ecological culture and society must incorporate three elements: (1) the rebuilding of primary and regional communities where people understand and take responsibility for the ecosystem of which they are a part; (2) just relations between humans where all have access to an equitable share in the means of subsistence; and (3) the creation of cultures of compassionate solidarity to replace the current culture of competitive alienation and domination.

In her final section Ruether examines the covenantal and sacramental traditions that emerge in biblical and Christian thought to see what resources may be reclaimed for an ecological spirituality. She notes that both were shaped under patriarchy and must be reinterpreted to be genuinely affirming of dominated women, men, and nature. Yet both have profound insights for ecological spirituality and practice, and can provide the grounding for the world's Christians to participate in healing the earth.

In her examination of the Hebraic roots of the covenantal tradition Ruether argues that biblical interpretation in recent centuries often was shaped by a strong history/nature dualism that made the Bible seem more antinature than it is. Instead the biblical writers had a holistic view that linked creation and redemption under a God who made both heaven and earth. While the Hebraic view of relationship to God is androcentric, anthropocentric, and ethnocentric, it does not preclude more inclusive perspectives. In fact much of the Bible testifies to a God who relates directly to nature. Hebrew teachings on dominion focus on the limits of human authority over nature, which is always a delegated authority, and the Levitical Jubilee laws provide a cyclical model of redemptive eco-justice. While much of this covenantal emphasis is diluted or lost in New Testament writings, it remains an important part of the Christian heritage.

An example of the contributions the covenantal tradition makes to ecological ethics is the debate on the rights of nature. Ruether draws on the covenantal insights of dominion to argue against the longing of some to return to a simpler world where humans are one species among many and animals can again live in the wild. Only by the human community asserting an enlightened guardianship over the diverse biota of the earth will it have a chance to be protected. She recognizes the intrinsic value of all things is part of this covenantal vision:

> A covenantal vision of the relation of humans to other life-forms acknowledges the special place of humans in this relationship as caretakers, caretakers who did not create and do not absolutely own the rest of life, but who are ultimately accountable for its welfare to the true source of life, God. This covenantal vision recognizes that humans and other life-forms are part of one family, sisters and brothers in one community of interdependence. Although we have limited rights of use of other life-forms, and also responsibilities of care and protection toward them, there is an ultimate thouness at the heart of every other living being, whether it be a great mountain lion or swaying bacteria, that declares its otherness from us.[50]

The covenantal vision overcomes fear of otherness to encounter all other life-forms simultaneously as other and kin. It carries with it the norms of justice, right relation, and restoration critical to healing the earth and maintaining an ecological society.

In examining the sacramental tradition, Ruether focuses in particular on the holistic vision of the cosmological understanding of Christ as both creator and redeemer of the cosmos. The framing question for her exploration is a central one for shifting from an anthropocentric to ecocentric frame of reference: how do our understanding of "God" and "human" change once the cosmos becomes the mediating context of theological reflection and spiritual experience? To address this Ruether explores the Christian synthesis of Hebraic, Oriental, and Greco-Roman thought, a synthesizing capacity she sees as one of the great strengths of Christianity. The synthesis of the Logos and the Christ as the divine person encountered in Jesus brings together the dramas of creation and redemption within a cosmological framework. Under different historical influences in medieval and later Christianity, this holistic vision becomes a minority tradition, but is never entirely lost. It reemerges in different contemporary ecological theologies such as those of Matthew Fox, Teilhard de Chardin, and process theology. Ruether builds on these models, insights from the new paradigms of science, and feminist theology to propose an ecofeminist theocosmology that flows from "God" understood as "the great Thou, the personal center of the universal process," reflected in the universe as "the ongoing creative Matrix of the whole."[51]

An ecological spirituality based in such an understanding must be built on three premises: "the transience of selves, the living interdependence of all things, and the value of the personal in communion."[52] Human kinship with other creatures spans space and time, linking us with the whole living Gaia today, as well as linking our material substance with all life that has gone before us and will follow us. Compassion for all living things replaces the illusion of otherness. The sacramental tradition provides a holistic cosmological spirituality necessary to envision and sustain healing the earth.

For Ruether, the covenantal and sacramental traditions are two voices of divinity from nature, the voices of God and of Gaia. Both of the voices are also our own voices, in the sense that they can only be experienced through the lenses of human experience. We need both of these voices—the voices of both politics and spirituality, of organized systems and ethical norms together with compassion and celebration—to form an ethic for ecological living. Ruether takes these two voices and returns to the themes of the earlier chapters to sort out helpful from harmful elements in constructing a new ethic. Looking at different understandings of evil in the Christian legacy, Ruether follows the Hebraic tradition in maintaining that evil lies in wrong relationship. From the wisdom of Gaia we learn that the good lies in respecting limits in our relationships with others, human and nonhuman.

Finally, from these insights Ruether lists several principles and guidelines involved in envisioning a good society. The vision begins with the principle of *equity*, between individual humans, regional communities, and across the globe, between humans and other species, and between generations now alive and those yet to come. Changing systems of domination and exploitation to ones of *biophilic mutuality* requires a fundamental restructuring at all levels. Efforts to change occur through *incremental steps guided by holistic vision*. New living patterns require *decentralization* and increased *local and regional control*. *True environmental costs* must be paid by producers and consumers. Needed efforts at population control and other social change must pay primary attention to *increasing women's moral agency* and examining *how women and their bodies are socially and culturally appropriated*. A *double transformation of women and men* in their relation to each other and to nature is needed. As part of this men must be fully integrated into the daily sustenance of life. Genuine, systemic *demilitarization* is required.

Ruether sees the basis for the struggle for ecological transformation to be rooted in creating and sustaining base communities of celebration and resistance. She sees three interrelated aspects of the work of these local, face-to-face communities: (1) In them the personal therapies, spiritualities, and corporate liturgies by which we foster a new biophilic consciousness are shaped. (2) They allow us to utilize local institutions over which we have some control—such as homes, schools, churches, farms, and local businesses—to make ecologically sustainable pilot projects. (3) They provide a base for building organizational networks at the regional, national, and international level to affect needed structural change. Amending the famous slogan, "Think globally, act locally," Ruether maintains that we must think both globally *and* locally, and act both locally *and* globally. Finally, Ruether sees *committed love* that remains committed to a vision and concrete communities of life regardless of the latest "trends," rather than optimism or pessimism as the stance that must guide our efforts.

Because social constructionist ecofeminism combines ecological insights with a feminist, liberationist methodology, it is not surprising that Rosemary Ruether's ecofeminist ethic has the closest affinity of the four models to the insights that emerge from discussion of lesbian, gay, and feminist hermeneutics of appropriation. Ruether's liberationist hermeneutical approach distinguishes her ethics from the previous three. She applies a feminist hermeneutics of suspicion and recovery to all her sources, working out of a praxis-based approach rooted in justice-centered communities of accountability. Like Cobb, Ruether provides an excellent example of critical engagement of both scientific and religious sources that incorporates an initial deconstructive stance but moves on to reclaim usable resources. In addition, Ruether insists on including and respecting ways of knowing other than scientific or critical knowledge, and explicitly opens her ethics to learning from religious sources outside the dominant biblical tradition. She embraces a communal hermeneutics that goes beyond Cobb and Hall to incorporate a liberationist epistemology in giving priority to historically excluded and subjugated perspectives as well as explicit accountability to communities engaged in the praxis of justice.

Since Ruether pays close attention to the socially constructed nature of human thought, she is particularly attuned to the ways modern science has constructed antiwomen and antiecological views. She takes scientific claims seriously, and uses methodological insights of new scientific paradigms, particularly the new physics, to level an internal critique at the still-dominant, mechanistic, "value-free" Enlightenment model of science. She notes that many parts of the new scientific paradigm are consistent with feminism and ecology, particularly its emphasis that the scientific observer is an internal part of the process who must take ethical responsibility for her or his actions. It also blurs the boundaries between objective and subjective, fact and value, and acknowledges that the standpoint of the observer influences both the description and constitution of reality. Finally, Ruether takes seriously other ways of knowing and argues that Westernized consciousness must heal itself of the divisions that have separated science as a way of knowing from wonder, reverence, and love—integral components of an ethics for healing and sustaining the earth. All of these perspectives on science are consistent with the liberationist insights outlined for an ethics of Gay and Gaia.

Ruether's approach to the Bible and tradition explicitly incorporates the four elements I have argued are necessary for the critical, deconstructive initial stance: critical-historical analysis, a hermeneutics of suspicion, explicit commitment to the well-being of the socially marginalized, and evaluation of the impact of sources on women and other subordinated groups. She seeks to engage the biblical narratives and tradition to sift out and recover usable resources from "the toxic waste of sacralized domination." Ruether embraces a contextual, relational understanding of authority with respect to the Bible. It has no a priori authority, no privileged or inherently normative character, but usable elements recovered through critical engagement with the text may be experienced authoritatively when they contribute to the praxis of earth healing. Hence her stance toward the Christian tradition is one of accountability to its resources and problems rather than uncritically privileging it as a source for ethics. As in the liberationist insights discussed previously, Ruether's approach embraces a communal hermeneutics, epistemological privileging of excluded and subjugated perspectives, and explicit accountability to communities engaged in the praxis of justice. She calls for an open canon of Scripture, including but not limited to Christianity's sacred texts.

One reservation with Ruether's ecofeminist approach may have more to do with this particular project than with Ruether's overall ethics. Because in *Gaia and God* Ruether focuses primarily on historical-critical analysis of the Western classical and Judeo-Christian tradition, there is little attention to or example of appropriation of other sources across lines of social difference. Two themes may indicate her stance on this: her social and historical analysis incorporates attention to gender, race, and class/slavery; and her emphasis on integrating politics and spirituality in a plurality of ecofeminisms suggests her awareness of the importance of liberationist insights in this area. Yet because to this point ecofeminism in the United States has been a movement predominantly of white feminist women, attention to appropriation of sources across lines of social and cultural difference must be looked at critically and incorporated explicitly into ecofeminist ethics.[53]

Another strength of Ruether's ecofeminist ecological ethics is that it is grounded in an explicitly ecocentric framework. In critiquing androcentric, anthropocentric, and ethnocentric perspectives, she calls for the cosmos and a sense of cosmic interrelatedness to become the mediating context for theological and ethical reflection. She combines ecological sensitivity to limits and interdependent biotic community with feminist emphasis on mutuality, relationality, and subjectivity for women, subjugated persons, and all other creatures to ground her underlying assumptions about the natural world.

In drawing on the deep forms of ecology and feminism, she pays attention to the ways cultural ideology and social structures are interconnected and in turn shape the way we perceive and construct our views of the world. Particularly important in this task is challenging hierarchical, dualistic ways of thinking that result in such hierarchies as male over female, humans over other creatures, mind over matter and body, spirituality over sexuality, and wealthy over poor.

As for Nash and Cobb, the starting point of anthropology for Ruether is recognition that human beings exist fully in nature, subject to the patterns and laws of nature. We can construct our existence within nature's patterns in constructive or destructive ways, but we cannot change the basic laws of the web of life themselves.[54] Yet humanity occupies a unique place in the natural world, what Ruether terms the "evolutionary growing edge" of an impulse to consciousness and kindness found in the cosmos. Human identity is shaped by the tension between these two "laws": the ongoing push of consciousness and kindness to move beyond what "is," and ecological laws that constrain the kinds of relationships and changes that are sustainable in the ecosystem.

While holding herself accountable to Christianity, Ruether makes no a priori privileging of her perspective or Christianity, calling instead for a plurality of ecofeminisms emerging from the particularity of different contexts. This frees her from some of the theological blinders noted in Hall and Nash that reproduce problematic theological views of humanity and sin by ignoring their socially constructed rather than presumed normative stance. Hence Ruether can recognize that human beings are latecomers to the planet, our arrival following over three billion years of prior evolutionary life. Drawing on both scientific insights from evolution and theological insights from Genesis, Ruether stresses repeatedly the kinship of human beings with the earth and other earth creatures. Against the metaphysical views of Greek tradition based on a dualism of eternal mind and mortal body, she embraces the Hebrew-Christian insistence that human beings are a psychophysical union. Hence evil lies in wrong relationship rather than in a metaphysical distortion of human identity for which we are culpable yet powerless to change.

Another contribution is Ruether's recognition that the psychocultural as well as socioeconomic roots of patriarchal structures must be addressed in order to transform patriarchal societal relations into structures built on equity. Her work thus serves to complement and deepen Cobb's attention to the relation between economics and community. More so than Hall, Nash, or even Cobb, Ruether casts a particularly critical eye on the ways human nature has been understood and constructed in the legacy of Western, Christian thought. A strong gender critique informs her analysis throughout. While recognizing different male and female biological roles in reproduction, she pays close attention to the ways gender roles are socially constructed and shape our views of nature and human nature. "Feminine" and "masculine" identities and values are socially constructed to reflect the values of particular cultures and societies rather than representing unchanging essential qualities grounded in the different biological sexes.

Yet while Ruether incorporates constructionist insights throughout her work, from the perspective of a lesbian and gay liberationist ethics, a problem with her ethics is the total absence and invisibility of those who differ from the heterosexual norm—lesbians, gay men, bisexual, and transgendered persons. Neither their experience nor sexuality and sexual orientation as tools of analysis seem to inform her work. This is a surprising, yet critical omission given her work elsewhere that explicitly addresses these issues,[55] and it leaves some clear gaps in revisioning an ecological ethics that is both liberating and inclusive. For example, in her examinations of the origins of human community, where are the contributions of "third gender" persons such as the *berdache* in Native American communities?[56] How does their presence or exclusion in society reflect the society's attitude toward the rest of the natural world? Can they help us to gauge the potential livability and sustainability of contemporary communities and cultures? How do they shape our understanding of human nature? With her focus on the centrality of matrifocal childrearing as the core of society, what is the role and place of nonheterosexual, nonprocreative persons to be constructed as we revision and rebuild communities today? The invisibility of those who do not fit into heterosexual gender constructs reflects a subtle but pervasive heterosexist bias in her work that must be addressed in drawing on it for a liberationist ecological ethics.

Yet because of its overall attention to constructionist critique of inherited traditions and forms of thinking, Ruether's ecofeminist ethics is the most conducive of the four models to rethinking the humanity/nature relation. Ruether pays close attention to how images of nature and the nature/culture boundary are constructed and how these shape human interaction with each other and nature.

Ruether's understanding of the humanity/nature relationship stems from her presumption that the earth is a living, interdependent system in which humanity is an inextricable part, interrelated with and kin to all other living creatures. She draws on the earth sciences to maintain that, far from beginning in paradise, the earth began as a molten globe enveloped in a poisonous atmosphere and bombarded with ultraviolet radiation. Hence life coevolved with its essential elements: water, air, soil, minerals, and other life-forms. Interdependency and interrelatedness are earmarks of life and humanity's relation to the rest of nature from the very beginning of life itself.

These fundamental insights frequently have been distorted by Western constructs of the "man/nature" relationship. Definitions of both sides of the concept have been problematic. While the male generic "man" has concealed the Western ruling-class male norm for human experience, "nature" also has been defined in at least three ways that distort our understanding: (1) as that which is "essential" to a being; (2) as the totality of the biophysical world apart from humans; and (3) theologically as the created world apart from God and divine grace.[57] Nature has been understood ahistorically, seen as unchanging apart from human influence. Nature has served as the foil over and against which humanity and divinity were defined. Nature was that which was nonhuman and nondivine.

In contrast to these concepts of the biophysical world, Ruether sees nature as the total earth system that constantly is undergoing its own process of adaptation and change. Human modification and reshaping of nonhuman nature takes place within and is a part of this process of continuous change. Humanly induced change is subject to the same ecological laws of Gaia that monitor other natural processes; changes that occur outside these limits produce deleterious effects for the biotic communities in which humans live, and ultimately for humans themselves. Hence human ethics concerned with the relation of humanity to the rest of nature must be shaped in part, at least, through correspondence to ecological laws and limits.

In understanding the principles and laws of Gaia, Ruether argues that cooperation and interdependency are the primary natural principles to which humanity must adapt its behavior. Stressing competition as the main "natural" impulse has led to a vastly distorted view of nature that manifests itself in ways such as exaggerating the place of meat eating in relation to the food chain, where the vast majority of food is supplied by plants.[58] It has also led to a severe distortion of humanity's "natural" relation to the rest of nature through stress on the "survival of the fittest" according to "laws of the jungle" that ignore our fundamental interrelatedness and interdependency with other life-forms.

In recovering insights from ecology for theology and ethics, Ruether pays close attention to the dissolution portion of the life cycle, where death and decay recycle vital elements to other parts of the ecosystem. She seeks to recover the early Hebrew understanding of mortality as natural rather than a result of sin or evil creatureliness to be avoided and fled. She is particularly sensitive to the ways ambivalence toward the status of nature as a source of both life and death, creatureliness and decay, has been projected onto women and other "excluded others." Too often these social distinctions in the human community have been projected back onto nature to give them ontological grounding. Hence she is explicitly critical of any forms of thinking that reproduce mind/matter dualisms, rigidly separating the "social" and "cultural" from the "natural" (and subjugating or repressing anyone or anything assigned to the latter).

Using this scientifically informed base to evaluate theological claims, Ruether opposes anthropocentric claims for dominion as an ontological right, noting instead that the earth appears to have gotten along quite well without human stewardship for more than four billion years. Yet given the current state of human proliferation and threat to the ecosystems of the earth, Ruether finds wisdom in some of the ethical implications of the Genesis teaching on dominion. Like Cobb, Ruether sees dominion primarily as a reality statement of humanity's current positioning in the web of life with important ethical wisdom about an enlightened guardianship of the earth's biodiversity. She cites the biblical narrative's nonexploitative intention, understanding of human authority as delegated authority subject to a higher authority and standard, a keen awareness of the limits of human power, and the responsibility for guardianship, all as critical insights for a contemporary ecological ethics in a world where the protection and preservation of the earth's biodiversity now lies primarily in human hands.

Two of Ruether's contributions are particularly important here. She draws attention to the ways our thinking about nature and other human beings have been shaped by the religious and other narratives of the tradition on themes such as creation, destruction, domination and deceit, healing and redemption. As in the case of Haraway, however, she also examines the material practices that shaped these stories and how they in turn enable and constrain our own understanding of humanity's relation to nature. An example of this is her exploration of the implications of the image of (male) God as artisan, one who forms things from dead, passive matter, rather than the earlier Goddess image of one who births creation. Hence Ruether calls for reincorporating other images of the divine that reflect a fuller range of the plurality of human experience.

This is rooted in Ruether's explicit awareness and examination of both sides of the nature/culture boundary as socially constructed. In naming the voices of God and Gaia as our voices as well, she acknowledges that all experiences of the divine and nature are always mediated through human experience. An example of the culturally mediated ways of experiencing nature are purity values and regulations in different cultures. Deconstructing purity values reveals the shifting and culturally constructed boundaries between what societies understand as nature and culture.

Hence Ruether's ethic offers constructive responses to some of the problems noted in the other models. In her attention to the constructed nature/culture boundary, she argues for a clear separation of questions of finitude and mortality from questions of sin and evil. In contrast to Nash, thus, she does not see physical death as a part of "natural" evil to be overcome eschatologically, but as a natural part of the earth as a living system (although like Nash she acknowledges a tragic dimension to death). Redemption lies within the limits of earth and in trusting surrender of the spirit in death back to God as the ongoing Matrix of life. Life on earth is not devalued through expectation of a future, nonearthly life where death is abolished.

Ruether goes beyond Cobb, Nash, and Hall, in stressing the need to pay attention to the human psyche in any ecological reshaping of our relations to each other and to nature. Integrating politics and spirituality must form the grounding of a liberationist ecological ethics, which must build on this insight to include specific attention to heterosexist understandings of the human psyche that fail to address anxiety around sexuality as critical ingredients in fostering patterns and structures of injustice and exploitation of human and nonhuman nature alike.

Ruether's failure to incorporate this should be noted in assessing her use of ecology for ethical guidance. Without a specific sensitivity to lesbian and gay insights to complement her feminist grounding, drawing on ecology for normative ethical guidance about the limits of science can blind one to and reproduce heterosexist constructions of science. Ruether's ethic has a built-in check on this with her insistence that values can no longer be abstracted from science and that the knower has responsibility for shaping reality in ways that can be benign or destructive. One sees this in her discussion of whether cooperation and interdependence or competition is seen as the primary form of relation in the natural world. Yet particularly with her focus on heterosexual gender roles in human community, it needs to be made explicit with respect to sexuality to avoid reproducing heterosexist assumptions, norms, and values in science and in ethics.

Given Rosemary Ruether's long history with feminist and liberation theologies, it is not surprising that the understanding of the divine that grounds her ecofeminist ethics is most congruent with the lesbian and gay theological insights in *Gay and Gaia*. Her views of the divine are rooted in a keen social-historical awareness of the constructed nature of all claims about God. She does not deny the validity of the sacred, but historicizes human understandings so that we may recognize the material and ideological practices that both shape and are justified by views of the divine. This grounding in turn enables us to draw critically from the traditions that have formed us while working to reshape our own views to empower our liberating praxis.

Much of Ruether's purpose in *Gaia and God* is to explore how the masculine dimensions of the divine have dominated and excluded feminine dimensions, resulting in distorted understandings of "God." The needed antidote to this historical process is not to throw out the masculine God in favor of a feminine Goddess, but to reclaim both voices of divinity as two inextricable and necessary dimensions of the divine. Ruether is most like Cobb and distinct from Hall and Nash in seeing these dimensions and hearing these voices of divinity as rooted in and coming from nature. Like Cobb and Birch's ecological understanding of God as Life, Ruether explicitly recovers the divine dimensions of the impulse to life in nature as Gaia.[59]

Ruether maintains we need both voices from nature: the masculine voice of God who speaks through power and law and the feminine voice of Gaia who speaks through the heart and relationality. Thus she is critical of the apocalyptic view of God who is unrelated to the earth, to bodies, to mortality as an unrelational God who silences Gaia. She argues that both divine voices are also our own voices, mediated by and reflective of human experience: "We need to claim them as our own, not in the sense that there is 'nothing' out there, but in the sense that what is 'out there' can only be experienced by us through the lenses of human experience."[60] Humans are not the source of life, in fact we are latecomers to life, but Ruether is critical of rigid demarcations between divine/human and divine/nature, particularly when these are reinforced and interpreted through a sexist dualism that assigns the weaker half to the female.

Ruether's ecofeminist ethics is unique among the four models in explicitly recovering feminine dimensions of the divine and integrating them with masculine elements to develop a more holistic understanding. Her view of the divine as both God and Gaia is consistent with the emphases of an ethic of Gay and Gaia of the divine as both just and indignant as well as tender and nurturing. It incorporates the centrality of both friendship and the erotic in describing the divine/human/nature relational matrix, and pays attention to the four attributes Hunt lifts up: attention, generativity, community, and justice.

Another of the constructive elements of Ruether's approach is her openness to learning about the divine from nonbiblical traditions while retaining connection with Christianity. This is congruent with the possibility and necessity of learning from lesbian and gay experiences of the divine that fall outside normative Christian tradition (although unfortunately Ruether does not do so). Consistent with a liberationist methodology, Ruether explicitly connects views of the divine and cosmology with the ethics and praxis they help enable. Hence she maintains that we must test our claims about the divine through our relationships: do they foster compassion and justice, or domination and enmity?

In light of these claims, Ruether talks about the divine through poetry, nonlinear images, and metaphors, as well as through classical understandings. She describes the divine as "the ongoing creative Matrix of the whole," "the great Thou," "the personal center of the universal process," and "the Matrix of life" who flowers in us as consciousness and the impulse toward kindness. This reality demands justice for the downtrodden yet is also our common mother, Gaia, who teaches us "patient passion."[61]

Ruether's view of the divine as nonhierarchical relational matrix calls into question any ontological dualisms based on hierarchy or rigid polarity such as male/female, spirit/matter, nature/culture, and divine/creation. She develops a relational panentheism that weaves together divine and human agency in relationship with all of nature. Overcoming the rigid God/creation boundary that characterizes the ethical models of Nash and Hall, Ruether develops a sensual epistemology to emphasize that the combined voices of divinity are rooted in and experienced within nature. Yet our experiences of the divine are always mediated by human experience. We need to reclaim these voices as also our own voices, divine agency as also our own agency. All of these elements are important in a liberationist ecological ethic.

With this understanding of the divine, Ruether stresses the cosmos as the mediating context for all theological reflection. Our capacity for ethical reflection on our relations with each other and nonhuman nature "is not rootless in the universe, but expresses this deeper source of life 'beyond' the biological."[62] Through our covenantal connection with the source of life, justice and right relation become the criteria by which we judge our relationships with each other and all of nature. Like Hall, Ruether stresses that in the covenantal relation we encounter each as both other and kin. She draws on the biblical concept of humans in the *imago Dei* to stress the unique responsibility of humans to be "accountable to the God of life to care for and protect the vast panoply of life-forms produced by millennia of earth's creativity."[63]

Creation narratives play a critical part in ethical reflection on our interrelatedness because they provide the blueprint for understanding the divine/humanity/nature pattern of relationship. In recovering the voices of both Gaia and God as dimensions of the divine, Ruether observes that societies with gender parity nearly always have creation stories that attribute creation either to a female or to a female and male together. A covenantal vision rooted in this multidimensional understanding of the divine recognizes that humans and other life-forms are part of one extended family in a community of interdependence. We need the full spectrum of human experience—including lesbian and gay experience—to more fully open ourselves to the divine in all its dimensions.

Methodologically, then, Ruether's views of the divine as God and Gaia could be developed further in two important areas. First, lesbian and gay voices as witnesses to the divine need to be acknowledged and incorporated. Without them, our understanding of God and Gaia is incomplete, and the ethics and praxis they support risks remaining heterosexist. Second, an ecofeminist ethics connected to the Christian community and tradition needs to develop better a Christological dimension to its understanding of the divine if it is to succeed in "luring" the more than one billion Christians of the world "into an ecological consciousness."[64] For example, Ruether concludes *Gaia and God* with a call to committed love, for "being rooted in love for our real communities of life and for our common mother, Gaia, can teach us patient passion."[65] Many, if not most Christians will want to know how this love of Gaia is connected to the love of Christ. This bridging of ecological and Christological issues is a strength of Cobb's ethics that is missing in Ruether's model. While Ruether has dealt with Christology at length in her other writings, it needs to be developed within an ecological and ecofeminist framework to ground an ethics intelligible and acceptable to members of the Christian community.[66]

Finally, Ruether develops a number of ecological and ethical guidelines helpful to a liberationist ethics. What distinguishes these is her pragmatic "critical both/and" approach: she refuses to be drawn into either/or thinking that suggests one approach to the exclusion of others, but instead looks critically at all sides for what can be reclaimed as most helpful. Hence, reflecting Christopher Stone's call for moral pluralism, she recognizes both the usefulness as well as the inadequacy of a rights-based approach to ecological ethics and calls for balancing it with approaches that stress relational responsibility within holistic biotic communities as the appropriate frame of reference. She locates herself critically within the Christian tradition while calling for critical openness to other perspectives and traditions. She pays attention to both the psychological and social dimensions of the ecological crisis while highlighting the need to reintegrate the political with the spiritual. Lastly, she outlines a praxis of ecojustice rooted in communities of resistance and celebration that can foster efforts toward healing and justice at the local, regional, national, and international levels. All of these guidelines are promising for a liberationist ecological ethics.

We now conclude our examination of the four models of Christian ecological ethics. The liberationist criteria of Gay and Gaia outlined in the opening chapters reveal both strengths and weaknesses in each of the four models. The liberal frameworks, epistemologies, and theological allegiances in Hall's stewardship ethic and Nash's ethic of integrity and responsibility are the main obstacles that need to be overcome in appropriating their insights and approaches. Cobb's process ethic of sustainability addresses many of these issues, and provides a theological grounding rooted in ecological insights, but still needs to be transformed through a liberationist hermeneutics that shifts the epistemological center of ecological ethics to the margins (both human and otherkind) of society. Ruether's ecofeminist ethic of earth healing is the most congruent model methodologically, and provides a wealth of theological and constructionist insights. The invisibility of lesbian, gay, bisexual, and transgendered experience and sexuality as an analytical tool in all of the models, however, must be addressed for them to provide an adequate grounding for a lesbian- and gay-sensitive liberationist ecological ethics. Grounded in the insights of this discussion, I turn now to outlining some of the features of a liberationist ecological ethic rooted in the paired criteria of Gay and Gaia.

Part 3

Gay and Gaia: Toward an Erotic Ethic of Ecojustice

10

Shifting Our Grounding: From Social Location to Ecological Location

All thinking worthy of the name must now be ecological, in the sense of appreciating and utilizing organic complexity, and in adapting every kind of change to the requirements not of man alone, or of any single generation, but of all his organic partners and every part of his habitat.

—Lewis Mumford, *The Pentagon of Power*

By my bedside sits a weathered copy of Esther Boulton Black's *Stories of Old Upland*, a chronicle of the early years of the town where I was raised and where my parents still live. Besides the fascinating stories of the earliest white settlers of the region that so intrigued me as a child, what is most striking to me now as I return to these stories and photographs is how much this area has changed in so little time. Fifteen hundred people made up the little citrus town Esther Boulton knew in 1910; by the time my parents arrived in California in the postwar boom of the early 1950s, Upland was a bustling town of 9,000. When I began kindergarten in 1962 the new Southern California freeway system already was beginning to transform Upland (then 12,000) and its neighboring towns from independent communities to a suburban bedroom sprawl enveloped by rapidly expanding Los Angeles, thirty-five miles to the west. By 1970 when my junior high class marched in the first Earth Day, my town's population had tripled to 33,000. Today Upland has topped 80,000 (and still growing!) and not a single remnant remains of the orange groves and vineyards I played in as a boy.

Aside from the grief and nostalgia I often feel for a childhood setting now lost (and I count grief and nostalgia as important sources for ecological ethics; as Donald Worster reminds us "Nostalgia runs all through society—fortunately, for it may be our only hope of salvation"),[1] reading Black's stories reminds me how differently she and I have learned to see the world while growing up in the same geographical place. She learned to see and "know" the world through a different set of social and ecological relations than did my parents forty years later or than I did in the 1960s and 1970s, and all of us knew a very different world than did the earlier indigenous Serrano and Cahuilla peoples we had displaced. The captions of her photographs record some of these changes. An early photo of the snowcapped San Gabriel mountains looming over citrus trees ready for harvest reads, "Upland was aptly advertised as *The Paradise of the Foothills.* This description may not have been too great an exaggeration for those of us who remember always blue skies and the heavenly fragrance of orange blossoms in May."[2] By the time of that first Earth Day, such days were rare, as Upland had achieved the dubious distinction of being the smoggiest community in all of California; then only infrequently could the mountains be seen on a summer day.

Southern California's natural beauty and climate have always drawn people; the worlds of those of us who were raised there were shaped by its distinctive combination of social and natural events. As a boy I learned early that "development" and "progress"—those unquestioned religious doctrines of California's twentieth-century prosperity—also meant the destruction of what had seemed natural to me: the surrounding orange groves and vineyards, the open hillsides and river washes. It was not until much later that I realized how much these in turn had resulted from the combination of human efforts and natural processes. Yet "nature" had a way of reminding us of ecological limits too often transgressed: earthquakes, fires, floods, and mudslides regularly, if only temporarily, humbled human efforts to reshape the landscape and reroute its natural processes of rain, wind, and fire. And the social upheavals of California in the 1960s—the Watts riots, the antiwar demonstrations, and drug culture—were early warnings of social limits too often transgressed. Both nature and the human society within it have been altered and changed as a result of these ongoing interactions.

I didn't realize how much this combination of the social and ecological had marked me and my ways of seeing the world and acting in it until college took me to the Midwest—and my first time ever away from mountains. Immersed in a new place where I was informed that what I thought looked like a tortilla was actually lefse, where the dense Minnesota woods in August were so lush and green they looked unnatural to me, and where I couldn't get my bearings without mountains and valley vistas to guide me, I had to learn new ways of seeing and acting in this new land of different topographies, seasons, human cultures, and ecological communities. Yet how many times since then have I had occasion to learn again how deeply embedded within me are the views of the world learned unconsciously in that early California mix of social and ecological patterns and interrelations.

In recent years, ethicists have highlighted the importance of paying attention to the social location of persons and communities as a way to reveal and examine power relations in society. Yet as my own upbringing illustrates, it is not only social patterns, but also biophysical and ecological ones that influence how we act and see in the world. If, as Lewis Mumford suggests, "all thinking worthy of the name must now be ecological," combining liberationist and ecological approaches suggests the need for an expanded concept of *ecological location*. This will offer a means to rethink within an ecocentric framework the liberationist attention to socially constructed categories of nature and human nature and the role of social location and identities, connecting them now with ecological theory and ethics. By *ecological location*, I mean enlarging the term *social location* to include both where human beings are located within human society and within the broader biotic community, as well as conceiving other members of the biotic community and the biotic community itself as locatable active agents that historically interact with and shape the other members of the ecological community, including human beings. Just as social location is an anthropocentric term that helps us to pay attention to how human identities are multiply formed with respect to various lines of human difference, ecological location is an ecocentric term that recognizes that human epistemologies—how we see and interpret the world—are also shaped by our relationship with the land and other creatures in our broader biotic environment.

The concept of ecological location is a logical outcome of liberationist efforts to deconstruct the nature/culture dichotomy that renders nature inert and invisible with respect to human affairs. Instead of the human social realm being seen as the only valued context, it must be understood as part of the broader ecological web of relations. Hence one's social location is a distinctive but interconnected part of a larger ecological location: where human and nonhuman creatures and communities are situated with respect to other members of the biotic community. How we are shaped to see and act in the world results from a complex interplay of physiological, social, cultural, *and environmental/ecological* factors. For ecological ethics (and, I would argue, the vast majority of social ethics), ecological location is the "relevant whole" or context that must be taken into consideration in ethical reflection.[3]

Stephanie Lahar captures the importance of thinking in terms of ecological location in her critique of the nature/humanity dichotomy in Western ways of thinking:

> The human/nature dualism is crucial to address and redress, since it is so fundamental, underlying and undermining our relations to the world around us and to that which is embodied and unmediated within ourselves. When we set ourselves apart from nature, we disembody human experience and sever it from an organic context. This means that we stop being aware of the shapings and natural containments that a particular environment places around human practices and social structures. *But of course environmental effects do not cease to exist.* Instead, society is shaped by a fractured relation to the ecosystem(s) it in-

> habits, losing both characteristic bioregional contours and a sensibility for natural limits. Additionally, I suggest that separating ourselves from our natural heritage, which has been a central project of human civilization, also has profound psychological and social implications as it supports our nonperception of others. When we cut off a part of ourselves that we share with all other human beings and, by extension, all of life, it is easier to deny that others, or a particular other, exists.[4]

In reconceiving ecological ethics as a liberationist ethic within an ecocentric framework, an ethic of Gay and Gaia sees ecological location as the necessary working tool and grounding for ecojustice. It provides a way to understand both the particularities and interconnections of intrahuman social relations and relations between humans and the surrounding ecology.

ENVIRONMENTAL FACTORS IN ECOLOGICAL LOCATION

Before looking at some of the characteristics and contributions of ecological location to ethics, it is helpful to note what things shape and mediate our interactions with our environments and thus influence the ways we are shaped by our ecological location. In his study of the relationship between environmental perception, attitudes, and values, Yi-Fu Tuan examines the interrelated questions of what affects our values of the environment, and how our values are in turn shaped by the environment.[5] Our responses to the natural environment result from a complex interplay between the sensory stimuli it provides, our individual temperament and purpose, and the cultural forces that provide meaning at a particular time.

Tuan argues that human interrelations with the environment are the result of an interplay of physiological, cultural, and environmental factors. Physiological factors such as sex, age, physical mobility, and differing abilities in the five senses of vision, hearing, touch, smell, and taste, all affect the way sensory stimuli from the environment are received and mediated in different persons; what is noticeable or pleasant to one person may be missed (such as a red flower to one who is color-blind) or found unpleasant (such as the sound of ocean waves to a light sleeper) to another.

Cultural factors strongly affect the meanings we give to received sensory stimuli. Tuan examines the vastly different historical responses of migratory and indigenous cultures to the identical physical setting to show how different prior experiences, socioeconomic background, and aim or purpose affect one's response to the environment. Northwestern New Mexico, for example, is home to five groups of people, two indigenous and three migratory. The native Navaho and Zuni populations, while differing in some aspects, each experience their environment as one of harmony and home, one with which to cooperate rather than subdue. Attitudes of the three migratory groups, Spanish-American, Mormon, and Texan, were shaped strongly by their previous experiences and the purposes each had in traveling to the new land. Early Spanish explorers, for example, came north from Mexico looking for gold. Finding few minerals and a climate and geography similar to the one they had left, their written records cite little remarkable in the new land, other than that it was colder in winter. In contrast, arriving from the humid east in search of suitable farmland, many Anglo-American explorers had a stronger reaction, describing the landscape as "barren," "sickening," or "disgusting." The Mormons and Texans who followed each came with a mission to subdue the land, which they saw in different ways as a rightful resource to be exploited.[6] The attitude of each of the five groups had been shaped by their prior experience and purpose in interacting with their environment.

Changes in attitude toward nature also occur within cultures through time. Tuan observes a general historical progression from religious to aesthetic appreciation as nature sheds its numinous aura and symbolic sacredness, to the more modern utilitarian view of nature as economic and recreational resource. Important for our discussion, however, is that all three attitudes are likely to be present among parties involved in environmental issues, often generating intense conflicts where people talk past each other in mutually nonintelligible ways. A contemporary example is the controversy over logging in old-growth forests in the Pacific Northwest. The forest is variously valued for economic, recreational, aesthetic, religious, and ecological reasons among proponents of the different perspectives; these attitudes depend in large part on the different social and ecological locations and histories of those involved with respect to the old-growth ecosystems. Attention to the historical, cultural, and environmental factors that shape our ecological locations can help to decipher some of the dynamics at stake in these conflicts.

Finally, the environment itself shapes our worldviews. Whether one is raised in an urban versus rural or temperate versus tropical environment will in part determine one's worldview. This is true from the smallest to grandest scale, including our underlying religious cosmologies. Tuan contrasts the vastly different religious cosmologies of peoples living in the tropical rain forest with the cosmologies of peoples living in open, arid environments. For the BaMbuti Pygmies nature is experienced as an all-encompassing rain forest reality that filters out the movements of the sun, moon, and stars, voids a concept of spatial distance or horizons, and has little seasonal variation. In contrast to the Genesis creation accounts, which, as Ruether noted, draw on earlier Mesopotamian creation stories that took shape in an open dusty land of occasional catastrophic floods and watery chaos, the BaMbuti religious cosmology lacks elements such as stars, seasons, the sky, and the earth and focuses instead on the all-sustaining forest. The effects of the surrounding environment on each group's sense of the way reality operates is profound and in turn provides the social and ecological blueprints for how they organize social relations and interactions with other parts of nature.

These religious worldviews in turn strongly influence moral norms for how one "ought" to interact with one's environment. This has important ramifications for an increasingly multiethnic global reality and a United States culture of mobility where the environments that unconsciously shaped one's worldview may have little or no relation to where one finds oneself located now. History is full of examples of migratory groups bringing with them and imposing views of the humanity/nature relation shaped by ecological locations very different than the ones in which they now find themselves, often with disastrous results for the indigenous peoples and ecosystems that find themselves supplanted.[7] By drawing attention to the ecological factors that shape us and our epistemologies, ecological location can help to identify and clarify such issues and social-ecological dynamics.

CONTRIBUTIONS OF ECOLOGICAL LOCATION TO THEORY AND ETHICS

In terms of ecological ethics and ethical theory, the concept of ecological location makes several contributions to the necessary rethinking of the humanity/nature relation within an ecocentric framework. These include recognizing the following: (1) nature's agency and humanity's unique location and agency within nature; (2) both differences within human communities and between humanity and other parts of nature; (3) the variation and particularity of human power and privilege vis-à-vis nature; (4) the historical dimension of ecology and nature; (5) recognizing the spiritual dimension of human interactions and histories with particular places, habitats, and geographies.

Critical to the concept of ecological location is that it acknowledges and places at the heart of ethical analysis the notion of nature's active agency, both as the whole ecosystem and biotic community, as well as its constituent parts of individual creatures, species, and niches. Human beings are a *uniquely positioned* agent in this matrix, but not unique in having agency, or even moral agency.[8] This is another way of building on the insights of process theology that all things are interconnected and relational; each affects the other and the whole. Some ramifications of this include building on Cobb's work connecting ecology and the economy by incorporating nature as active agent rather than as passive resource into socioeconomic analysis and policy development.

In her efforts to develop a model for understanding the relationship between ecological and social transformations, Carolyn Merchant stresses the importance of considering nature's agency:

> Nonhuman nature, therefore, is not passive, but an active complex that participates in change over time and responds to hu-

> man-induced change. Nature is a whole of which humans are only one part. We interact with plants, animals, and soils in ways that sustain or deplete local habitats. Through science and technology, we have great power to alter the whole in short periods of time. The relation between human beings and the nonhuman world is thus reciprocal. Humans adapt to nature's environmental conditions; but when humans alter their surroundings, nature responds through ecological changes.[9]

It is important to remember that, like the mechanical view of nature they replace, ecological paradigms are still social constructions. Like all social constructs, they will reflect the values, norms, concepts, and presuppositions of the individuals and communities that form them. Yet an ecological perspective differs from other viewpoints in that it requires one to take account of the ramifications of human *and nonhuman* actions throughout the ecosphere: "Ecological thinking constructs nature as an active partner. The 'nature' that science claims to represent is active, unstable, and constantly changing. As parts of the whole, humans have the power to alter the networks in which they are embedded. Nature as active partner acquiesces to human interventions through resilience and adaptation or 'resists' human actions through mutation and evolution. Nonhuman nature is an actor; human and nonhuman interactions constitute the drama."[10] Hence ecological location provides the methodological grounding for understanding how human (and nonhuman) epistemologies are shaped in response to their *total* environment, which includes but is not limited to the human environment.

Ecological location can help us recognize simultaneously power differences within the human community and in humanity/nature relations. Attention to power in ecological location can show how both the dynamics of intrahuman community and the human/nature biotic community are built on either cooperation or domination—that is, relations that either sustain or deplete us socially and ecologically. Thus it can help to better integrate understandings of ecological and social justice as right relation, a critical component of a liberationist ecological ethics. Keeping the intrahuman dynamic of social location is an integral part of ecological location. It retains the concern of liberationist thinking about how human social differences are constructed into social relations of domination and exclusion. Expanding this to ecological location shows both how these human social relations affect the wider biotic community as well as how they are affected *by* the limits and makeup of the biotic community.

Ecological location draws on the ecological insight that all creatures affect their environment, but it is primarily humans who consciously change it in decisive ways. The conscious character of human alterations of nature means that they can be evaluated ethically as well as ecologically. In her analysis of ecological transformations in colonialist and capitalist New England, Merchant provides a helpful model for a liberationist ethics through extending the categories of a socialist feminist analysis into an ecological context. She argues that power in the human/nature relation is mediated through the relation of ecology to production, reproduction, and consciousness:

> The relations among animals (including humans), plants, minerals, and climatic forces constitute the *ecological core* of a particular habitat at a particular historical time. Through *production* (or the extraction, processing, and exchange of resources for subsistence or profit), human actions have their most direct and immediate impact on nonhuman nature. *Human reproduction*, both biological and social, is one step removed from immediate impact on nature: the effects of the biological reproduction of human beings are mediated through a particular form of production (hunting-gathering, subsistence agriculture, industrial capitalism, and so on). Population does not press on the land and its resources directly, but on the mode of production. Two steps removed from immediate impact on the habitat are *the modes through which a society knows and explains the natural world*—science, religion, and myths. Ideas must be translated into social and economic actions in order to affect the nonhuman world.[11]

While some have criticized Merchant's emphasis on production and reproduction as too anthropocentric and implicitly partaking of a nature/culture duality,[12] the advantage of her approach for a liberationist ecological ethics is that it provides a framework for analyzing the effects of human material practices within both the social and ecological realms. Not only the effects of human production, reproduction, and consciousness on nature, but also nature's (changing) reciprocal effect on production, reproduction, and consciousness can be accounted for. Ecojustice criteria such as sustainability, participation, and just distribution can then be employed to evaluate these relations. As a critical starting point, the category of ecological location is broad enough to include each of these factors and relationships in examining how our worldviews are shaped by our social and ecological contexts.

An example of how ecological location can help us to make sense of the interplay of social and ecological justice issues is the ongoing conflict over access to the land in Central America. Rodolfo and Teresa Hernández are campesinos (peasants) and community leaders in the small village of La Mora, a "repopulation zone" on the northern slopes of the Guazapa volcano in central El Salvador. I first met Teresa in 1986 in a refugee camp in San Salvador during the height of El Salvador's twelve-year civil war, a war generated by a complex interplay of social and ecological conflicts. Rodolfo and Teresa's social and ecological locations had been shaped by five centuries of social and ecological exploitation that had determined the patterns of human interaction with each other and the land in El Salvador. Spanish colonialization first destroyed the culture and ecologically sustainable farming and communal land patterns of the indigenous Pipil in order to create an agroexport economy that relied on cheap human labor and privatization of land. Successive waves of different export monocrops and land expropriations left the vast majority of the rural population in poverty; by 1970 over 70 percent of El Salvador's cropland was controlled by less than 2 percent of the population. Forced from their traditional fertile farmlands in the highlands, campesinos like Rodolfo and Teresa moved onto the steep slopes of El Salvador's many mountains and volcanoes to try to subsist on the poorer farmland there. The resulting ecological deterioration of these now deforested lands only deepened their poverty.

When in the early 1980s Rodolfo and Teresa decided to join the thousands of campesinos who had begun to take up arms to demand access to land and more just social conditions, their ecological location high on the slopes of the Guazapa volcano had been shaped largely by these centuries of poverty, social injustice, and ecological deterioration. Two additional factors have since reshaped their ecological location and the consciousness that comes from it: twelve years of civil war and subsequent efforts to repopulate their former lands.

Fearing it was losing the war after its brutal campaigns of the early 1980s failed to dislodge the campesino guerrillas of the Farabundo Martí National Liberation Front (FMLN) from their territory, in the mid-1980s the Salvadoran military (on the advice of its U.S. military advisors and financiers) began a program of "ecocide"—a deliberate attempt to destroy the ecology that sustained life in areas controlled by or sympathetic to the guerrillas. January 1986 saw the unleashing of Operation Phoenix on the guerrilla-controlled Guazapa volcano: a scorched-earth campaign modeled on earlier U.S. campaigns in Vietnam that combined bombing with military groundsweeps to completely deforest and depopulate Guazapa.[13] Following several months of flight on foot, Teresa and her children ended up in the refugee camp where I met her, and Rodolfo escaped with the FMLN guerrillas. The Guazapa region was totally devoid of human life, and much of its ecological balance shattered.

With the end of the war in 1992, Rodolfo and Teresa returned to Guazapa with thousands of other campesinos to begin to rebuild their lives with each other and the land. Organized into sixty repopulation villages, the communities are building on years of experience at organizing to restructure their social and ecological relations. Yet their ecological location has shifted in the intervening fifteen years of social and ecological conflict. Dramatic changes are apparent at all three levels of the humanity/nature relation that

Merchant describes. In terms of production, the communities have abandoned the ecologically most devastated lands and steep slopes and have created a reforestation program to begin to rehabilitate and restore some of the shattered ecology. To reduce dependence on outside chemicals, organic farming techniques are being introduced. At the level of human reproduction, the communities have reorganized themselves socially, recovering some of the communal relation to the land and working cooperatively. At the level of consciousness, there is increasingly a deepened awareness that as Salvadoran environmentalist Ricardo Navarro says, "The struggle for the environment is the struggle for life."[14] Environmental issues are seen now as important social, political, and economic issues that affect each of these at all levels. Finally, nature itself has changed. The Guazapa region to which the campesinos returned is not the same land they left; it has been damaged significantly, which has affected many of its natural patterns, such as rainfall and runoff patterns.

As this brief overview of Rodolfo and Teresa's situation in El Salvador demonstrates, looking at the ecological locations that have shaped individuals, communities, and contexts can help to identify and clarify the complex interplay of social and ecological factors involved in situations of social and ecological injustice. In Guazapa, the shift in ecological location as a result of the war and repopulation means that both the people and the nature that surrounds them have changed, as has the humanity/nature relationship of the region. Examining their ecological locations draws attention to the historical dimension of the ecological as well as the social factors in this shift, and the different forms of power and agency involved. Each of these in turn provides a critical component to good ethical reflection on the moral dimensions of this conflict and its resolution.

Ecological location can help us to recognize the current privileged position of human beings as a species vis-à-vis the rest of nature without collapsing that location and identity into a generalized, universalized human posture. For example, middle-class suburban North Americans, poor urban Mexicans in Mexico City, and rural farmers in Zimbabwe or El Salvador all may have some degree of power and privilege—as well as vulnerability—with respect to "nature," but it is very different in kind and degree. Their ecological locations and power relations within them are vastly different, and the larger ecosocial analysis is distorted in simply universalizing the discussion as the relationship between "humanity and nature" without taking this into account. Paying attention to their different locations in both the human community and the ecosystem may be one way of keeping in mind both the human justice and ecological issues at stake, and how they intersect.

Wendell Berry's analysis of the dynamics of racism in his native Kentucky and its impact on the human relations with the land provides an illustration of how ecological location can incorporate analysis of the historical effects of human power differences such as those found in racism with their ecological consequences. While these human differences have historically generated roots going back generations to the slave period, Berry shows how they have resulted in contemporary ecological effects that continue to shape different attitudes toward and interactions with the land structured along race and class lines. Berry is a white farmer and writer whose reflections on the connection of racism to how we interact with the land began in the midst of the civil rights and black power movements in the 1960s. As he worked on his small Kentucky farm, he thought of all those—black, white, and Indian—who had worked that land before him. In noting the different attitudes among Indian, white, and black people toward working the land, he was struck by how white racism had influenced and altered the relational patterns of each of the three racial groups:

> I became thoughtful of all the work that had been done there on my home ground either by despised men [*sic*] or by men who secretly despised themselves for doing the work of despised men—so many of the necessary acts of my history, neither valued nor understood, wasted in the process of wasting

> the earth. And I thought back to the time before the brief violent spasm of my people's history there, to the thousands of years when the Shawnees and their forebears lived in the country in its maidenhood, familiar with it as they were with their own bodies, as much at home in it as the plants and animals, wedded to it by an exquisite awareness of its life. They were native as no one has been since. And I began to understand how the racism of my people has been a barrier not just between us and our land but between us and our exemplary predecessors.[15]

Berry identifies the crucial link between racism and mistreating the land as the white slave-owners' belief that manual labor on the land was "nigger work," appropriate only for racial inferiors. Hence, when whites found themselves forced to do the manual work they had identified with "despised" (black) people, they found themselves consciously or unconsciously humiliated and ended up despising themselves—and deepening the cycle of racial alienation and alienation from the land. The denigration of blacks and the land was thus linked, and both were valued only for their (related) utilitarian value to the white owner. Yet this set in place a racially destructive human/land relationship where members of both races were deprived of critical components needed for an ecologically sustainable relation: divorced from manual labor, whites could never know the land intimately and thus were ignorant of many of their effects on the rhythms of the land, while blacks who gained an intimacy with the land through manual labor were denied of any sense of a permanent relationship with or responsibility for the land.

It is worth citing Berry's eloquent analysis at length here. He begins by arguing that only when white people seriously confront their racist heritage and their own racism, rather than relying on governmental legislation to do the work for them, can the destructive heritage and current reality of racism begin to be healed. Yet it is only as we confront and work to end racism and its association of "nigger work" with living closely to and working with the land that we can heal our broken relation to the land. It should come as no surprise that a race of people that thought it could "own" another race of people should end up alienated from the land as well.

> No man [*sic*] will ever be whole and dignified and free except in the knowledge that the men around him are whole and dignified and free, and that the world itself is free of contempt and misuse.
>
> For want of the sense of such freedom, even as an ideal, the white race in America has marketed and destroyed more of the fertility of the earth in less time than any other race that ever lived. In my part of the country, at least, this is largely to be accounted for by the racial division of the *experience* of the landscape. The white man, preoccupied with the abstractions of the economic exploitation and ownership of the land, necessarily has lived on the country as a destructive force, an ecological catastrophe, because he assigned the hand labor, and in that the possibility of intimate knowledge of the land, to a people he considered racially inferior; in thus debasing labor, he destroyed the possibility of a meaningful contact with the earth. He was literally blinded by his presuppositions and prejudices. Because he did not know the land, it was inevitable that he would squander its natural bounty, deplete its richness, corrupt and pollute it, or destroy it altogether. The history of the white man's use of the earth in America is a scandal. The history of his effort to build here what Allen Tate calls "a great culture of European pattern" is a farce. To farm here, as we have done for centuries, as if the land and the climate were European, has been ruinous, ecologically and agriculturally, and no doubt culturally as well. . . .
>
> The notion that one is too good to do what is necessary for *somebody* to do is always weakening. The unwillingness, or the inability, to dirty one's hands in one's own service is a serious flaw of character. But in a society that sense of superiority can cut off a whole class or a whole race from its most necessary experience. . . .
>
> The abstractness of the white man's relation to the land has forced the black man to develop resources of character and reli-

gion and art that have some resemblance to the peasant cultures of the old world . . . but at the same time it has denied him the peasant's sense of permanent relation to the earth. He has wandered off the land into the cities in the hope of being better treated, only to be scorned as before. And on the land his place has been taken by machines—and we are more estranged from our land now than ever we were.

For examples of a whole and indigenous American society, functioning in full meaning and good health within the ecology of this continent, we will have to look back to the cultures of the Indians. That we failed to learn from them how to live in this land is a stupidity—a *racial* stupidity—that will corrode the heart of our society until the day comes, if it ever does, when we do turn back to learn from them. Inheriting the cultural growth of thousands of years, they had a responsible sense of living within creation—which is to say that they had, among much else, an ecological morality—and a complex awareness of the life of their land which we have hardly begun to have. They had a cultural and spiritual wholeness of which the white and black races have so far had only the divided halves. . . .

Empowered by technology, the abstractions of the white man's domination of the continent threaten now to annihilate the specific characteristics of all races, virtues and vices alike, absorbing them as neutral components into a machine society. It is, then, not simply a question of black power or white power, but of how meaningfully to reenfranchise *human* power. This, as I think Martin Luther King understood, is the real point, the real gift to America, of the struggle of the black people. In accepting the humanity of the black people, the white race will not be giving accommodation to an alien people; it will be receiving into itself half of its own experience, vital and indispensable to it, which it has so far denied at great cost.

As soon as we have fulfilled the hollow in our culture, the silence into our speech, with the fully realized humanity of the black man—and it follows, of the American Indian—then there will appear over the horizon of our consciousness another fig-

> ure as well: that of the American white man, our *own* humanity, lost to us these three and a half centuries, the time of all our life on this continent.[16]

Berry's essay reveals several contributions of the concept of ecological location for discerning the variation of human power and privilege in our relations with nature and the ecological ramifications of these. In paying attention to the different racial ecological locations—their different power relations to the land—-ecological aspects of racism that might otherwise remain invisible are revealed. Different power relations between blacks and whites led to different relations to and locations on the land; these distorted human power relations in turn resulted in distorted and damaging ecological relations for both whites and blacks. These distorted ecological relations resulted in and contributed to different worldviews along racial lines that reinforced and exacerbated the racist and ecologically damaging cycle. Paying attention to race in ecological locations reveals the complex interconnections of cultural and ecological values and how these shape human relations with each other and the land. Finally, as Berry notes, examining how we are located and interact ecologically reveals a culture's "ecological morality," that is, its ability to integrate its spiritual, cultural, and material practices into an ecologically sustainable and socially just whole.

This same dynamic permeates what today has come to be known as "environmental racism": the fact that people of color in the United States bear a disproportionate share of the nation's pollution problem without participating fully in either the economic benefits or the decision-making procedures that set environmental policy. The Rev. Benjamin F. Chavis Jr., who first coined the term *environmental racism,* describes its multifaceted character as follows:

> Environmental racism is racial discrimination in environmental policymaking. It is racial discrimination in the enforcement of regulations and laws. It is racial discrimination in the deliberate

> targeting of communities of color for toxic waste disposal and the siting of polluting industries. It is racial discrimination in the official sanctioning of the life-threatening presence of poisons and pollutants in communities of color. And, it is racial discrimination in the history of excluding people of color from the mainstream environmental groups, decisionmaking boards, commissions, and regulatory bodies.[17]

Ecological location can help to pinpoint the historical, social, and ecological factors and the differential power relations between human groups and between these different human groups and the rest of nature. Power differences along lines of race and class in the human community affect ecological reality and the dynamics of different humanity/nature relations in fundamental ways, whether it is the disproportionate siting of nuclear waste dumps on or adjacent to Native American reservations, the flight of chemical plants to the cheap labor and lax environmental regulations of the Mexico maquiladora border industry, or the selection of Warren County, North Carolina, a predominantly poor, African American county, for a PCB landfill. *Where* we are located affects *how* we see. The not-in-my-backyard attitude that has characterized much of white suburbia (and the mainstream environmental movement)—and has been enforced with white political and economic power—has led to ecological locations differentiated largely by race and class. It has meant that we see and know nature differently, in largely unacknowledged and unconsciously racist and classist ways. Ecological location draws attention to these social and ecological dynamics.

Ecological location integrates a *historical* dimension into the relationship of humanity to the broader biotic community. It recognizes the historical agency of nature that changes both independently of and in response to changes in human society, while at the same time influencing and interacting with human history. As we saw in Tuan's analysis, it provides a framework for analyzing how human communities have been shaped differently in their view of the human/nature relationship depending on how and where they have been located. Important for lesbian and gay realities, and life in general in the late twentieth century, ecological location can help to look at the effects of human mobility and dislocation, both voluntary and involuntary, on this relation, as well as our historic impacts on and change in the biotic community.

Critical to this aspect of ecological location is developing an understanding of history as ecological and ecology as historical. In his study of the interconnections of ecological and historical changes in New England, William Cronon notes, "The replacement of Indians by predominantly European populations in New England was as much an ecological as a cultural revolution, and the human side of that revolution cannot be fully understood until it is embedded in the ecological one. Doing so requires a history, not only of human actors, conflicts, and economies, but of ecosystems as well."[18] This replacement of human populations led to fundamental changes in nature—both to how human beings interacted with the rest of nature, as well as in nature's very ecosystems themselves. Some species became extinct, others flourished, but none were unaffected.

One goal of such an understanding of history is to overcome the nature/culture dichotomy that has characterized the vast majority of historical studies: "Our project must be to locate a nature which is within rather than without history, for only by so doing can we find human communities which are inside rather than outside nature."[19] Nature, no less than human culture, has a historical dimension that has coevolved over time with human cultural development, sometimes changed by and sometimes changing independently of human influences. As Donald Worster observes, every part of the globe has "a story to tell not only of human history but also of the history of nature—a history of environmental transformation."[20]

In the last decade an interdisciplinary field of environmental history has begun to develop that can provide important tools for ecological ethics in learning how to incorporate a historical dimension to ecological location. Such recent books as Clive Ponting's *A Green History of the World: The Environment and the Collapse of*

Great Civilizations, Alfred Crosby's *Ecological Imperialism: The Biological Expansion of Europe, 900–1900,* and Donald Worster's *The Wealth of Nature: Environmental History and the Ecological Imagination* provide a much needed broad ecohistorical analysis and context for ethical analysis, particularly of the ecological ramifications of human migrations and degrees of ecological sustainability and depletion of different cultural patterns. More regional and local studies, such as Cronon's and Merchant's works on New England, include the specifics of particular ecosystems that can furnish the necessary specificity for examining our current ecological locations and what has shaped them.

Good ecohistorical analysis can help us to better understand the (often contradictory) mix of attitudes, values, and practices that shape our current ecological locations. Two examples may illustrate this. Carolyn Merchant understands "ecological revolutions" to be major transformations in human relations with nonhuman nature which radically affect both sets of actors by altering the local ecology, human society, and human consciousness. She sees two such transformations as having occurred in New England since the first permanent contact with Europeans in the 1600s, with our contemporary society possibly standing on the brink of a third. The first, colonial ecological revolution led to the collapse of indigenous Indian ecologies and the implantation and adaptation of a European ecological complex of animals, plants, pathogens, and people built around a mercantile economy that saw indigenous plants and animals primarily as commodities. Merchant argues that this ecological revolution "was legitimated by a set of symbols that placed cultured European humans above wild nature, other animals, and 'beastlike savages.' It substituted a visual for an oral consciousness and an image of nature as female and subservient to a transcendent male God for the Indians' animistic fabric of symbolic exchanges between people and nature." The second, capitalist ecological revolution occurred between the American revolution and the Civil War with the onset of the industrial revolution and created air and water pollution and resource depletions as externalities outside the market economy.

Human consciousness became "split into a disembodied analytic mind and a romantic emotional sensibility."[21]

Crosby's analysis of European migrations is broader in scope but similar in perspective to Merchant's analysis of New England. He sees these migrations as ecological phenomena in which the interaction of people, animals, plants, weeds, pathogens, and microorganisms resulted in enormous transformations of both cultures and ecologies. Crosby's work exposes the false dichotomy between the biological and the social, and his ecological frame of reference provides a necessary complement to more prevalent social analyses of the movements of European peoples. It reveals the enormous ecological as well as cultural ramifications of mobile cultures where practices and values formed in one ecological location are transferred to another, often with disregard or contempt for indigenous natural and human ecologies and cultures—and with disastrous results for both. Crosby's and Merchant's studies reveal the complex interplay of ecological and cultural influences in history and provide critical resources for employing the historical dimension of ecological location in ethics.

In addition to drawing attention to dynamics of power and historical elements in humanity's relationships with other parts of nature, ecological location can help us to recover for ethical analysis what many indigenous religious traditions have long recognized: the spiritual dimensions of the humanity/nature relation that develop for many and consciously or unconsciously influence the ways we see and act in the world. At least initially, human beings develop attachments not with nature or the biosphere in an abstract or universalized sense, but rather with particular places, particular communities of animals, plants, bodies of water, weather patterns, rock formations, seasonal rhythms.[22] Many believe there is little chance for ecological ethics to succeed in helping to reverse the ecological crisis without human beings developing a renewed sense of spiritual connection to the land and all its creatures and parts. Ecological location can be one part of building up what Mitchell Thomashow has called an "ecological identity"[23] by drawing attention to the particularity of our spiritual and aesthetic relations with nature, and how this affects our way of acting and seeing in the world.

I believe this last aspect, the spiritual dimension to ecological location, has important ramifications for a cultural and socio-economic structure built largely on mobility and what I term "ecological *dis*location." Alfred Crosby's analysis of the massive European migrations in recent centuries, for example, demonstrates that these migrations are ecological as well as social phenomena in which the interaction of people, animals, plants, weeds, pathogens, and microorganisms resulted in enormous transformations of both cultures and ecologies. They are also *religious* and *ethical* phenomena. The Native American cultures and societies Europeans largely displaced and replaced included complex religious and spiritual ties to their local ecosystems—religious cosmologies, beliefs, and practices that could not simply be uprooted and transferred to a reservation without dramatic impact on both the Native communities and the ecosystems that sustained them. The transplanted religious and ethical values and practices of the European migrant communities that replaced them have contributed to the rapid and largely deleterious ecological transformation of much of North America and its diverse ecosystems. Without developing new religious, spiritual, and ethical ties to the land, our ecological efforts in other areas will be incomplete.

ECOLOGICAL LOCATION AND LESBIAN AND GAY REALITIES

Drawing on these insights, ecological location can help to identify components of lesbian and gay realities that either may foster or obstruct developing an ecological perspective, spirituality, and praxis. For example, viewed ecologically, an ambiguous element of many lesbian and gay locations is our nearly ubiquitous experience of displacement and mobility, especially among the lesbians and gay men who have contributed most to our predominantly urban cultures. While many of us were nurtured with positive experiences of our "eco-habitats of origin"—for me personally, it was sustained contact with the prairies and mountains of Colorado that helped me to survive the initial battering I experienced in coming out—we often find ourselves removed and cut off from direct contact with these eco-habitats in order to have the relative safety and security of lesbian and gay urban subcultures. The positive side of this is the wide diversity of perspectives and resources we bring from the variety of ecological locations that nurtured us: we come from all walks of life and environmental niches.[24] Out of the intentionality of effort required to find each other and form community, lesbians and gay men have built on this diversity to create a cultural richness in our communities and wider culture that finds expression in diverse and colorful ways. Particularly important to an ecojustice ethics are the ways we have found and reclaimed pleasure and beauty in our body-selves and communities.

Yet this very mobility and experience of displacement can extract a high ecological cost when it precludes or hinders the development of a close, intimate, and sustained relation to and knowledge of the land and all its inhabitants. Wendell Berry states clearly the damaging ecological ramifications of this: "A nation of urban nomads, such as we have become, may simply be unable to be enough disturbed by its destruction of the ecological health of the land, because the people's dependence on the land, though it has been *expounded* to them over and over again in general terms, is not immediate to their feelings."[25] For this reason it is that much more important that we learn to reintegrate our erotic and ecological selves. Violations of *any* form of right relation—whether our most intimate, personal relationships or of the ecological web in which we live—damage all forms of right relation.

A similarly ambiguous reality is the location of much of the gay community—and certainly its most visible and active part—in concentrated urban settings or "gay ghettos." From the perspective of discovering, creating, and preserving our identities as lesbians, gay men, and bisexual and transgendered persons, and thus adding to the richness of diversity of the human community, the ghetto has been a vital and indispensable resource. It has provided us with safe space to find each other and create and value our diverse cultures and communities, which would not have been possible elsewhere given the pervasive homophobia and heterosexism of mainstream society. In serving to connect us with each other and nourish the life of our communities, gay and lesbian ghettos have a distinctly ecological character as an important "niche" in the human and broader biotic community.

Yet ghettos can become unecological when they foster in us a solely inward focus, cut off from or refusing to recognize our wider interdependence on other portions of the human community and on the broader earth community.[26] When ghettos become rigid and isolate us physically and psychologically from others to whom we are related—even when these relationships may be distorted through social dynamics of domination and subordination—they can become dysfunctional and unecological. In saying this, I am not arguing that those who have been victims of violence or abuse need to remain in relationship with their abusers. Ghettos as safe space provide one of the most important resources for beginning the healing process from abuse precisely because they shield those who have been victimized. I am suggesting, however, that ghettos need to help foster a process of moving beyond a victim stance to reclaiming agency as one who has survived and may now even hope to thrive, and this entails recognizing a broader interrelationship and reconnection with others, no longer operating out of a victim stance.

Rather, just as an ethic of Gay and Gaia calls into question rigid boundaries of "us/them" in the nature/culture dichotomy, it insists on relational understanding of the ghetto as primary social-ecological location. The ghetto as location should be seen in terms of dialectical movement, an ebb and flow of connection with and withdrawal from other parts of the human community. At times one's energy must be focused inward on dynamics in the gay community; other times we focus outward as we join in solidarity and celebration with others. An ethic of Gay and Gaia reminds us that in both moments we need to maintain an ecological perspective that keeps in mind the earth and other subordinated peoples and creatures, and the impact our material practices have on them.

Related to this is the predominance of urban locations reflected in the discussion and theorizing of lesbian and gay male realities. The danger of exclusively urban locations comes from being largely cut off from the land and its vital cycles and seasonal rhythms. Yet as J. Michael Clark has argued, reclaiming our right to "home"—whether to the homes from which we have been displaced or to the new urban homes we create—can give us a stake in the ecological health of urban locations as well as connect us with others who are similarly situated. Urban locations have the potential to connect us with a variety of social groups in common struggle for justice and ecological well-being. This must be seen as an extension of rather than as an alternative to important liberationist issues indigenous to our communities:

> Even with our considerable in-house agenda, *which absolutely must not be forsaken*, groups such as the various "faerie circles" and Gays United Against Nuclear Arms have pursued ecological concerns, while individuals have worked within local neighborhood groups on similar issues. In Atlanta, for example, gay men and lesbians have been active in fighting unnecessary freeways and trucking facilities which threaten our neighborhoods and their immediate ecosystems. Leaving our psychological or epistemological ghettos does not mean relinquishing our very important gay liberation agenda; it does mean developing a broader vision which sees the connections among *all* forms of oppression, exploitation, and disvaluation and which thereby facilitates liaisons to confront all of these.[27]

Attention to ecological location also can help to recall the important place nature and our surrounding environments play for many of us in coming to terms with our sexuality and coming out. I have talked to countless lesbians and gay men for whom the solace of the woods, the rhythms of the sea, or the breezes on the prairie provided the sustenance that saw them through these difficult periods. I recall my own healing walks on the prairies of Colorado when suicide seemed the only viable alternative to discovering my gay sexuality in the confines of a homophobic boys' boarding school that reflected and focused my church's and society's contempt for and hatred of gay men and lesbians. In the mountains and on the prairies I felt at home, that there was one place I fit in; returning to those places still brings back to me the confusing mixture of fear and exhilaration I associate with that critical period of my life. Without a cultural ritual of discernment, such as that provided by the Native American vision quest, intimate contact with the natural world for many of us gave us our first bearings in a homophobic world.

These themes connecting discovery of sexual identity in and with nature often are reflected in lesbian and gay male literature. In his novel *Native*, for example, William Henderson explores how his relationship with nature mediates the growing gay sexual awareness of Blue, a twenty-three-year-old cowboy in Wyoming. As Blue contemplates the loss of home, occupation, and friends—virtually his entire social world—he initially turns to nature to find his bearing. In a society that has largely constructed a view of nature that excludes homosexuality as unnatural, this connectedness to nature has allowed countless lesbians and gay men to intuit our rightful place in the diversity of the natural world.

Finally, attention to ecological location in lesbian and gay realities can help to reveal and clarify the regional and geographical diversity within our communities. There are ecological as well as cultural reasons why lesbian and gay communities in Minneapolis and Denver, in San José, Costa Rica, or Johannesburg, South Africa, differ in important ways from the more dominant and visible communities on the East and West coasts of the United States. And much of the diversity *within* each of these communities reflects the diversity of the different ecological as well as social locations of their members.

In its exploration of lesbian and gay locations, an ethic of Gay and Gaia needs to explore ways to extend lesbian and gay emphasis on embodied relationships to include our ecological relationships, so that we *feel* as well as *know* the ecological ramifications of our actions and attitudes. It can enable us to see the diversity of knowledge we hold from our many ecological locations as a strength and to build on these to overcome the alienation from the particular social and ecological locations in which we find ourselves that too often come from displacement and mobility. While acknowledging that displacement for many people is often not a choice, or is a choice necessitated by survival demands, an ethic of Gay and Gaia nevertheless calls into question our cultures' mobility and dislocations in favor of forming long and sustainable relations within our particular communities and locations.

11 Erotic Ecology: Interconnection and Right Relation at All Levels

[Lesbians and gay men] have helped me to embrace a larger vision of divine and human loving. Dante found eros in the Love that moves the sun and the other stars, and many of you have also. I have come to see divine eros as that fundamental energy of the universe that is the passion for connection and hence the hunger for justice and the yearning for life-giving communion.

—James Nelson, *Body Theology*

As we have seen, an ecojustice ethic of Gay and Gaia roots ecological moral agency in reclaiming and reintegrating the connections between the erotic and the ecological. The insights of lesbian and gay theo-ethical reflection provide important components for rethinking the theoretical grounding of a liberationist ecological ethic. In this chapter I explore the possibilities of grounding an ecological ethic in such areas as revisioning the erotic as ecological and ecology as erotic, interpreting ecojustice as right relation at all levels of human and nonhuman nature, developing an ecological liberationist hermeneutics, and deepening ethics by learning cross-culturally from other ecologically oriented value systems. In the next chapter, I give attention to specific features of an ethic of Gay and Gaia, and what these reveal about the possibilities of constructing a liberating ecological praxis in lesbian and gay communities and in wider society.

TOWARD A LIBERATIONIST ECOLOGICAL HERMENEUTICS: THE EROTIC AS ECOLOGICAL AND ECOLOGY AS EROTIC

Reclaiming and revisioning the erotic provides the key starting point for moving from an intuitive sense of the connectedness of all things to theorizing the erotic as ecological and ecology as erotic. This leads to the possibility of grounding a liberationist ecological hermeneutics in the erotic. Viewed theologically, the erotic is that part of sacred power that moves us toward connection, with ourselves, with each other, with all of nature, and with the divine. With James Nelson we affirm that divine eros is a "fundamental energy of the universe that is the passion for connection."[1] The way Nelson conceives God in an embodied theology is equally appropriate for a liberationist ecological ethic: "We need to recapture a vision of the divine eros as intrinsic to God's energy, God's own passion for connection, and hence also our own yearning for life-giving communion and hunger for relationships of justice which make such fulfillment possible."[2]

A deep sense of interconnection forms the starting point and grounding of a liberationist ecological ethic. This interconnection rooted in the erotic is what Carter Heyward affirms when she writes about the shared work of justice between lesbians and gay men: "We need to help each other realize that our desire for intimacy, connection, and touch—whether explicitly sexual touching or a more diffuse sensual yearning—is erotic. It is good. And it is sacred." She goes on to claim that, "Our primary moral agency—our capacity to do what is more or less right in any given situation—is to help create the yearning and conditions for non-violent, non-abusive life together."[3] A liberationist ecological ethic extends this understanding of moral agency to humanity's *ecological* moral agency: creating the positive scenario (the yearning) and material conditions for ecologically sustainable, "non-abusive life together" within our ecosystems and biotic communities.

Writers outside lesbian and gay communities also are connecting the erotic and the ecological in ways important to an ecological ethics. Writing from an African American context, Alice Walker makes explicit this sense of sacred power in erotic/ecological interconnection in many of her writings. Returning to the scene in *The Color Purple* I examined in the opening chapter, Shug explains to Celie how she came to believe that God was not white, male, or "out there," but "is inside you and inside everybody else":

> My first step from the old white man was trees. Then air. Then birds. Then other people. But one day when I was sitting quiet and feeling like a motherless child, which I was, it come to me: that feeling of being part of everything, not separate at all. I knew that if I cut a tree, my arm would bleed. And I laughed and I cried and I run all round the house. I knew just what it was. In fact, when it happen, you can't miss it. It sort of like you know what, she say, grinning and rubbing high up on my thigh.[4]

In the context of an erotic relationship between two women in an oppressed community, Shug ties feeling interconnected with all things in nature to erotic interconnection with another woman by debunking a racist patriarchal concept of God that fosters disconnection and relations of power-over. These connections empower both Celie's and Shug's sense of moral agency (symbolized for Celie in her decision to no longer write letters to an oppressive image of God) and their connection to all parts of the earth.

This attention to the power of the erotic in fostering ecological relations of mutuality, integrity, and sustainability highlights the possibility of extending Mary Hunt's work to view ecological ethics as an ethic of friendship, of actively befriending creation. This does not mean romanticizing the natural world or denying the destructive and harmful (particularly to human individuals) elements present in nature.[5] Rather it means drawing on the erotic energy of connection to develop relations of care-filled affection and respect with other parts of the natural world, recognizing as we do in human friendship both the relationships that connect us and the differences that need to be observed and respected. Just as authentic friendship requires discerning and respecting appropriate boundaries and limits, grounding an ecological sensitivity in the erotic provides a powerful and empowering basis in connection from which to discern and respect appropriate ecological limits—what Ruether terms "the laws of Gaia." This will mean deepening an erotic ecological spirituality to undergird our ethics, one that fosters connection to and intimate knowledge of the diversity of our ecological world, its creatures and processes—a spirituality that may learn from while respecting the boundaries of the spiritualities of those indigenous peoples who continue to live more closely with the land.

ECOJUSTICE AS RIGHT RELATION AT ALL LEVELS OF HUMAN AND NONHUMAN NATURE

Grounding a liberationist ecological ethics in lesbian and gay insights facilitates the integration of ecological and social justice concerns by understanding them both as manifestations of right relation. Conceiving ecojustice as right relation among all the earth's creatures both gives us guidelines by which to assess our relations with each other and otherkind, as well as provides a foundation on which to construct a positive scenario of ecological and social integrity and sustainability toward which we can strive. While to this point most lesbian and gay liberationist thought has focused on right relation among human beings and communities, a liberationist ethic of ecojustice expands this to focus both on the ecological integrity of the relationships between human communities and the rest of nature, as well as the ecological impact of intrahuman relations on nonhuman creatures and the ecosystem. J. Michael Clark makes these connections explicit in his work on gay ecotheology: "Our urge for justice as right relation for our spouses and friends logically broadens to a concern for right relation for all people (social justice) and for all the earth (ecojustice)."[6]

Hence ecologically damaging patterns such as Nash documents under the categories of "The Pollution Complex" (poisoning our neighbors) and "Exceeding the Limits" can be understood as violations of right relation between human and otherkind. And ecological (as well as human) damage resulting from unjust intrahuman activities such as environmental racism or militaristic counterinsurgency "ecocide" can be seen as violations of right relation at many levels, both within human communities along boundaries of social and cultural difference, and within our broader biotic communities.

Yet a liberationist ecological ethics must probe deeper in examining right relation within an integrated social-ecological context. Central to a focus on ecojustice as right relation at both social and ecological levels is attention to power dynamics in relation. Lesbian, gay, and feminist insights on power pay attention to both the material dynamics of domination and subordination manifest in our relations as well as the cultural and ideological values that enable and sustain them. In addition to examining the problem of anthropocentrism in the ecological crisis as deep ecologists have done, we must go beyond this to ask how androcentric, heterosexist, and racist values and practices in human communities shape our relations to each other *and* to our biotic communities in ecologically nonsustainable and damaging ways.

This means reconceiving power in our social and ecological relations, moving away from hierarchical patterns of "power over" to "power with," enabling an ethics and pattern of care and respect.[7] Here lesbian, gay, and feminist attention to eros and the erotic as the grounding for understandings of power that are mutually enabling rather than coercive has a critical contribution to make to ecological ethics. Ecojustice understood as right relation is grounded in the ecological and erotic insight that all of nature, including its human component, is intimately and inseparably connected. Reclaiming our erotic experience as the basis for knowing and connecting to others can ground an ecological ethic of interconnection in right relation. Clark notes this when he argues that "our sexuality is an urging into relationship with another person and, implicitly through that relationship, also an urging for justice in all relationships, including our relationship with the earth itself."[8]

Developing the interconnections of erotic and ecological relationality as right relation brings with it a commitment to examining issues of human sexuality from an ecological perspective. We need to re-theorize our understanding of human sexuality as issues of both social *and* ecological right relation. How do societal constructions and arrangements of sexuality either reinforce patterns of hierarchical domination and subordination or enable more mutual, interdependent relations among humans, and how are these in turn reflected or translated into human relations with otherkind and the land? What are the connections between eroticized violations of human beings and ecological violations of the earth and its nonhuman inhabitants? Is it mere coincidence that descriptive language from the one realm is used metaphorically to describe abuse in the other, such as in "the rape of the land"?

These last questions suggest two cautions that must be observed in positing the erotic as the appropriate grounding for ecological ethics. With the pervasive and increasing eroticization of violence and sexual abuse of women and children in twentieth-century capitalist societies, any liberationist ethic must pay close attention to the ways the erotic serves as a vehicle of violence and oppression rather than connection and empowerment.[9] Indeed some feminist theologians have argued compellingly that the eroticization of domination and subordination is so pervasive and its impact on women and children so devastating that the erotic cannot function as the source of identity or moral agency in women's lives.[10] While I believe that such arguments finally serve to deny out-of-hand an important source of moral agency to women and gay men, they make a critical point. Any liberationist ecological ethic grounded in the erotic that does not take seriously social constructions of the erotic as a vehicle of violence against others, particularly women and children (as well as some socially marginalized men and some nonhuman creatures) risks reproducing the deadly combination of misogyny and eros.

Second, while attention to the erotic provides an important grounding for ecological ethics, a constant challenge is to make the language, concepts, and relations it describes more concrete. For example, lesbian, gay, and feminist ecological writings have called attention—appropriately, I believe—to the problems of reproducing anthropocentric hierarchical relations through applying hierarchies of intrinsic value that place human beings at the top. Yet what can we offer instead to help adjudicate the very real conflicts between human communities and other parts of nature, such as Nash and Cobb point to in the ecology/economics dilemma? I address some of these issues below, but they remain an important challenge to an ethics of ecojustice grounded in the erotic as right relation.

ECOLOGICAL HERMENEUTICS AND LESBIAN AND GAY HERMENEUTICS OF SUSPICION

Like all liberation theologies, lesbian and gay hermeneutics incorporate as a critical first step the suspicion that dominant portrayals of reality and the world do not tell the whole story, and the story they do tell is distorted, often at our expense. From concrete experiences of either being totally invisible in the human ecology, or portrayed as dangerous or exotic aberrations, lesbians and gay men have learned to identify and question those values and practices that lead to our being ignored, dismissed, or actively exploited and repressed. Even our close allies, such as ecofeminists, too often limit their analysis to the connections between the domination of women and nature, missing or ignoring the links between *hetero*-sexism and ecological exploitation, and thus further reproducing lesbian and gay invisibility.[11] Since many of these same values and practices lie at the heart of the ecological crisis where the well-being of the environment has been ignored, dismissed, or actively exploited, lesbian and gay hermeneutics of suspicion have an important contribution to make toward developing a more inclusive liberationist ecological hermeneutics.

Clark argues that common to lesbian and gay male experiences and the ecological crisis is not only being *devalued*, valued instrumentally in relation to our worth to a society defined by white heteromale capitalist norms, but also being totally *disvalued*, judged as valueless or having negative value in a society of compulsory heterosexuality. Hence critiques that locate the problem of the ecological crisis in worldviews that are false because they are anthropocentric (the critique made by deep ecology) or androcentric (the critique made by ecofeminism) are incomplete, for they ignore lesbian and gay exclusion *as homosexuals* from a predominantly Western, white, heteromasculinist world view.[12] Clark notes, "Not only are women, nature, and sexuality *de*valued, but heteropatriarchy's hierarchy of values and categories *dis*values diversity; reductionism exploits and destroys anyone and anything designated as 'other.' . . . What we see is not just a *de*valuing (or *lowering* of value) which leads to domination and exploitation, but a *dis*valuing which *strips away* all value and which thereby leads to exclusion, to being disposable, to being acceptable for extinction."[13]

Lesbian and gay sensitivities to the dynamics of both devaluing and disvaluing all those deemed "other" has important contributions to make to an ecological hermeneutics that connects oppression of people with oppression of nature. Clark cites one ecological example to illustrate:

> Not only has our society and its governmental representatives *de*valued the old growth forests of the Pacific northwest (*lowering* their value from something intrinsically valuable in their own right to something valuable only as a natural resource to be exploited for financial gain), but it has also *dis*valued the spotted owl whose habitat these forests provide (assigning it *no* utilitarian value and thereby deeming it expendable). A mind-set which understands value only in terms of dollar signs readily accepts the exploitation of the forests for short-term gain as well as the virtual extinction of one species of owl, an extinction which would further weaken an already *dis*valued diversity of life.[14]

To recognize and overcome social and ecological relations of devaluation and disvaluation, a liberationist ecological ethics follows Cobb and others in affirming intrinsic value in all creatures as the starting point for ecological ethics.[15] Operating from suspicions generated by the particularities of our locations as lesbians and gay men, an ecojustice ethic of Gay and Gaia collaborates with, reshapes, critiques, and extends the analyses of radical forms of ecology such as ecofeminism and deep ecology, while it also exposes the inadequacy of reformist ecological ethics that do not examine critically the worldviews that shape our ecological practices.

HERMENEUTICS OF SUSPICION, HIERARCHY, AND INTRINSIC VALUE

Affirming intrinsic value in all things, however, poses the question of what to do when unavoidable conflicts arise between creatures, species, and habitats. Because of our negative experiences with hierarchical thinking that places us either at the bottom or entirely off scales of human value, many lesbians and gay men have joined ecofeminists in casting a suspicious eye on hierarchical rankings of intrinsic value (especially when humanity is the norm and conveniently is placed at the top). This suspicion of hierarchical solutions forms the starting point of our exploration of the relation between affirming the intrinsic value of all things and suggesting guidelines for how to resolve conflicts between them.

Clark joins many ecofeminists in trying to resolve this by arguing for the *absolute* and *equal* intrinsic value of all things. He criticizes ethicists like Nash, who Clark maintains posit an anthropocentric hierarchy of rights of nature by projecting human values onto nature and then giving them ontological grounding by distinguishing different levels of intrinsic value. Against this Clark argues, "[The] fundamental reality of interdependence and interconnectedness—of becoming-in-relation—means we can no longer accept any hierarchical evaluation of human over nonhuman, animal over plant, biosphere over geosphere."[16] He ties this explicitly to lesbian and gay suspicion of hierarchical ways of thinking that result in reductionism rather than affirmation of diversity: "The hope that connects gay eco-theology with gay liberation theology is the belief that if people can learn to value diversity throughout all life, then they will also appreciate diversity in human life. Then will homophobia disappear; then will no one and no thing ever be expendable again."[17]

Clark recognizes that refusing to see differences in intrinsic value among living things does result in a pragmatic dilemma in a world where human beings must necessarily make instrumental use of other living things, at a minimum for food and shelter. Yet while humans may be forced to *act* in a graded or hierarchical manner, "these self-preferring actions *should not* be taken to imply that any other organism is inherently less valuable in the cosmic scheme of things."[18] To avoid ecological ethics resulting in a "futile idealism," Clark suggests that the degree of human necessity combined with a grateful prudence guide human use of animal life.

While I share Clark's suspicion of hierarchical patterns of thinking that perpetuate relations of domination and subordination in human and ecological communities, I do not find his solution of an absolutely equal level of intrinsic value throughout the biosphere and geosphere entirely persuasive or ultimately very useful.[19] Clark too quickly jumps to the assumption that recognizing differences in intrinsic value between things necessarily implies a hierarchical ordering of relationships of domination and subordination and leaves human domination unchallenged. I argue that an alternative is to view ecological reality as a web of relationships where differences between nodes (organisms or species) in the web may reflect differences in intrinsic value without portraying these relationships hierarchically and nonrelationally, taking them out of their ecological context.

The problem with positing an absolute and equal intrinsic value in all things in ecological ethics is that it provides no basis other than arbitrary human whim from which to actually practice a biocentric ecological ethics. As Ron Long argues in his critique of Clark, "Without some prioritization, some hierarchialization, our choices come to seem purely arbitrary. . . . Clark's fear of absolute hierarchicalization threatens to render 'love of nature' ethically vacuous."[20] In Clark's case this results in the irony that in trying to overcome anthropocentrism in intrinsic valuing, he is forced to posit an anthropocentric criterion open to the vagaries of arbitrariness—degree of human necessity—for guiding human use of animal life.[21]

A weakness in Clark's position is the lack of discussion of what he means by intrinsic value. Here the perspectives of process theologians such as John Cobb and Jay McDaniel are helpful. They agree with Clark that everything has intrinsic value, but because they more carefully consider what they mean by intrinsic value and how we recognize it in ourselves and in others, they also recognize different grades of intrinsic value. McDaniel, for example, defines intrinsic value as the ability of something to be a subject, that is, the degree of "experiential richness and self-concern" an organism has.[22] Because the experience and capacity for experience of some organisms is richer and their self-concern greater than others, we can discern a greater intrinsic value in them.

McDaniel goes on to maintain that an understanding of intrinsic value must incorporate three elements: first, intrinsic value as the subjectivity of creatures is something that is recognized rather than assigned or ascribed—that is, subjectivity has an objective dimension;[23] second, intrinsic value is also relational in that subjectivity and richness of experience require interaction with one's environment and the creatures in it; and third, in addition to intrinsic value, all organisms can be of instrumental value to others, which may cause damage or death to the creature. Clark affirms the first two of these elements, but seems to ignore the basic biological fact that *all* organisms simultaneously have intrinsic and instrumental value and thus conflict between organisms, including human beings, is inevitable.

Human/nonhuman conflicts can be clarified ethically by locating and highlighting the moral dimensions. McDaniel argues that while all things have intrinsic value, not all things have "moral patience or considerability,"[24] that is, the required status of receiving *ethical* (as opposed to, say, aesthetic) regard by humans. As organisms with psyches and a high degree of sentience, many animals have a claim on human moral agency in a way rocks or even plants do not. To illustrate this, McDaniel uses the example of a dog infected with ringworm. On what basis may human beings intervene to rid the dog of ringworm, when that will destroy the ringworm fungus? An ethics of absolutely equal intrinsic value can give no guidance for human action here. Recognizing a greater degree of subjectivity, capacity for experience, and sentience in suffering—

that is, greater intrinsic value—in the dog than the ringworm, however, allows human beings to take the ethical action of healing the dog by destroying the ringworm.

Against the charge that using criteria such as psyche and sentience to understand gradations in intrinsic value is anthropocentric and therefore will always reflect human self-interest, McDaniel argues that the only option we have in moral deliberation is to generalize from criteria we discover in our own human experience. It is inevitable that there will be an anthropocentric bent, as we have no experience apart from *human* experience filtered through human bodies, sense, and cultures to guide us. This is justified in addition because human experience is in continuity with animal and other experience; those qualities that contribute to human richness of experience are likely to contribute to nonhuman experience as well.

McDaniel's examination reveals two qualities that human beings find rich in their experience: *harmony*, "the general feeling of attunement, balance, accord, and affinity," and *intensity*, "zest or energetic vitality in relation to other things."[25] These qualities are present in other creatures and can guide our recognition of differing degrees of intrinsic value. We can guard against mere anthropocentrism by divorcing harmony and intensity from aspects and associations that are uniquely human, and applying them with imagination, humility, and tentativeness, always recognizing our own limits.

Finally, McDaniel addresses some of Clark's other concerns by recognizing that "the need for judgment on the basis of degrees of value must be complemented by reverence for life, and this reverence must itself involve empathy for organisms on their own terms."[26] He stresses that measure of an organism's intrinsic value is done *as a last resort*, not as a first resort, to give guidance in conflict situations where a choice must be made.[27]

I believe that an approach of graded understandings of intrinsic value ultimately is a better grounding for an ecojustice ethic of Gay and Gaia than Clark's solution of seeing absolute and equal intrinsic value in all things. At the same time I agree with Clark that taking seriously lesbian and gay hermeneutics of suspicion about hierarchical valuings and rankings means critically examining the ways differential gradings of intrinsic value function in an ecological ethics. When taken out of an ecological context, viewed reductionistically as one-on-one conflict without regard to impact on other creatures and relations in the ecosystem, differential valuing can only result in continued human domination and mastery. However, viewed within an interconnected web of relationships where conflict between any members affects the other members, recognizing difference in intrinsic value becomes one factor, but not the only one, in resolving conflicts.

Hence recognizing a difference in intrinsic value between the spotted owl and human loggers in the Pacific Northwest is only one factor in an ethical analysis of that conflict. Viewed ecologically, the fact that one of the actors faces extinction, as well as the broader context of the threat to the whole ecosystem, must also be brought into play in the debate. Our experience with AIDS in the gay male community itself gives strong reason for recognizing differences in intrinsic value. Rather than maintaining that the AIDS virus and persons with AIDS have equal intrinsic value, we acknowledge the greater value of the human lives at stake, focusing on living with AIDS until the virus can be destroyed.[28]

LESBIAN AND GAY INSIGHTS FOR ECOLOGICAL LEARNING CROSS-CULTURALLY FROM OTHERS

Attention to the connections between affirming the erotic in its diversity of expressions, and valuing social and ecological diversity leads an ecojustice ethic of Gay and Gaia to learn from other cultures where these connections have been stronger than in the Western tradition. Lesbian and gay appropriations of cross-cultural insights in other areas suggests ways to do this consistent with a liberationist ethic.

Since the earliest days of the lesbian and gay liberation movement in Europe and then the United States, suspicions about dominant Western portrayals of a fixed, "natural," and bipolar dichotomy of gender and sexuality around the norm of compulsory heterosexuality has led lesbians and gay men to listen to and learn from non-Western cultures where homosexual behaviors were affirmed and included and which thus provided alternatives to the Western model. In the process many lesbians and gay men have noted the greater degree of integration of sexuality and spirituality in these cultures and how these are grounded in and contribute to a more respectful and friendlier attitude toward the earth.

This may be a distinctive contribution that an ecojustice ethic rooted in lesbian and gay insights has to make to ecological ethics. In the United States, for example, much attention in the environmental movement has been paid to Native American spiritualities because of their affinity to ecological values and practices. A lesbian and gay hermeneutics asks explicitly what is the place of the erotic and sexuality in this. How do native cultures understand sexuality in relation to spirituality and their attitudes toward the earth? What insights might this give us for formulating a liberationist ecological ethics?

Underlying this approach is the belief that Clark articulated above: affirming and respecting ecological integrity and diversity is integrally connected to affirming human diversity, including human sexual diversity. An ethic of Gay and Gaia moves beyond affirming erotic and ecological diversity to claiming that an adequate ecological moral agency must be grounded in an integration of the two. It thus looks to human cultures that affirm both natural and human diversity for the insights they may offer a liberationist ecological ethic.

To explore further the insights of Native American ecological wisdom and its connections to sexuality and spirituality, North American gay men and lesbians have long focused on the place of the *berdache* in many North American Indian cultures.[29] Existing as an alternative "third gender," the berdache in Indian cultures is a morphological male with an androgynous, nonmasculine character who often held a place of esteem and respect in Native American society. In his groundbreaking study of the berdache, Walter L. Williams emphasizes the integration of spirituality with gender and sexuality in Indian understandings of the berdache (and, indeed, of all persons) as a reflection of nature in that person. While the sexual behavior of the berdache was nearly always homosexual and much of his activity involved work normally reserved for women, attention in the Indian communities Williams studied was less on gender or sexuality than on the sacred power believed to have been endowed on the berdache by the spirits. The high esteem for the berdache lies in his spiritual power to mediate between men and women and between the physical and the spiritual for the benefit of the community: "Generosity and spirituality, not homosexual behavior, are what underlie the social prestige of the berdache from the Indian viewpoint, but these qualities are emphasized without denying the sexuality of the berdache. Spirituality, androgyny, woman's work, and sex with men are equally important indicators of berdache status. They are all seen as reflections of the same basic character of a person; this is what Indians mean when they talk about berdaches being 'spiritually different.' "[30]

Important for an ecological ethic, Native American acceptance of ambiguity and human diversity, including sexual diversity, flows from their reverence and respect for nature. As lesbians and gay men have emphasized, the erotic and the ecological are intimately intertwined as an interweaving of the physical and the spiritual. Respect for nature implies respect for human diversity, and vice versa. Equally important, where the berdache was respected in Native American cultures, the status of women's roles usually was high. Hence the berdache's involvement with feminine tasks and roles was not interpreted as a betrayal by men of a "superior" masculinity, but as a (spiritually powerful) variant within the natural spectrum of human diversity on gender and sexuality.[31] The explicitly profeminist stance of an ethic of Gay and Gaia echoes this understanding.

Overcoming rigid dichotomized thinking, whether between humanity and nature, or between male and female, is basic to a liberationist ecological ethic. While his focus is on Native American societies, Williams cites examples across the globe of berdache- and amazonlike gender and sexuality patterns.[32] Not coincidentally many of these indigenous cultures are among the most ecologically sustainable and earth-friendly, holding nature in high esteem and central to their spirituality and understanding of human identity and culture. Conspicuously absent is the othering mechanism of associating certain groups and classes with a "lower" nature as a justification for domination both. This correlation lies behind my critique of Ruether's examination of male and female roles in culture in *Gaia and God*. By ignoring alternative gender and sexuality constructs found cross-culturally around the globe, she both reproduces heterosexist patterns of fixed bi-gender complementarity as well as overlooks an important correlation between respect and affirmation of human sexual and gender diversity and diversity in nature.

Yet as I emphasized in chapter 2, a liberationist ecological ethic must pay attention to the danger of misappropriating the wisdom of others across lines of social and cultural difference. This is especially the case when some of us are part of and have benefited from a historical relationship of domination and oppression as those of us who are of European ancestry have with respect to Native Americans. In using the insights of the berdache and Native American ecological spirituality as mirrors for self-knowledge that highlight parts of our own lesbian and gay experience, we must be attuned to the oppressive as well as liberating aspects it exposes.

In his own appropriation of the berdache tradition for a gay ecotheology, Clark has noted this danger. He argues that the location of gay men and lesbians in Western society between socially acceptable dichotomized poles of gender and sexuality means that "we share at least an empathic connection with the berdache."[33] Yet Clark is clear that we "must be exceedingly careful not to co-opt or misappropriate native American wisdom as if it represented our own insights."[34] We may avoid this, in part, by learning from both Indian wisdom and the history of European oppression from which we have benefited: "If any of us are to discern life-shaping ecological wisdom from native American experience—and particularly if those of us who find ourselves at some intermediate, 'third gender' place between the gender and sexuality opposites intend to use that standpoint to reinvigorate the role of the berdache, not in usurpation, but as an ecologically healing role—then we must synthesize both the wisdom and the tragic history of native America, hearing both its prophetic judgment and its prophetic demand."[35] For Clark the goal of such an "ecological berdache" is the joint valuing of human- and biodiversity, integrating the spiritual and the physical, the erotic and the ecological, to engage in healing and restoring *all* of the earth.

For those of us continuing to work within the Christian tradition, drawing on the berdache tradition means paying particularly close attention to those aspects of Christianity that were instrumental in justifying European repression and eradication of whole Native American cultures, as well as of the berdache and his place in Native American spirituality and society. Williams points out that Christian condemnations of sodomy as "the abominable sin" served as one of the primary justifications of the Spanish conquest of Native Americans.[36] The Spanish, and later other European groups, undertook a concerted effort to wipe out berdachism through a campaign of suppression that included torture and execution of those suspected of this "crime." Missions, in particular, were effective in destroying or driving underground the berdache tradition by separating Native American children from their communities in mission boarding schools. Williams has observed that where native peoples have become most acculturated in Christianity they also have absorbed Christian homophobia and either condemn or refuse to acknowledge the berdache tradition and gays and lesbians in their midst.[37]

Those who seek a liberating praxis from within the Christian tradition must see challenging and overturning the churches' traditional condemnatory stance toward homosexuality as central to ecojustice work, not only for lesbians and gay men in our communities, but also for non-Christian peoples. It must form an integral part of reshaping an ecological worldview: affirming and advocating for human diversity—including human sexual, gender, cultural, and religious diversity—must be seen as inseparably linked to affirming and protecting ecological diversity. One particular challenge to this is the absence of positive references to same-sex identities and relations in the Christian scriptures. Williams cites many examples of Native American creation myths that show an important and positive role for the berdache in Native American society. These religious stories function to instill respect and esteem both for the berdache and for human diversity. He notes, "Gender egalitarian societies often have creation stories which give important roles to women. Without the active explanation in myth, there is no ideological underpinning for a high female status. The same may be true for the berdache. In cultures where berdaches have high status, there is usually mythological justification for the practice. It is not enough that the religion be neutral or tolerant. It must actively explain the phenomenon in a positive manner."[38] The biblical Genesis creation accounts, in their bias toward maleness and heterosexual complementarity, do not offer a similar flexibility. This suggests again the importance of opening up the canon of Christian scripture to other voices that reflect a broader array of human and ecological diversity.

In conclusion, this limited exploration of some of the methodological aspects of an ecojustice ethic of Gay and Gaia reveals a number of resources for a liberationist ecological ethic. Grounding ecological ethics in lesbian and gay liberationist insights allows us to revision the erotic as ecological and join together ecological and social justice as interconnected manifestations of right relation. Attention to cross-cultural learnings about sexuality reveal connections between sexuality, spirituality, and nature critical to forming ecologically sustainable and just societies. In the next chapter I build on this to explore how seeing Gay and Gaia as paired criteria of ecojustice facilitates prophetic critique of mainstream society, the environmental movement, and lesbian and gay communities from the integrated vantage point of liberationist and ecological concerns.

12

Gay and Gaia: Features of an Erotic Ethic of Ecojustice

All over town crab
apple trees have come out
blooming in the wrong season
the wrong climate
People have noticed
yet no one can account for it
so some have begun
to consult specialists
Others hold that an early frost
last spring set the trees
on a different course
But in any event
it is nature, true nature
at her surprisingest
The way I see it
the only thing to do
is sit back, enjoy
or get out of the way
of their blooming
in our front yards
along roadways
and in the far distance
But I would urge you
to stop long enough to attend
closely to their beauty
the fragileness of flower
Study the blooms & tight buds
the leaves coming out all
around us
as most other midwestern trees
turn gray.

—Karen Herseth Wee, "Coming Out"

Like the untimely blossoming of the crabapple trees in Karen Herseth Wee's poem "Coming Out," lesbians and gay men are claiming our place within "true nature at her surprisingest." Though many continue to see our coming out as "blooming in the wrong season and wrong climate,"[1] my contention here is that by attending to the fragileness and beauty of recently gained lesbian and gay insights, ecological ethics may gain a richer and surer grounding, a more inclusive and liberating character.

In addition to the promising possibilities explored in the previous chapter of grounding ecological hermeneutics in the reintegration of the erotic and the ecological, several specific themes important to lesbian and gay theo-ethical reflection provide important components for an ethic of Gay and Gaia. These include *the centrality of embodiment* in an ecological ethic, *valuing diversity at all levels* of human and nonhuman ecologies, overcoming attitudes and practices of *disposability and dispensability*, and attention to *the danger of appropriating resources from others without reciprocity*.

Attention to these constructive resources in turn highlights areas of alienating ecological praxis. Three primary audiences come into play in drawing on Gay and Gaia as ecojustice criteria of prophetic critique grounded in lesbian and gay liberationist commitments. Here I sketch initially how the ethics looks outward to critique mainstream society and the environmentalist community for the ways both ignore lesbian and gay experiences and perpetuate the interconnected oppression of lesbians and gay men and the earth. What features of Gay and Gaia are needed to complement, critique, and deepen other efforts at ecological ethics such as those explored earlier? I then show how the ethic looks inward to provide tools for self-critique of lesbian and gay male communities from both ecological and social justice guidelines, taking seriously our own liberationist commitments while learning from and incorporating ecological critiques. How, where, and in what ways do we need to rethink our relations with each other and with all of nature in light of the criteria of Gay and Gaia?

EMBODIMENT: CONNECTING BODILY INTEGRITY TO ECOLOGICAL INTEGRITY

Central to a liberationist focus centered in lesbian and gay realities is embodiment, that is, paying attention to our health and integrity as body-selves and how this intersects with the health and integrity of our relations with others. In our efforts to reclaim sexual and bodily wholeness in our lives and relationships, lesbians and gay men have discovered the critical importance of beginning with our embodied realities to ground and empower our moral agency and efforts at wholeness and liberation. These insights provide an important grounding for a liberationist ecological ethic.

Embodiment as bodily integrity also requires attention to respecting physical limits and meeting physical needs within our own body-selves and within our web of relationships. Extending embodiment into an ecological context means first recognizing that who we are as body-selves is inseparably enmeshed in a web of ecological relations, each rooted in embodied, material, creaturely reality. Respecting our own bodily integrity, our own material embodiment as living creatures, means respecting the bodily integrity of other creatures and the ecological integrity of the interweaving of relations. Embodiment as both bodily and ecological integrity thus serves as the key to beginning with our own bodily experience as that which mediates reality and seeing this inseparably connected to the bodily integrity of other parts of the ecology.[2]

Within a Christian context, embodiment reminds us of the importance of incarnation as a central dimension of Christian understanding. The divine is known through embodiment, taking on material, bodily flesh. All efforts to view the sacred in disembodied ways, perpetuating dualistic splits between spirit and body, mind and matter, are called into question through an understanding of sacred reality as embodied reality. Viewed ecologically, divine embodiment/incarnation invites a panentheistic understanding of the divinity/humanity/nature relation: the divine Spirit infuses all reality, weaving all things together in embodied interrelation and imbuing the material world with a sense of the sacred.

LESBIAN AND GAY EXPERIENCE AND VALUING DIVERSITY

Building from this understanding of ecological embodiment, incorporating lesbian and gay experience into ecological ethics can also provide models for better valuing and integrating ecological diversity and human diversity. Because of our own experience of exclusion from the "natural" human diversity, lesbians and gay men have developed a sensitivity to diversity and patterns of inclusion and exclusion, expendability and disposability that can aid us in recognizing similar practices toward other parts of the earth community.

A corollary of this is insistence on valuing diversity in the human community as an important part of ecological ethics. While ecological ethics has placed great emphasis on the need to value diversity within the ecosystem, especially the biodiversity of species, too often it treats humankind as one undifferentiated species within a web of interconnected species.[3] To the extent that this contributes to keeping certain sectors of the human community invisible—such as lesbians and gay men, people of color, and women—this ecological reductionism furthers mechanisms of oppression.

Because we experience lesbian, gay male, bisexual, and transgendered lives as expressions of valued diversity, intrinsically valuable in ourselves *and* as part of the human and "natural" ecology, a liberationist ecological ethic looks to our communities (as well as other communities) for resources in recognizing and protecting human and ecological diversity and caring relations. Within our communities, for example, we have developed webs of relationships and friendships that offer constructive new models to the wider society while providing us with critical resources of strength, caring, and courage to draw upon in times of crisis, such as the AIDS pandemic currently engulfing us. Other examples include lesbian- and gay-parented families that operate outside socially sanctioned models of heterosexual complementarity to contribute to the richness of experience and diversity of our own communities as well as to the broader social makeup.

This lesbian- and gay-rooted attention to diversity within the human social realm and its broader ecological ramifications makes a much-needed addition to and deepening of one of the central points made by Rosemary Ruether: to address the pattern of human domination over nature we must address the psychosocial roots of patterns of domination within human culture. Ruether argues that only a fundamental reconstruction of the primary roots of cultures based in "hierarchies of exploitation and control that emanate out of the family pattern of female mothering and domestic labor" can provide the basis for liberating *ecological* as well as social structures.[4] An ethic of Gay and Gaia agrees wholeheartedly with this attention to the psychosocial roots of social and ecological domination, but adds that this reconstruction will be incomplete without close attention to the role played by homophobia and heterosexism (which are linked to, but not subsumed within sexism) in reproducing these hierarchies and repressing valuable forms of human diversity.

In the collection *Homophobia: How We All Pay the Price*, the authors address a wide spectrum of the hidden costs of homophobia and heterosexism in family relationships, religious institutions, social policy, and other aspects of personal and social life.[5] Among the costs of socializing the population in homophobia and structuring society according to the logic of heterosexism, they include the following:

- Homophobia inhibits appreciation of other types of diversity, making it unsafe for everyone because each person has unique traits not considered mainstream or dominant.
- Homophobia locks all people into rigid gender-based roles that inhibit creativity and self-expression. Homophobia conditioning compromises the integrity of heterosexual people by pressuring them to treat others badly, further reifying social structures based on hierarchy and domination.
- Homophobia prevents heterosexuals from accepting the benefits and gifts offered by sexual minorities: theoretical insights, social and spiritual visions and options, contributions to the arts and culture, to religion, to family life, to all facets of society.
- Homophobia inhibits one's ability to form close, intimate relationships with members of one's own sex. It generally restricts communication with a significant portion of the population, and particularly strains family relationships.

An ethic of Gay and Gaia examines the psychosocial processes by which homophobia and heterosexism are maintained, particularly for the wider ecological ramifications of the assumptions about nature and "animalistic" and "unnatural" homosexuals that are presupposed. Sensitivity to maintaining human as well as ecological diversity requires incorporating the critique and deconstruction of homophobia and heterosexism in the social-ecological reconstruction of culture and society presupposed by a liberationist ecological ethics.

This attention to the value of social diversity can add an important contribution to the critique that John Cobb and others have made of current national and global economic practices built on capitalistic market models. Not only do the so-called free-trade economic agreements that currently dominate within and between wealthy and poor nations threaten the *ecological* diversity of these lands,[6] they also threaten the cultural and ethnic diversity. When a market *economic* mentality also becomes the dominant *moral* mentality, diversity becomes an obstacle to profit as all things, human and otherkind, are commodified as resources to be used instrumentally for economic growth.[7] As ecofeminist Vandana Shiva explains, "The linear reductionist view superimposes the roles and forms of power of the western male on women, all non-western peoples, and even on nature. Based on these . . . values and concepts, nature, women, and indigenous third world peoples become 'deficient,' in need of development. Diversity—and the unity and harmony of diversity—become epistemologically unattainable."[8] Or, as J. Michael Clark has argued in observing the variety of popular resistance movements that have emerged connecting social justice and environmental issues in the face of free-market exploitation, "Just like gay men and lesbians, people everywhere are beginning to realize and claim their own intrinsic value and to celebrate their own difference and the larger diversity that their difference enriches. As the disempowered peoples of the earth come to appreciate the intrinsic value and diversity of the human community, *those* values will be superimposed upon nature."[9] Here an ethic of Gay and Gaia builds on and contributes to proponents of environmental justice and social ecology who have linked the social and ecological threats to diversity resulting from current capitalist economic practices.[10]

Recognizing the diversity lesbian and gay experiences bring to the human community means also recognizing and valuing other expressions of human diversity that contribute constructively to the richness of the human and ecological fabric, whether racial, sexual, religious, or cultural. Hence a critique of all forms of religious or cultural exclusivism and triumphalism forms an explicit part of a liberationist ecological ethic.

DISPOSABILITY AND DISPENSABILITY

Ecological ethics has rightly critiqued the consumerist attitude of disposability that pervades Western societies' attitudes toward nature; too often nature is seen as a "resource" at human disposal to be used and discarded with little regard for the cost to the ecosystem and human communities of exploiting its components and generating mountains of (often toxic) refuse that are dumped back into the ecosystem. In a liberationist ethic this attitude must be connected to the practice of disposing of entire segments of the human community seen as valueless under mainstream standards.

Gay and lesbian sensitivity to attitudes and practices of disposability and dispensability stems from our long experience of being devalued and disvalued in mainstream society. Perhaps nowhere has this been more starkly revealed than in the AIDS crisis. In reflecting on the churches' early response to AIDS, gay theologian Chris Glaser observed the contradictory messages he received from dominant societal values, his upbringing as a Christian, and the practice of the churches towards those with AIDS: "One of my first employers tried to teach me a business principle: 'Everyone's dispensable.' I was a poor learner because the lesson was at odds with everything I'd been taught as a Christian. I was one of God's children. God loved me personally and intimately. According to an old Jewish saying, God created us because God loves stories. I believed God was as passionately interested in my story as in everyone else's."[11] Yet as a gay man who works with persons with AIDS,

Glaser found the gospel value of the sacredness and indispensability of all God's children regularly contradicted by the practice of the churches. He noted that this gospel value became most clear to him in his conversations with people with AIDS who suddenly found themselves dispensable and disposable in many of the contexts that previously had provided them with meaning and security: their professions, families, and religious communities.

At a church-sponsored AIDS consultation, the group Glaser worked with made this contradiction explicit, titling its paper "We Are Not a Throw-Away People": "Our statement affirmed that no person affected by AIDS—homosexual, racial minority, female, IV drug user, hemophiliac, prostitute, Third World—is dispensable. Every life is of value. At the time I did not care for the negative implications of our title, but now I value its prophetic stance, which calls into question the heresy that *anyone* is dispensable."[12]

With the ascendance of the Christian right, efforts to institutionalize the disposability of gay persons have gained politically powerful advocates in the federal government. In justifying his efforts to reduce federal AIDS spending, for example, Senator Jesse Helms of North Carolina argues that people with AIDS do not deserve treatment because it is their "deliberate, disgusting, revolting conduct" that is responsible for the disease. "We've got to have some common sense about disease transmitted by people deliberately engaging in unnatural acts."[13] Clark connects the dispensability of people with AIDS with our attitudes toward the earth:

> In the history of the gay and lesbian communities, never has our own expendability been so evident as in the rising incidence of anti-gay/lesbian violence and particularly in the AIDS health crisis. The same heteropatriarchal value hierarchy that insists that nature is reducible to expendable resources also insists on dichotomizing innocent and not innocent (read: expendable) victims of AIDS. . . . We are being treated as expendable objects to be used or found useless and then discarded. And our experience of our expendability becomes the symbol or paradigmatic metaphor of western culture's attitudes toward all the earth.

> Hence, our gay eco-theology must adamantly oppose *any* disvaluation and exclusion which leads to dispensing with diversity and disposing of life. Neither gay men and lesbians, nor the biosphere, nor the geosphere, nor any of the great diversity that god/ess creates and delights in is expendable. Inclusively defending the value of diversity and opposing disvaluation and expendability become the foundation of a gay eco-theology.[14]

Explicitly opposing the disposability of any portion of the human community must form a central part of a liberationist ecological ethics. Paying attention to the experience of *lesbian and gay* disposability and dispensability has two important components. First, it can help make visible the connections between lesbian and gay oppression and the oppression of other disposable segments of human society, such as African American youth and men, Native Americans, homeless and mentally ill persons, and all others deemed dispensable under the logic of white supremacist consumerist capitalism. Second, it draws attention to the ecological costs of violations of bodily integrity. Whether it is the integrity of individual lesbian and gay bodies or of the wider lesbian and gay body politic that is violated, there is a high cost to the larger society that in turn is passed on to our biotic communities. This is another reminder that the health of the (social and ecological) whole depends on the health of *all* of the parts.

APPROPRIATION WITHOUT RECIPROCITY: CRITIQUING SOCIAL AND ECOLOGICAL EXPLOITATION

I have argued previously that critical to an ethic of Gay and Gaia is attention to how one appropriates resources across lines of social difference in an effort to break down social relations of exploitation and domination. Lesbian and gay experiences of enforced invisibility and silence within the dominant heterosexual culture can be helpful here in recognizing some of the dynamics of exploitative appropriation and working toward an alternative, mutually liberating praxis.

The social dynamics of the closet, constructed and maintained by attitudes of homophobia and structures of heterosexism, mean that the resources and gifts of lesbians and gay men constantly are drawn upon without acknowledgment of the locations and experiences that helped to produce them. This is typically the case whether the persons are historical figures who were homoerotically oriented, such as the artist Michelangelo and the writer Willa Cather, or contemporary gay and lesbian figures, such as Howard Ashman, who delighted millions of children with his lyrics for such films as *Beauty and the Beast, The Little Mermaid*, and *Aladdin*. In the churches it certainly has been the case for countless pastors, musicians, layworkers, and theologians whose gifts have had to go unmarked by the erotic energy and sexuality that helped shape them. A liberationist ecological ethic recognizes this as exploitation, perpetuated by homophobia and heterosexism. It is analogous to being ecologically unaware, such as when we refuse to acknowledge our mutual interdependency with other creatures, using other parts of nature instrumentally as resources without acknowledgment or reciprocity through sustainable practices.

An ecojustice ethic of Gay and Gaia calls especially on the churches to reverse this dynamic of exploitation, whether of lesbians and gay men or of the earth and its members. The church as an agent of ecojustice needs to be about acknowledging and celebrating its diversity, by acknowledging and celebrating openly, where it can, the contributions of its lesbian and gay members, and elsewhere working to create the safe conditions for its closeted members to be able to live and work openly. By doing so it would reveal a truer picture of both church and society, rather than one based on lies, denial, and repression. Similarly, the church must recognize its ecological dependency on the earth, acknowledge the ways it unecologically has appropriated and exploited the earth's nonhuman members, and work to reverse this process.

APPLYING AN ETHIC OF GAY AND GAIA IN LESBIAN AND GAY COMMUNITIES

In the preceding pages I have argued for the numerous positive ecological contributions that emerge from lesbian and gay communities and insights. Yet developing an ethic of Gay and Gaia also helps to draw attention to ongoing areas of alienating ecological praxis *within* lesbian and gay communities. When the ethics developed here is applied internally, it highlights certain practices and assumptions for discussion and evaluation from a liberationist ecological perspective. Three areas considered here include the dynamics of gay and lesbian participation in consumerism, elements in our relational patterns, and our relations to animals as animal products. Reflection within an ecological framework suggests more liberating alternatives in each case.

Rethinking Our Patterns of Consumption

As suggested above, the experiences of lesbians and gay men in working for our survival and liberation provide numerous constructive resources to the larger society for developing an ecologically liberating ethic. One area where we too often mimic the ecological destructive habits of mainstream society, however, is in our participation in its patterns of consumption.

Clark writes, "As gay men and lesbians look out upon our disposable society of planned obsolescence and throw-things-away consumerism, we cannot help but be aware of the growing trash heap, the over-burdened land fills, the industrially polluted waters, and the wastelands of deforestation."[15] Yet many of us continue to be all too unaware. Clark is correct that for *ecologically aware* lesbians and gay men, the connections between mainstream society's view of our disposability and its "throw-things-away consumerism" is clear. Unfortunately, however, this consumerism is one area that much of the gay male community, at least, shares in common with mainstream society. A glance at the advertising that fills the pages of catalogs, newspapers, and magazines aimed at gay men (and increasingly also at lesbians) and a walk through the stores in our neighborhoods reveals that much of gay male culture is highly consumer-oriented with little or no awareness of the negative ecological consequences.

This consumerism is the negative side of the enormous cultural richness we have created in our communities that finds expression in diverse and colorful ways. It is important to emphasize that a liberationist ecological ethic is not a call to asceticism. It reaffirms the ways gay culture has found pleasure and beauty in our bodyselves and communities, and acknowledges the central role gay men and lesbians always have played and continue to play in the rich diversity of human culture, particularly in music and the arts.[16] An ecojustice ethic of Gay and Gaia calls gay men and lesbians to examine, however, how these rich resources and our relationships themselves are distorted through consumerism and exploitative relational dynamics that reflect the dominant society's utilitarian "use it and toss it" attitude, with destructive ecological and social consequences.[17]

It is also important to highlight the role of the closet in facilitating gay and lesbian participation in our consumerist society. For white gay men in particular, remaining closeted can facilitate the generation of a large income and provide us with access to economic privilege through our participation in a (white heteromale-dominated) political economy based on high rates of ecologically nonsustainable consumption and defended through an ecologically disastrous militarism. Hence the closet can function by providing access to heterosexual privilege to facilitate our participation in economic gain at the cost of ecological destruction. Passing as heterosexual men leaves white gay men unmarked by race and gender in ways that gay men of color and lesbians are not. It opens up possibilities of collaborating in the ecologically destructive practices of our consumerist society less available to those who either are out of the closet or who differ from the white heteromale norm by gender, race, or class.[18]

On the other hand, it is important not to blame gay and lesbian people for the existence of the closet. We did not invent or institute homophobia and heterosexism, the primary reasons nearly all lesbians and gay men find the need to pass as heterosexual in at least some periods or circumstances of our lives. Nevertheless, an ecojustice ethic of Gay and Gaia calls us to account for how we relate to the closet, particularly when it facilitates unecological practices and unjust social relations. Our liberationist ecological ethics strongly encourages the creativity and imagination that characterize gay and lesbian cultures while insisting they not be constructed or practiced in ecologically destructive ways: Creativity and affirmation of material, bodily reality, yes; consumerism and exploitative relationships, no.

Rethinking Limits in Sexuality and Relationship

Central to gay and lesbian liberation has been affirming the intrinsic goodness of sexuality and sensuality, yet the brutal reality of AIDS and other sexually transmitted diseases has forced gay men in particular to confront the ecological limits of the sexual dimensions of our relationships. It has been an extremely painful lesson to learn to recognize bodily limits in taking responsibility for the health of our sexual actions. Yet it is critical to locate accurately where these ecological limits lie: in the unchecked spreading of bodily fluids (between any human beings) that transmit viruses and parasites between our bodies. Spreading these fluids has been shown to be harmful to our bodily health and integrity, and hence damaging to the human ecology. In this sense safe sex is a primary ecological practice that recognizes ecological limits in our relationships.

It is equally critical to assert what these ecological limits *do not* proscribe: they do not proscribe the erotic and sexual dimensions of our relationships. Instead they remind us of the importance of taking responsibility for the ecological, social, and personal ramifications of *all* our actions—especially those that involve the most embodied and intimate parts of ourselves, as sexuality does. Against the ecological fascists and the Christian right-wingers who see AIDS as either a "natural check" or "divine punishment" of gay sexual acts and identities, an ethic of Gay and Gaia insists on locating ecological limits and moral responsibility in their proper place. In the midst of tragedy and terrible suffering, gay men have led the way in taking responsibility for learning the appropriate (ecological) limits of our relationships without sacrificing what we have gained in reclaiming our erotic and sexual dimensions. "Safe Sex Is Hot Sex" keeps the ecological and the erotic united as it reaffirms our stance as active agents, not passive victims in the midst of tragedy. Moral agency, not moralism, is the lesson of the gay and lesbian response to AIDS.

An ethic of Gay and Gaia combines ongoing evaluation of gay sexual practices from ecological and justice perspectives by acknowledging and affirming the constructive changes that have come about. In response to AIDS, new relational models are emerging that emphasize taking responsibility, talking about sex, our bodies, and values, and our interconnectedness. The late Michael Callen noted some of these themes in his analysis of what he termed "the Second Gay Sexual Revolution" among gay men. He argued that there is

> a much more realistic sense about the body and its physical limitations. A decade of physical devastation wrought by AIDS has shattered once and for all the naive, hubristic illusion we had in the '70s that one's body was merely a plaything, an indestructible pleasure machine requiring little more than a monthly shot of penicillin to keep it in good running repair. These days, when gay men touch, there is a sober, unspoken awareness of just how high the stakes are: (Unsafe) sex can kill you. In a heart-breaking sense, all gay male sex these days is group sex: the two men having sex, and the almost palpable presence of AIDS and death. The sexual ethics of the '70s was basically "Every man for himself." In sharp contrast, the Second Sexual Gay Revolution is built upon *a hard-won awareness of the essential sexual, emotional, and microbiological interconnectedness of gay men.*[19]

Callen's observations reflect the critical ecological insight that just as in the other dimensions of our lives, our sexual relationships are never merely private, but always also interconnected with the lives of others—sexually, emotionally, and biologically. Embodiment always includes both bodily and relational integrity.

Rethinking Lesbian and Gay Relationships with Animals

Long consigned to the "animalistic" side of the nature/culture dichotomy, lesbians and gay men may have a particular sensitivity to classifying something as an animal or animalistic in order to justify exploiting it. In deconstructing the nature/culture boundary, an ethic of Gay and Gaia follows process theology in recognizing the subjectivity and agency of animals, and calls all members of society, including lesbians and gay men, to reexamine our relations with them.

Here we have much to learn from process theologians, such as John Cobb and Jay McDaniel, and animal rights activists who have led the way in focusing on the utilitarian and often inhumane ways animals are used in the food and fashion industries to satisfy human wants—many of which are either unnecessary or ecologically damaging.[20] Starting from the assumption of animal subjectivity, the recognition that animals experience feelings, pain, and emotions, these writers argue that our relations to animals that are dominated by treating them as commodities (eating animal flesh and wearing animal fur and skins) must be seen as political acts that depend on the general invisibility of the material practices that support them and inflict suffering on animals.[21] They often depend on and reproduce the subject/object and humanity/nature dichotomies that have distorted both intrahuman and humanity/nature relations in ecologically and socially damaging ways.

A particular difficulty facing many gay men and lesbians whose primary locations are urban is that beyond contact with pets and "pests" (the ubiquitous rats and roaches of city life), most of us are distanced from daily experience of either animals or segments of the animal industry. Many of us have little firsthand input into how animals are treated in our society. This is particularly ironic in that invisibility of animals and the animal products industry functions in ways similar to the invisibility of lesbians and gay men to hide—and thus perpetuate—oppressive practices.

Nevertheless, in at least two important areas we can make concrete personal and political choices that affect these practices: in what we eat and what we wear. As a liberationist approach, an ethic of Gay and Gaia begins with a social and ecological analysis of the reality and impact of the animal industry on animal, human, and ecological communities. From there several of the principles articulated above can give guidance for how we rethink our relations to animals.

Rethinking Eating Animals: A Liberationist Case for Vegetarianism. An ethic of Gay and Gaia grounded in lesbian and gay liberationist concerns sees these as inseparably linked with the liberation of other oppressed groups and the well-being of the earth as a living ecological system. Reducing our consumption of meat and other animal products in affluent societies through a vegetarian diet can link these concerns and incorporate liberationist insights concerning social justice, ecological justice, embodiment, and justice for animals. There are at least four ways that a concern for animals expressed through a commitment not to eat them can complement a liberationist concern for other people and the earth.

First is the issue of *social justice.* It takes far more energy, land, and plant foods to support an animal-based diet than one based on plants. This inefficient and unjust use of resources contributes directly to the growing problem of the inequitable distribution of wealth resulting in increased poverty and world hunger. As McDaniel notes,

> A billion people could be fed with the amount of grain and soybeans that now go to feed U.S. livestock. Ninety percent of the protein, ninety-nine percent of the carbohydrates, and one hundred percent of the fiber available for direct human consumption in grains and vegetables are wasted when cycled through livestock. . . . Given population projections, available land is used much more efficiently as cropland than as pasture for grazing. On the average, land used to supply grains and vegetables can feed twenty times more people than can an equivalent amount of land used as pasture to supply meat. Thus, if we want to contribute to a long-term solution to hunger, we ought to refrain from eating meat. This is an option, at least, for those of us in industrial nations, whose survival does not depend on the eating of meat.[22]

Attention to animal rights and pressure to eat fewer or no animals often are portrayed as "bourgeois" First World concerns that are peripheral to the "real" demands of social justice. Yet how we consume and eat in the affluent First World directly affects the lives and possibilities for justice of poor people around the globe. To cite just one example, the United States alone imports nearly two hundred million pounds of meat annually from the countries of Central America. The region's rapidly diminishing rain forest is being destroyed to provide pasture to raise cattle for export, in the process also depriving the local population of cropland to support themselves.[23] Far from being peripheral, attention to changing our relations to animals can complement and strengthen our commitment to social justice for all human beings.

Second, *ecological justice* must be considered. Many of us grew up with the nursery rhyme "Old MacDonald's Farm" and storybooks set on farms that gave us early on an image of domestic animals as a necessary and integral part of the rural farm ecology. When most people lived on farms or in rural communities, this well may have been the case; economies were much more local and farms and rural towns functioned more as independent and interdependent local social and ecological systems.[24] With the dramatic shift this century toward urbanization, the stimulation of a predominantly meat-based diet modeled after the (previously unattainable) diet of the European aristocracy, and rapid population growth in the United States, animal production has had to be radically reorganized away from the family farm to much more intensive factory-style methods to meet consumer demands.

Yet supporting this broad animal industry for human consumption has had widespread deleterious ecological consequences: destruction of tropical rain forests for pasture land, groundwater and river pollution from feedlots and slaughterhouses, compaction and loss of topsoil from overgrazing, depletion of resources from the land through not recycling animal wastes, a possible contribution to global warming through methane production from cattle flatulence—the list goes on and on.[25] While the damaging effects of some of this can be reduced through better practices, many are the inevitable and unavoidable result of restructuring our relationship with the land and with domestic animals in order to meet consumer demand. We support and participate in this ecological destruction through buying and consuming animal products. It has become increasingly clear that with the growth in human population and the increased dietary expectations of affluent populations, an animal-based diet is a luxury the planet can no longer afford.[26]

The issue of *embodiment* means that the bodily integrity of both our own bodies and those of animals is at stake in the kind of diet we choose. While much of the focus on embodiment in lesbian and gay male experience has been on sexual wholeness and protection from physical and emotional abuse, increasing attention is being paid to good nutrition as a critical component of embodied integrity.[27] Lesbian and gay liberationist thought joins feminist thinking in conceiving attention to nutrition in our communities as a justice issue of resistance to the self-destruction of addictive eating disorders that result from the internalized self-hatred fostered by wider cultural forces. As we reclaim the sacredness of our bodies, more attention must be given to the physiological and spiritual effects of animal protein-based diets.

Both widespread nutritional studies and anthropological studies of predominantly vegetarian cultures suggest not only that all our nutritional needs can be met through a diet of fruits and vegetables, but that it is far healthier than heavily meat- and dairy-based diets.[28] In addition, despite the popularity of the "man the hunter" thesis of human origins, paleoanthropological studies indicate the great predominance of plants in the diets of human beings as we evolved. Eating large quantities of meat has not been a necessary part of a healthy human diet. It has been only in the past two hundred years, in fact, that most people in the West have had the opportunity to eat meat on a daily basis; such a diet is still out of reach for the large majority of the earth's human population.[29]

Finally, we consider the issue of *justice for animals.* In addition to the reasons cited above, an ecojustice ethic of Gay and Gaia argues that we should consider being vegetarian for the sake of the animals themselves. Taking seriously the ethical guideline of embodiment in an ecological context draws attention to the widespread abuse of animals in the animal industry that causes unnecessary suffering of fellow sentient creatures, beings whose subjectivity and right to exist for their own sake we recognize. Given our history, lesbians and gay men need to be particularly sensitive to the utilitarian exploitation and disposability of other creatures consigned to the "passive resource" side of the nature/culture dichotomy and their invisibility that results in the process. At the very least we need to become aware of the actual material practices of animal production so that we can assess them and our participation in them from a moral standpoint. We need to employ our hermeneutics of suspicion in examining such seemingly neutral terms as "meat" that function to naturalize social and political constructs of animals as consumable objects while obscuring that the mass category of "meat" is actually made up of individual animals that once had lives of their own and experienced pain and pleasure.[30]

While ethicists and activists who take seriously the subjectivity of animals differ on whether the killing of animals per se is wrong—human beings, after all, are part of an ecological web that includes animal predation—they agree that the killing of animals in the current structure of agrianimal industry is morally untenable. McDaniel argues that killing animals may be morally justified if they are allowed some kind of quality of life prior to slaughter, and if their killing is done instantly, minimizing both physical and emotional pain. While such conditions were once widespread on small family farms where animals were an integral part of the farm ecology and economy, they have rapidly been replaced through agribusiness as the only viable way to meet consumer demand for animal products, so that today nearly all poultry products and over half the milk and red meat consumed in the United States come from large-scale indoor factory farms, all of which is subsidized by government aid.[31]

Within an ecojustice ethic of Gay and Gaia, attention to the combined mandate of each of these arguments—embodiment and bodily integrity, and the demands of social, ecological, and animal justice—presents a compelling argument for vegetarianism, or at least reduced meat consumption in affluent societies as an integral part of a liberationist approach to ecology. As McDaniel maintains, "To say 'no' to meat is to say 'yes' to animals, other people, ourselves, and the earth. A preferential option for animals is simultaneously a preferential option for the earth and the poor."[32]

Rethinking the Place of Leather in Gay Eroticism. Wearing animal products such as leather is another area a liberationist ecological ethic calls us to reconsider. In our efforts to reclaim the erotic and reintegrate sexuality with spirituality, the use of leather has played an important role in much of lesbian and especially gay male culture. In addition, gay male practices such as cross-dressing (drag) and leathersexuality—ritualized exploration of domination and subordination using leather clothing and objects in sexual relationships—have served to blur, bend, and extend gender boundaries and roles that uphold compulsory heterosexual complementarity. These practices have been critically important to many gay men in helping to break through years of heteromale socialization and internalized self-hatred for not conforming to expected masculine roles. They also have generated controversy and debate within gay male and lesbian circles. Examining these practices from an ecological perspective and commitment to the rights of animals reveals a further ambiguous dimension.

Most critiques of gay male leathersexuality have seen it as inhumane and misogynist in developing and acting out ritualized practices of "hypermasculinity" drawn from patriarchal gender roles and values. Clark, in contrast, argues that leathersexuality has a constructive role to play in its potential to subvert unconscious patriarchal power constructs through reflective ritual play: "Ritualistic leathersex shifts dominance and submission, power and pain, from the unreflective realm of social interactions and power hierarchies into a controlled arena, at once turning roles inside out and shattering the facades of everyday human power abuses. . . . The rituals of leathersex employ assumed and exaggerated roles to disconnect participants from dayworld reality, not as an escape, but as a route to deepened understanding."[33]

Carter Heyward has picked up on this in suggesting that profeminist gay men can teach lesbians "about how playing sexually with power is *not* necessarily to collude with patriarchal principles of domination and control. It can be a way of embodying some profoundly sacred tensions."[34] Key for Heyward and Clark is combining gay male erotic power play with a feminist sensitivity and commitment to examine the consequences of gay male sexual practices on women. This requires listening deeply to each other about the experiences and interpretations of meaning gay men and lesbians furnish as we seek to find ways to work together. For gay men it means an ongoing search to find ways to affirm the erotic in our lives, to avoid scapegoating both masculinity and women while we create new models of masculinity that honor and affirm both men's and women's experience.

Yet other questions about using leather to integrate gay sexuality and spirituality arise from an ecological perspective. Missing from Clark's and Heyward's discussion is any consideration of the perspective of animals (let alone the other ecological and social justice aspects raised above in connection with buying products of the livestock industry). If gay men and lesbians take seriously the ecological demand that we expand our community of accountability to include animals and the whole earth community, it may also mean questioning the connection of our integrated erotic spirituality and sexuality to wearing leather. What does it mean, for example, to have a central part of our reclaiming the erotic and reintegrating it with spirituality involve "trafficking in leather"—that is, dead animal hides—appropriated in purely utilitarian ways where we by and large are ignorant of the material practices that convert them from living animal to a sensuous second skin covering our bodies?[35] As in the case of the animal food industry, the material practices that produce leather products have shifted dramatically from earlier indigenous cultures oriented around the ritualized hunt and subsequent killing and preparation of animal skins, or even from family farms where the animals had some quality of life and relationship with their human consumers.

As in questioning the eating of meat, exploring ecological and ethical issues related to wearing leather simultaneously raises difficult questions about identity. For many, eating meat tastes good and feels "natural." Giving it up can leave a large void. Similarly, for many gay men (and lesbians), wearing leather feels and smells erotic. Are we "by nature" drawn to eating meat and/or wearing leather? Do these practices flow from an unconscious side that connects us in powerful ways with our early tribal origins? How much are our erotic preferences and values constructed? Is it enough, as some have suggested, to be aware and grateful of the sacrifices animals make to provide us with leather—using all of the animal as Native Americans practiced[36]—or does awareness of our very different material relations with animals today than our forbears or members of other cultures had, and the ecological damage of sustaining a large animal industry require us to forswear wearing leather products altogether? Does the fact that leather is a "durable good" that can be used for years make it differ from meat that is consumed repeatedly and thus creates a much higher demand for animals make a difference in the ethical evaluation of the two issues?

As in the case of food, these are simultaneously political and very personal questions and issues, for me, and, I would venture, for many gay men. In my closet hang a leather jacket and vest that feel like old friends and make me feel more connected to myself and other men when I wear them; alongside them are my leather boots that I wear two-stepping when I whirl the dance floor with other men and women. They have contributed to my identity as a gay man, helping me to reconnect with my erotic and sexual self as I connect with others. I feel a sense of emptiness, of loss of richness of experience when I anticipate giving them up. And what of other gay men who continue to wear leather (or eat meat)? If I give up wearing leather and incorporate a critique of it into my liberationist ecological stance, will my at-times fragile sense of connection with them, often strained already by the homophobic, sexist, and racist pressures we live under, be further stretched?

While there are no easy answers, an ecojustice ethic of Gay and Gaia compels us to raise and debate these questions within the gay male and lesbian communities. As a now-included part of our community of accountability, animals must be given moral consideration in the debates we engage in within our communities. Whether it is the food we eat, the clothes we wear, or the products we buy (which may be tested on animals), the consideration of the well-being of animals must form an integral part of our reflections with each other and the broader human community.

In this chapter I have tried to outline some of the central features of an ethic of Gay and Gaia that complement, critique, and deepen the insights and approaches being developed in ecological ethics. If sensuality, embodiment, and erotic connection in right relation, so central to lesbian and gay liberationist thinking, are also central to an ecological moral agency, what are some of the implications of this for how we conceive ecological ethics and apply it in our communities? While I make no claims to have the definitive answer, I do believe that rethinking and revisioning our ecological ethics to make them inclusive and liberationist along the lines suggested here in *Gay and Gaia* can give us, in the words of a gay ecotheologian, "a place to start."[37]

Conclusion

Teach us
To continue the Creation
To help the seeds
To multiply,
Giving food
For the people
And for the beasts

Teach us
to further the joy
You never tire of offering
When weary travellers find you,
A signpost to their home.

Teach us
To make the horizon
Become a beautiful image
Of Creation's grandeur.
—Helder Camara, "Earth, Sister Earth," in *Sister Earth*

James Nash concludes his ethical analysis of the ecological crisis in *Loving Nature* with these words: "A summary at this point would be absurd. But a note of hope is theologically and politically reasonable. . . . The best we can do is hustle and hope."[1] Hustle and hope. There is wisdom in this admonition, in its sense of urgency, in its theological grounding. Any moral response to the "earth and its distress"[2] must be rooted in hustle and hope.

Yet what nourishes these roots of "hustle and hope"? What guides and sustains the vision that enables our moral response to be more than a series of steps in crisis management, but rather a life-giving, life-sustaining reorientation to a sense of the sacred that permeates the earth? Throughout *Gay and Gaia* I have argued that our moral responses to the ecological and social crises that challenge us on every front today need to be rooted in and nourished by both the erotic and the ecological as two distinctive yet interwoven sacred powers of right relation. To begin to do so means rethinking and reimagining our most fundamental assumptions about the world, the sacred, and our human location with and within each.

Larry Rasmussen has argued that what is most needed today is *conversion* to the earth—a turning to earth in both orientation and allegiance.[3] I believe that Rasmussen is right, and that conversion to the earth comes through mutual recovery of the erotic and the ecological, a rediscovery and reappreciation of the interweaving of the spiritual and the material. Conversion to earth as a positive, energizing, joyful moral response comes in part, at least, through linking the joy and intimacy of the erotic in our lives with the earth intimacy and joy that comes from reconnecting with the wider web of life.

Dom Helder Camara, former archbishop of Olinda and Recife, Brazil, gives some sense of what this might look like in his poem "Earth, Sister Earth" cited in the epigraph. Whether through an ecologically understood Judeo-Christian sense of dominion as limitation of human power exercised in ecologically responsible ways, or a recovery of earth-based religious teachings that stress humanity's place within the broader web, we need to look to the Earth to teach us how to continue the Creation for the well-being of the whole life community. Camara links this with a second invocation, that the Earth "teach us to further the joy you never tire of offering,"[4] providing the earthy, erotic energy to sustain our efforts. The result of such a moral response may well be a horizon that reflects the beauty of "Creation's grandeur."

Let me finish by returning to the story of Celie's erotic and ecological transformation in *The Color Purple*. A story that begins in the violent act of rape, in such erotic brokenness and alienation that Celie is able to pray only to a distorted image of God, ends in a prayer of erotic and ecological joy: "Dear God. Dear stars, dear trees, dear sky, dear peoples. Dear Everything. Dear God."[5] The pages in between these two prayers reveal a lifelong journey of movement toward healing and reconnecting sexuality and spirituality in the context of both human community and the wider earth. The personal and the communal, the human and the whole earth, the erotic and the ecological—the well-being of each depends on the interdependent well-being of each other and the whole. This, finally, is what an ethic of Gay and Gaia seeks to embrace.

Gay and Gaia. A challenge to envision a world of reintegration in right relation at all levels of life:

- where humanity is reintegrated into nature's web;
- where persons of all sexualities, genders, and colors are reintegrated as vital, necessary, enriching parts of a diverse human ecology in a diverse natural world;
- where the diverse ways of knowing each other and all creatures in right relation expand our awareness and acknowledgment of the diverse ways of knowing the divine.

Gay and Gaia. A challenge to move from dislocation to relocation at all levels.

- where nature and all nature's creatures no longer suffer dislocation to fulfill utilitarian human desires and greeds;
- where lesbians and gay men, bisexual and transgendered persons are relocated within nature and nature is relocated in our communities;
- where erotic, sensual, embodied ways of knowing, relating, and experiencing are relocated at the center of moral experience and our experiences of the divine.

Gay and Gaia. Glimpses and intuitions of a hopeful future in our shared commitments and practices for ecological and human justice and integrity throughout the planet today. To honor these intuitions, to expand and build on these fragile glimpses is to envision and relocate—to re-incarnate—the divine in our midst, for the healing and wholeness of the planet and all who dwell within.

Gay and Gaia. Remember our roots in struggle and grace, nourish the signs of life and commitments to justice around and within us, plant seeds of hope and vision for the future.

Gay and Gaia. Hustle and Hope. Continue the Creation, further the joy, and make the horizon become a beautiful image of Creation's grandeur.

Appendix: A Postscript on Method

The social and ecological dislocations described in *Gay and Gaia* are the result of pervasive social, political, economic, cultural, and ecological changes that have marked recent decades in the United States and throughout much of the world. As we approach the twenty-first century, the expanding global ecological crisis has shifted and reshaped fundamentally the context in which all social change, and thus ethical reflection and praxis take place. This has critical implications for ethical methodology. It is now imperative that the framework for theological and ethical reflection expand to include within its community of accountability the entire earth community whose very sustainability and survival are increasingly threatened by human physical, industrial, and technological activity. This, in turn, requires a fundamental rethinking of the assumptions that have shaped understandings of humanity's relationship to the rest of the natural order.

The ecological crisis and subsequent necessary shift away from a solely anthropocentric ethical framework have coincided with the development and articulation of liberation theologies, which have brought renewed emphasis on the centrality of human struggles for justice and liberation in Christian faith and ethics. Many in these struggles are justifiably suspicious of calls from the same societal elites perceived as responsible for the social injustices they experience to now shift their attention from these critical social issues to the ecological crisis. It is increasingly clear, however, that struggles for human liberation are contingent on a healthy and sustainable natural ecology that can provide an environment supportive of human life. Destruction of the earth's ecological systems has accelerated to the point it now threatens to engulf not only efforts for justice and liberation, but life itself. As the Salvadoran ecologist Ricardo Navarro puts it, "The struggle for the environment is the struggle for life."[1]

Until recently there has been little interchange at the level of ethical reflection between those working to articulate and live out a constructive ecological ethics rooted in the Christian community and those working on human liberation issues. Christian ecological ethics in particular often has been blind to or ignorant of the way societal constructions along lines of race, class, sexuality, and gender influence the way we understand the earth, diagnose ecological problems, and propose solutions that inevitably affect the earth community as a whole. Yet when divorced from liberationist concerns, the necessary shift in ecological ethics away from a predominantly anthropocentric framework risks obscuring already existing, inequitable societal power relations. Such ethical praxis may thereby inadvertently reproduce or aggravate existing social contradictions in the human community while seeking to attend to the urgent question of humanity's place and role within the larger earth community.

A primary goal of *Gay and Gaia* is to provide a methodological grounding for a liberationist ecological ethic that takes seriously lesbian and gay male experiences and insights. A working assumption that I explore throughout the book is that epistemology and theory in ethics are shaped by our combined social and ecological locations: where we are located in the web of social and ecological relations in large part influences how we see the world and understand its problems and our responses. Hence attention to the paired criteria of Gay and Gaia as markers of social and ecological justice can reveal important connections about the links between oppression of lesbians and gay men and oppression of nature, as well as sources for our shared renewal and liberation.

A central focus of *Gay and Gaia*, therefore, is on methodological issues, particularly the epistemological and hermeneutical assumptions and issues that inform different approaches to ecological ethics. Assumptions about *what* we know about nature, humanity, the divine, and the relations between them, and *how* we know provide a fundamental grounding for Christian ecological ethics that too rarely is investigated critically for the ways it affects intrahuman as well as interspecies relations. This is especially critical for those committed to lesbian and gay liberationist concerns, given our history of being defined outside the "natural" human and broader biological realms under implicit and explicit assumptions regarding heterosexual "normalcy" that in turn may be reproduced unconsciously in ecological ethics.

The contribution of lesbian and gay male insights to constructing ecological ethics is particularly important in two areas. First, working with the tools of constructionist theory, lesbian and gay theorizing have focused attention on how societies construct views of what is natural and unnatural that reflect cultural biases posing as empirical reality in claims made about human nature and human relationships to the rest of the natural world. In particular, gay and lesbian theorists have questioned what may be termed "heterosexual naturalism": views of human nature and the natural world that assume heterosexuality to be the normative and natural state for all biologic life.

Second, those working within lesbian and gay male liberationist frameworks have examined critical hermeneutical and methodological issues with respect to the ways theology and ethics appropriate insights from other sources, especially from the natural and human sciences, from Western and Christian tradition, and across lines of cultural, racial, ethnic, economic, and sexual difference. Both of these areas—the critical examination of culturally constructed views of the natural world and human nature, and the appropriation of insights from other sources—must form part of a liberationist ecological ethic that focuses with urgency on the ecological crisis without sacrificing aspirations for human liberation.

Because the vast majority of the liberationist theorizing on sexuality that has emerged thus far has come from explicitly lesbian and gay male frames of reference, I ground *Gay and Gaia* primarily in gay male and lesbian experiences and theorizing. Yet recent activism by bisexual and transgendered (transsexual and transvestite) persons has drawn attention to the need to expand this grounding to include bisexual and transgendered experience in theorizing sexual experience and identity. Wherever possible such experience and insights are incorporated into the book. I also include bisexual and transgendered persons, experience, and theorizing within the primary community of accountability of this project; any liberationist ecological ethics must be sensitive to their valid place in the human and natural ecologies.

The danger of grounding *Gay and Gaia* primarily in the insights of lesbian and gay experience and theorizing rather than expanding this to lesbian, gay, bisexual and transgendered experience and theorizing—is that it reproduces bisexual and transgendered invisibility and reinforces a binary framework of either homo- or heterosexual identity and experience, rather than blurring the boundaries as much recent constructionist theory has emphasized. But the other alternative—simply adding bisexual and transgendered to the list without having a sufficient body of literature and theorizing to accompany them—is problematic as well, for it leads to the illusion that one is engaging seriously bisexual and transgendered experience when one is not. The result is the collapse of the particularity of bisexual and transgendered experience into lesbian and gay experience and theorizing, reproducing a different but equally problematic form of invisibility. Hence I ground *Gay and Gaia* explicitly in the insights lesbian and gay experience and theory, drawing where possible on the insights of other sexual experience and identities.

Most broadly, then, the methodology for *Gay and Gaia* is grounded in the combined contemporary concerns for the well-being of socially marginalized persons and the entire earth community. Liberationist approaches acknowledge that all perspectives operate with biases, whether implicit or explicit, that are shaped in part by our social locations. Hence where we position our claims and to whom we are accountable in making them become important methodological considerations. My primary accountability in *Gay and Gaia* is shaped by the paired concerns of Gay and Gaia:

to lesbians, gay men, bisexual and transgendered persons of all colors, and to the ecological well-being of our threatened planet, itself a living system of interconnected parts. The primary audiences to which this project is directed are three: to lesbian and gay male, bisexual and transgendered individuals and communities, to the mainstream church and society in North America, and to the various parts of the environmental movement, both religious and secular.

To derive my methodological approach, I critically appropriate the combined methodological insights and approaches of four positions that have been important to gay and lesbian theorizing.[2] While I develop each position in depth in the book, a brief review may be helpful here. Because ecological ethics depends so greatly on insights from science, first is emphasis on a critical use of the natural sciences, especially ecology and biology, that avoids positivistic or reductionistic approaches rooted in mechanistic Enlightenment assumptions. This approach is central to both ecological and liberationist perspectives. Second and third are appreciation for the insights of identity politics and deconstructionism that call for a critical approach to the sciences as they in turn critique and balance each other's problematic tendencies. Identity politics grounded in the liberation struggles of lesbians and gay men, feminists, and antiracism and anticolonialism/imperialism movements have shown how a strong sense of group identity can empower liberation struggles and reveal aspects of social reality obscured by the dominant social practices and discourses. Their tendency toward rigid boundaries and fixed social identities is problematic, however, and can result in new patterns of exclusion and invisibility.

Social constructionist theorizing and postmodern deconstructionism have called attention to the socially constructed nature of these social boundaries and identities and plays a critical role in the liberationist effort to deconstruct dominant social categories such as race, gender, and sexuality that are portrayed as inevitable and rooted in nature. By itself, however, deconstructionism can lead to a moral and political relativism that can undercut liberationist efforts on behalf of marginalized communities. This is partially addressed by the fourth source, Marxist-feminist insights on standpoint theory that identify and emphasize certain social locations with respect to the material and ideological relations of society. These locations or standpoints can generate perspectives that are given greater weight in liberationist work because they allow us to see more clearly how social power operates along axes of social and cultural difference.

All four of these methodological positions—a critical scientific realism, identity politics and insights, social constructionism and deconstructionism, and standpoint privileging of perspectives—contribute to a liberationist ecological ethic. Taken together they help to generate a social-ecological hermeneutics of suspicion that takes seriously both scientific and social claims but deals with them critically. To be liberationist, an ecological ethic must take seriously, for example, the knowledge claims of the natural sciences without being tyrannized by them. Similarly the other three sources from the human social realm must be held accountable to the ways they operate within the larger ecological realm. These four sources combined provide the necessary methodological grounding for a liberationist approach to ecological ethics.

Because of its central role in *Gay and Gaia* and because of the often impassioned if not acrimonious debate that has engulfed lesbian and gay communities over origins of our sexualities and sexual identities, I need to say a further word about my use of social constructionist theory and insights. In the past decade, lesbian and gay male scholars have tended to divide into oppositional camps on the questions of origins and characteristics of human sexuality, particularly homosexuality. Those grounded in essentialism have generally theorized that a homosexual orientation is an innate, biological characteristic that predetermines a minority of persons to be physically and emotionally drawn to persons of their own biological sex. Followers of constructionism, on the other hand, have theorized a more fluid understanding of sexuality, emphasizing that social practices and values rather than biological predispositions give meaning to our sexual identities, and thus sexuality, far from being an ahistorical, transcultural biological phenomenon, changes and evolves with culture across geography and time.[3]

Increasingly scholars in both camps are seeing the essentialism/constructionism standoff as a sterile, nonproductive debate that does little to advance lesbian and gay male political and theoretical interests. Both have played important roles strategically in lesbian and gay male politics, yet both have shortcomings. The essentialist position underlies many of the strategies focused on attaining civil rights for lesbians and gay men as a recognizable and definable social minority. Given the increase in antilesbian/gay violence, harassment, and expulsion of lesbian and gay male soldiers in the military, and ongoing discrimination in such areas as housing, employment, insurance, and spousal rights, this approach remains critically important. Yet by itself it leaves the constructed nature of heterosexuality unchallenged as the dominant societal norm. It also implies static and unchanging realities when the diversity of cultural forms of sexual experience needs to be recognized and respected. Constructionism is important in this because it draws attention to the ways all sexualities—homo-, hetero-, and in-between—are constructed and reflect social values.

The affinity for constructionist theory in much lesbian and gay studies has grown out of concern that because essentialist understandings read gender, race, and sexuality as rooted in nature, biology often is read to equal social destiny. As many feminists have argued, however, the political and ethical ramifications of essentialist views too often are to reify and eternalize historically constructed unequal social relations as rooted in nature. Hence men are seen as "naturally" superior to women, whites are "naturally" superior to people of color, and heterosexuality is "naturally" superior to homosexuality. Constructionist insights help to destabilize these social relations of domination by exposing the ways they have been constructed historically and culturally to reflect the political and social values of certain segments of humanity over other groups. If unequal relations grounded in different social identities are an outcome of historical processes rather than expressions of biologically determined "nature," they can be challenged and reshaped within historical processes. Constructionism does not deny the contributions of biological factors to identities such as gender and sexuality, but insists that their meanings are never self-interpreting and can only be understood accurately in light of the social, cultural, and historical factors that have shaped them.

Constructionist insights have important implications for ecological ethics with its assumptions about nature, human nature, and their relation. There are realities that will be perceived inaccurately in any theory that extends heterosexual naturalism into revisioning ecological ethics. Formulating an ecological ethics without attention to constructionist insights grounded in lesbian and gay experience risks perpetuating heterosexist social relations that are damaging both to lesbians and gay men and to the wider human and nonhuman ecology.

Examining claims about nature are therefore critical to any ecological ethics that seeks to be liberationist. Historically, justification for the persecution and exclusion of many groups, including women, black males, and lesbians and gay men, often has been grounded in essentialist understandings of sexuality and claims about nature. With respect to nature, for example, African Americans and lesbians and gay men have been oppressed for being both "unnatural" and "too natural," in the sense of animalistic. The self-contradictory logic informing this stance results from the conflicting ways Western society has understood and symbolized nature as both nurturing mother and as threatening, uncontrollable chaos. When nature is viewed primarily as procreative mother, lesbians and gay men have been seen as the essence of unnatural because of the allegedly nonprocreative nature of our relationships. Heterosexual complementarity and procreation are posited as unchanging essential norms of nature and human nature; whatever falls outside this is therefore perverted or "unnatural." When nature is associated with sexuality as nonrational and noncontrollable forces that threaten a rational and ordered culture, gay men and lesbians are seen to represent the epitome of "animalistic" sexual otherness whose sexual expression escapes the bounds of culture and threatens the "natural" heterosexual social order.

Unnatural or "too" natural—in either case Western constructs and practices have excluded people with homoerotic feelings and behavior from the ecology of both human culture and nature. Homosexuality has been seen as a threat to a social order understood as rooted either in nature or threatened by nature. To avoid simply reproducing heterosexist values and social relations, therefore, a key question in revisioning ecological ethics becomes, how do current models of ecological ethics respond to epistemological assumptions of heterosexual naturalism that ground the categories of thought we inherit? Do they actively challenge these assumptions in order to transform them, or do they perpetuate them through silence or uncritically repeating them?

To understand the ways we construct categories of natural and unnatural I focus on the three rubrics considered previously in each model: views of humanity, nature and the nature/humanity relation, and divinity and the divinity/humanity/nature relation. I approach these questions with the paired criteria of Gay and Gaia outlined in chapter 1. At one level this can be understood as a lesbian- and gay-rooted ethic of ecojustice. Hence Gay is a justice criterion that critiques heterosexism: any ecological ethic that explicitly or implicitly excludes lesbians and gay men from the ecology of human community and the broader ecosystem is unjust and therefore inadequate. Gaia is an ecological criterion that insists we view ourselves, our communities, and the world ecologically, that is, holistically and relationally. The community of accountability for our ethics is Gaia, the living earth, and it is to all the earth that we are accountable.

At a deeper, ecological level, these two criteria in turn mutually inform each other. Hence Gay is also an ecological criterion. In insisting on lesbian and gay inclusion, it recognizes the intrinsic value of each member of human communities and the broader ecosystems, apart from perceived utilitarian values such as procreation. While recognizing the unique positioning of gay and lesbian people—as well as of non-gay peoples—it critiques any understanding of positionings or identities that are nonrelational or not connected to the broader whole. And Gaia is a justice criteria. It insists that all understandings of justice in the human community be framed ecologically and extended to the entire earth community. It draws from its lesbian and gay members as much as from its non-gay members in constructing an understanding of human nature and humanity's relation to the rest of the earth and to God.

Given the contested nature of much of the discourse used in this project, it is helpful to define some of the key terms and place them in their interpretive context. Studies of sexuality and sexual identity inevitably draw on other categories and concepts such as sex and gender, which traditionally have been defined as complementary pairs of a nature/culture dichotomy. *Sex* has been understood as a biological classification dividing organisms into male or female categories according to purely physical characteristics such as genitalia or chromosomal makeup. *Gender* has been understood as a cultural or sociological classification, a complex interaction of biogenetic and sociocultural systems that defines sex roles for human beings, usually along lines of sex, into masculine and feminine roles and identities.[4] While much of the deconstructive work of feminist analysis in recent decades has focused on gender, increasingly feminists and others are calling into question the nature/culture dichotomy itself as a social construct, examining the ways human beings and cultures construct their understandings of nature. Examination of the nature/culture boundary in ecological ethics forms an important part of this project.

Sexuality involves both sex and gender (as well as other constructs, such as race and class) to describe the relational context in which we are drawn emotionally and physically to relate sexually to ourselves and other persons.[5] It is that "array of acts, expectations, narratives, pleasures, identity-formations, and knowledges, in both women and men, that tends to cluster most densely around certain genital sensations but is not adequately defined by them."[6] *Sexual identity* refers to the meaning and sense of personal and corporate identity one gains from or gives to the expression of one's sexuality at any particular point in time. As noted above, much of the debate between essentialists and constructionists is how much, if at all, the content of one's sexuality is shaped by, and shifts in response to, the surrounding web of societal-cultural meanings and practices.

The oppression of lesbians and gay men stems from structures and attitudes of *compulsory heterosexuality*, the tool of social control that mandates that all expressions of sexuality be channeled through patriarchal, heterosexual patterns.[7] *Homophobia*, the irrational fear and hatred of or prejudice toward lesbians, gay men, and homosexual behavior, operates with societal and religious sanction to keep in place *heterosexism*, the structuring of society to value and favor heterosexual men and their procreative sexuality.

Feminists and profeminist gay men insist that lesbian and gay oppression must be understood within the broader framework of patriarchal sexism: "Gay oppression is part of the very fabric of our society, reinforcing both male socialization to fear homosexuality and female socialization to remain passively attached to a traditional, dominant male. This combination of homophobia and sexism is the very core of heterosexism."[8] As a part of *sexism*, the domination of women by men, and valuing maleness and masculinity over femaleness and femininity, heterosexism "maintains the subservience of women to men by punishing homosexuality and any deviance from the currently accepted range of masculine and feminine heterosexual roles."[9] Homophobia and heterosexism are the result of anti-body and antisexual attitudes and dualisms that pervade traditional Christian theology and teaching, and have in turn shaped Western attitudes and societal structures. These anti-body dualisms intersect with the fundamental patriarchal male/female dualism to provide patriarchal societies with a powerful and pervasive method of control over anything and anyone designated as "other."[10]

Many of these attitudes in turn are embedded in beliefs and assumptions of how nature operates, and therefore what is "natural" human behavior. Appeals for sanctions against lesbians and gay men, such as Colorado's infamous Amendment 2 that prohibited equal treatment under the law for homosexuals, often are grounded in appeals to nature and "natural behavior" (that is, heterosexual) for their justification. Because much of the current focus on ecology is centered on concern for rethinking humanity's "natural" place within the environment, persons committed to equality and justice for lesbians and gay men must address the claims and assumptions emerging from the environmental movement about human nature.

Working definitions of key ecological terms also are helpful. *Ecology* is both the branch of biology that deals with interrelationships, the science of the development of natural communities, as well as the web of organic relations that make up the earth as a living system. As a science, ecology explores how ecosystems and biotic communities function as interrelated and interdependent webs of life, and how they become disrupted. When ecology focuses on the effects of human disruption on the sustainability of biotic and human communities, it functions in the expanded sense of a combined socioeconomic and biological science. *Ecological* refers to a philosophical awareness of the interdependence of living things, and as such functions as both a scientific and philosophical modifier as in the phrases "thinking ecologically" and an "ecological point of view."[11]

Defined most simply, *nature* is the biophysical world, of which humans are a part. Hence human culture exists within nature as a uniquely human product. Nature is a living system that undergoes a constant process of adaptation, evolution, and change, with and without human participation. It is in this sense that I use the term *nature* in this work. Yet in Western tradition nature has been constructed to hold several other senses, including the following: the biophysical world apart from human beings; that which is the essence of something, as in their "nature"; and, theologically, the created world apart from God.[12] Investigating the different understandings and constructs of nature that inform ecological ethics is central to this project.

Location refers to where a creature (human or nonhuman) is situated in the web of human and nonhuman relations that make up biotic and human communities. *Social location* restricts this to the human realm and focuses on the power differentials that exist in these relations as structured along lines of social and cultural difference such as race, gender, class, and sexuality. *Ecological location* includes social location but expands it to the biotic realm to focus on the effects of where humans and other creatures are situated ecologically, how these relations are structured in terms of power and sustainability, and the relation between intrahuman relations in the social realm and interspecies relations that include but are not limited to interaction with humans.

It is this methodological outline that I develop to give grounding to an emerging ethic of Gay and Gaia. The liberationist (and social justice) dimensions of lesbian and gay struggle are reconfigured and expanded into an ecological context, while ecological ethics is rooted firmly within a justice framework. This methodology allows us to revision the erotic as an important component of and grounding for ecology, as well as placing the erotic dimensions of right relation within an ecological context. In the interrelation of the erotic and the ecological we find the grounding and inspiration for an ecojustice ethic of Gay and Gaia.

Notes

PART 1: TOWARD AN ETHIC OF GAY AND GAIA: GUIDES FOR THE JOURNEY

1. Starting the Journey: Initial Reflections

1. Alice Walker, *The Color Purple* (New York: Harcourt Brace Jovanovich, 1982), 166–68.

2. Chief Seattle, "You Will Suffocate: A Chief's Lament," *Wildlife* 18, no. 13 (1976): 115. Cited in Aline Amon, *The Earth Is Sore: Native Americans on Nature* (New York: Atheneum, 1981), 53.

3. Sallie McFague, "Imaging a Theology of God's Nature: The World as God's Body," in *Liberation Theology: An Introductory Reader*, ed. Curt Cadorette, Marie Giblin, Marilyn Legge, and Mary H. Snyder (Maryknoll, N.Y.: Orbis, 1992), 285.

4. Walker, *The Color Purple*, 187.

5. In the summer 1995 debates around the efforts of the Republican Contract with America to roll back environmental legislation, Representative Randy Cunningham (R-Calif.) made this association clear when he decried those supporting continued environmental protection as the same people "who would put homos in the military."

6. Audre Lorde, "Uses of the Erotic: The Erotic as Power," in *Sister Outsider: Essays and Speeches* (Freedom, Calif.: The Crossing Press, 1984), 57.

7. Carter Heyward, "Sexuality, Love, and Justice," in *Our Passion for Justice: Images of Power, Sexuality, and Liberation* (New York: Pilgrim Press, 1984), 83–93. For a more detailed discussion of the methodological issues and commitments that ground *Gay and Gaia*, see the appendix.

8. See Beverly W. Harrison, "The Power of Anger in the Work of Love," in *Making the Connections: Essays in Feminist Social Ethics*, ed. Carol S. Robb (Boston: Beacon Press, 1985), 3–21.

9. For helpful discussions of the epistemological privilege of the poor and oppressed in Latin American liberation theology, see José Míguez Bonino, *Toward a Christian Political Ethics* (Philadelphia: Fortress Press,

1983), 42–44; Robert McAfee Brown, *Gustavo Gutiérrez: An Introduction to Liberation Theology* (Maryknoll, N.Y.: Orbis, 1990), 107–8.

10. James E. Lovelock, *Gaia: A New Look at Life on Earth* (Oxford: Oxford University Press, 1979); James E. Lovelock, *The Ages of Gaia: A Biography of Our Living Earth* (New York: Norton, 1988); Lynn Margulis and Dorian Sagan, *Microcosmos: Four Billion Years of Evolution from Our Microbian Ancestors* (New York: Summit Books, 1987). See also Lawrence E. Joseph, *Gaia: The Growth of an Idea* (New York: St. Martin's Press, 1990).

11. For analysis of the ambiguities of the closet for gay men and lesbians, see Eve Kosofsky Sedgwick, *Epistemology of the Closet* (Berkeley and Los Angeles: University of California Press, 1990), esp. 1–4.

12. For the relation of this methodological grounding to recent insights emerging from bisexual and transgendered activism and theory, see the appendix.

13. Conspicuously absent from this list are Christian ecological ethics formulated outside a North Atlantic context, particularly within increasingly ecologically and economically devastated areas of Latin America, Africa, and Asia. These conversations are well under way, as evidenced by the spring 1993 course on ecology and theology in Central American at the Seminario Bíblico Latinoamericano in San José, Costa Rica, and the Brazilian theologian Leonardo Boff's recently released *Ecology and Liberation* (Maryknoll, N.Y.: Orbis, 1995). Apart from Boff's book and some preliminary work formulated by Ingemar Hedström of the Departamento Ecuménico de Investigaciones in San José, Costa Rica—see in particular Ingemar Hedström, *Volverán las Golondrinas? La Integración de la Creación desde una Perspectiva Latinoamericana* (San José, Costa Rica: DEI, 1990), and three articles in the collection *Liberating Life: Contemporary Approaches to Ecological Theology*, ed. Charles Birch, William Eakin, and Jay B. McDaniel (Maryknoll, N.Y.: Orbis, 1990)—there is little available in print on Christian ecological theo-ethical reflection from this part of the world.

2. Crossing Bridges: Hermeneutics and Connection from the Margins

1. J. Michael Clark, *A Place to Start: Toward an Unapologetic Gay Theology* (Dallas: Monument Press, 1989), 16.

2. Debate still exists as to how much sexual identities are in-born versus constructed. Here I focus on the constructed element, which I be-

lieve predominates, because it highlights the area of ethical agency and the choices we make. For an extended discussion of lesbian and gay identities, including the essentialist/constructionist debate, see my essay, "How Shall We Name Ourselves? The Ethical and Political Implications of (De)Constructing Gay and Lesbian Identities in the Current Ecclesial-Political Context," unpublished field examination (New York: Union Theological Seminary, 1991).

3. See Carolyn Merchant, *The Death of Nature: Women, Ecology and the Scientific Revolution* (San Francisco: HarperCollins, 1989); Carolyn Merchant, *Ecological Revolutions: Nature, Gender, and Science in New England* (Chapel Hill, N.C.: University of North Carolina Press, 1989); Donald Worster, *The Wealth of Nature: Environmental History and the Ecological Imagination* (New York: Oxford University Press, 1993).

4. Lynn White Jr., "The Historical Roots of Our Ecological Crisis," *Science* 155, no. 3767 (10 March 1967), 1203–7. For a thoughtful assessment of the Christian tradition's attitudes toward nature and a response to White's thesis, see Paul Santmire's excellent book, *The Travail of Nature: The Ambiguous Ecological Promise of Christian Theology* (Philadelphia: Fortress Press, 1985).

5. The term *hermeneutics of suspicion* was first articulated in liberation theology by Juan Luís Segundo in *The Liberation of Theology* (Maryknoll, N.Y.: Orbis, 1976). It refers to the ideological suspicion that persons who traditionally have been excluded from the dominant realm of experience bring to the task of interpreting texts. It is rooted in the particular experience of the excluded community and seeks to uncover the ideological interests at play that have covered up liberating aspects in the interpretation of the text. A feminist hermeneutics goes further to ask what interests were involved in silencing the voices of women and other excluded peoples before the text was even written down. This approach is also taken from a Marxist perspective articulated by the black South African biblical scholar Itumeleng J. Mosala in his book *Biblical Hermeneutics and Black Theology in South Africa* (Grand Rapids, Mich.: Wm. B. Eerdmans, 1989). Mosala draws on a materialist hermeneutics to identify different layers of power interests involved in formulating and redacting texts in the Hebrew scriptures and the strengths and limitations of drawing on these texts by the oppressed black community in South Africa.

6. For early theorizing on "standpoints" as a critical positioning that must be achieved through praxis rather than inherited with skin color,

sexuality, or gender, see Nancy Hartsock, "The Feminist Standpoint: Developing the Ground for a Specifically Feminist Historical Materialism," in *Discovering Reality: Feminist Perspectives on Epistemology, Metaphysics, Methodology, and Philosophy of Science*, ed. Sandra Harding and Merrill Hintikka (Dordrecht, Holland: D. Reidel Publishing Company, 1983), 283–310. This concept is developed further below.

7. Worster, *The Wealth of Nature*, 27.

8. For example, Simon LeVay's research on the size of the male hypothalamus as a biological factor in sexual orientation made no claims that this was the *only* factor, but when it was published in 1991, the media was quick to pick it up as proof that homosexuality is "biological." See Simon LeVay, *The Sexual Brain* (Cambridge, Mass.: The MIT Press, 1993), xii–xiii, 137–38; and Joe Dolce, "And How Big Is Yours? An Interview with Simon LeVay," *The Advocate*, June 1, 1993, 38–44. See Sedgwick, *Epistemology of the Closet*, for an excellent discussion of the complexity of experiences around sexual orientation and identity, esp. 22–27.

9. Donna Haraway, *Primate Visions: Gender, Race, and Nature in the World of Modern Science* (New York and London: Routledge, 1989), 3. See also Sandra Harding, *The Science Question in Feminism* (Ithaca, N.Y.: Cornell University Press, 1986).

10. Ursula K. Le Guin, "Why Are Americans Afraid of Dragons?" in *The Language of the Night: Essays on Fantasy and Science Fiction* (New York: HarperCollins, 1989), 39–40, cited in Ann K. Wetherilt, *That They May Be Many: Voices of Women, Echoes of God* (New York: Continuum, 1994), 100.

11. See Thomas Kuhn, *The Structure of Scientific Revolutions* (Chicago: University of Chicago Press, 1962).

12. Haraway, *Primate Visions*, 4.

13. Ibid., 8.

14. Ibid., 6.

15. For further discussion of the problem of the social construction of the "other" and the use of "othering" as a means of exploitation, see chapter 4.

16. Haraway, *Primate Visions*, 6.

17. These perspectives are privileged not because the oppressed are somehow inherently or morally superior, but because reality viewed from their locations is more likely to generate critical knowledge: "The standpoints of the subjugated are not 'innocent' positions. On the contrary,

they are preferred because in principle they are least likely to allow denial of the critical and interpretative core of all knowledge. . . . 'Subjugated' standpoints are preferred because they seem to promise more adequate, sustained, objective, transforming accounts of the world." Donna Haraway, "Situated Knowledges: The Science Question in Feminism and the Privilege of Partial Perspective," in *Simians, Cyborgs, and Women: The Reinvention of Nature* (New York and London: Routledge, 1991), 191.

18. Sandra Harding, "Rethinking Standpoint Epistemology: 'What Is Strong Objectivity?'" in *Feminist Epistemologies*, ed. Linda Alcoff and Elizabeth Potter (New York and London: Routledge, 1993); cited in Wetherilt, *That They May Be Many*, 230.

19. Haraway, *Primate Visions*, 8.

20. Ibid., 13 (emphasis added).

21. Ibid., 8.

22. For an excellent overview of this, see Santmire, *The Travail of Nature*.

23. See, for example, Elisabeth Schüssler-Fiorenza, *In Memory of Her: A Feminist Theological Reconstruction of Christian Origins* (New York: Crossroad, 1983), xiii–xxv.

24. For example, John Boswell, *Christianity, Social Tolerance, and Homosexuality: Gay People in Western Europe from the Beginning of the Christian Era to the Fourteenth Century* (Chicago: University of Chicago Press, 1980); L. William Countryman, *Dirt, Greed, and Sex: Sexual Ethics in the New Testament and Their Implications for Today* (Philadelphia: Fortress, 1988); John McNeill, *The Church and the Homosexual* (Boston: Beacon Press, 1988).

25. I draw on two sources for this discussion of Brooten's hermeneutics: Bernadette J. Brooten, "Paul's Views on the Nature of Women and Female Homoeroticism" in *Immaculate and Powerful*, ed. Clarissa Atkinson, Constance Buchanan, and Margaret Miles (Boston: Beacon Press, 1985), 61–86; and "Why Did Early Christians Condemn Sexual Relations Between Women?" paper given at the annual meeting of the Society for Biblical Literature, San Francisco, November 22, 1992. See also the discussion of Brooten in chapter 3.

26. For an excellent example of this, see Countryman, *Dirt, Greed, and Sex*.

27. Wetherilt, *That They May Be Many*, 36–37. Wetherilt is a white lesbian feminist theologian whose work centers on engaging the voices of women across multiple differences and locations in theology.

28. Carter Heyward, *Touching Our Strength: The Erotic as Power and the Love of God* (San Francisco: Harper and Row, 1989), 74.

29. J. Michael Clark, "Erotic Empowerment, Ecology, and Eschatology, or Sex, Earth, and Death in the Age of AIDS," paper delivered to the 1993 annual meeting of the American Academy of Religion, Washington, D.C., November 1993, 3–4.

30. Heyward, *Touching Our Strength*, 75.

31. Mary E. Hunt, *Fierce Tenderness: A Feminist Theology of Friendship* (New York: Crossroad, 1992), 9.

32. Ibid., 19.

33. For example, in *Touching Our Strength* (p. 7), Carter Heyward states,

> I am writing to serve the interests of lesbians and gaymen, especially those who are coming into a sense of pride and delight in an embodied spirituality that, as yet, they may only intuit as a resource of empowerment and liberation. I am writing also in the service of feminists and womanists—women and men of different colors; lesbian/gay, heterosexual, bisexual, celibate and sexually active—who are committed to making connections between sexuality, spirituality, and the ongoing struggle for justice *for all*. I am writing finally on behalf of the interests of progressive and radical religious folks—especially, though not exclusively, christians and former christians—who know the debilitating, violent character of the dualistic splits that break our collective body and shatter us, one by one. My people hold me accountable—responsible for what I say.

34. Clark, "Erotic Empowerment, Ecology, and Eschatology," 3.

35. Gary D. Comstock, *Gay Theology Without Apology* (Cleveland: Pilgrim Press, 1993), 4.

36. Ibid., 6.

37. Ibid., 58.

38. Wetherilt, *That They May Be Many*, 38.

39. Clark, *A Place to Start*, 19–20.

40. Comstock, *Gay Theology Without Apology*, 39.

41. See Harrison, "The Power of Anger in the Work of Love," 3–8.

42. In a particularly insightful examination of the Jonathan and David relationship, Comstock makes this point when he concludes: "The body of scholarship that interprets Jonathan and David's covenant as Jonathan's abdication to David's kingship may be an attempt to rational-

ize the disbelief and annoyance that many feel, but cannot reasonably explain, when confronted by noncompetitive, intimate, loving relationships between men. Gay men, especially those who have risked coming out in some or all aspects of their lives, more readily recognize and credit human love and not politics as the arena of David and Jonathan's relationship; and we do not flinch when Jonathan risks social security and familial approval for personal affection." *Gay Theology Without Apology*, 87.

43. Clark, *A Place to Start*, 23.

44. Wetherilt, *That They May Be Many*, 38.

45. Heyward, *Touching Our Strength*, 81 (emphasis added).

46. Comstock, *Gay Theology Without Apology*, 108.

47. Wetherilt, *That They May Be Many*, 38.

48. An excellent example of this is Merchant's *The Death of Nature*. Merchant documents how feminine images of nature that prevented or ameliorated human exploitation of the natural world during the medieval period in turn provided the justification for exploitation during and after the Enlightenment, when cultural assumptions about women and nature shifted but the feminine images of nature remained. See especially 1–41.

49. This insight is central to extending of the category of analysis of social location to ecological location which I explore in more depth in chapter 10.

50. Joanna Macy, together with other ecological activists, has created a ritual liturgy, *Thinking Like a Mountain: The Council of All Beings*, to encourage and facilitate people beginning to imagine the world through the experience of nonhuman creatures. See Joanna Macy, Arne Naess, Pat Fleming, and John Seed, *Thinking Like a Mountain: Towards a Council of All Beings* (Philadelphia: New Society, 1988). Adams brings this "epistemological empathy" closer to home in urging us to consider our actions through the experience of creatures who regularly adorn our tables. "Why," she asks, "do we welcome animals to be *on* our tables, but not *at* them?" What does this reveal about our attitudes, assumptions, and relations toward nonhuman nature? See Carol Adams, *The Sexual Politics of Meat: A Feminist-Vegetarian Critical Theory* (New York: Continuum, 1990).

51. Anne Primavesi, *From Apocalypse to Genesis: Ecology, Feminism, and Christianity* (Minneapolis: Fortress Press, 1991), 20–22.

52. I contend that one can affirm a socially constructed dimension to sexual identities without dismissing the contribution of biogenetics, as gay scholar Jonathan Katz appears to do in arguing that "nature had noth-

ing to do with it." Constructionist views that dismiss out of hand the place of "nature" or biology in sexual identity risk perpetuating a culture/nature dualism and result in a distorted understanding of sexuality. See Jonathan Ned Katz, *Gay/Lesbian Almanac: A New Documentary History* (New York: Harper and Row, 1983), 149.

53. Toinette M. Eugene et al., "Appropriation and Reciprocity in Womanist/Mujerista/Feminist Work," *Journal of Feminist Studies in Religion* 8, no. 2 (fall 1992): 91–122; some of the quotes here are taken from the taped recording of this meeting.

54. The previous two gatherings addressed the themes "Inheriting Our Mother's Gardens" and "Troubling the Waters."

55. Joan M. Martin, presider, opening comments of the 1991 Women and Religion/American Academy of Religion session on "Appropriation and Reciprocity in Womanist/Mujerista/Feminist Work." Some comments by the presiders and participants do not appear in the written article in the *Journal of Feminist Studies in Religion*, but are preserved on the taped recording of the original presentation. Where I draw on those comments I cite the tape; otherwise note references are to the written text.

56. Comments of Judith Plaskow, in Eugene et al., "Appropriation and Reciprocity," 105.

57. Ibid., 107.

58. Hunt, from transcript of tape of "Appropriation and Reciprocity."

59. Toinette Eugene, from transcript of tape of "Appropriation and Reciprocity."

60. Eugene et al., "Appropriation and Reciprocity," 100.

61. Ada María Isasi-Díaz, from transcript of tape of "Appropriation and Reciprocity." See also Eugene et al., "Appropriation and Reciprocity," 101.

62. Eugene et al., "Appropriation and Reciprocity," 100. The Lugones citation is from María Lugones, "On the Logic of Pluralist Feminism," in *Feminist Ethics*, ed. Claudia Card (Lawrence: University of Kansas Press, 1991), 35–44.

63. Isasi-Díaz, from transcript of tape of "Appropriation and Reciprocity."

64. Eugene et al., "Appropriation and Reciprocity," 108.

65. Ibid., 117–18.

66. Ibid., 102.

67. Ibid., 103.

68. Ibid., 115–16 (emphasis added).

69. Ibid., 103.

70. Ibid., 103; and from transcript of tape of "Appropriation and Reciprocity."

71. Eugene et al., "Appropriation and Reciprocity," 108. Similarly Ann Wetherilt remarks: "While we cannot attempt to engage the works of, for example, African-American women without attention to questions of race, including our relationship to it, reduction of the voice and expertise of African-American women to the issue of race is but a form of tokenism." Wetherilt, *That They May Be Many*, 22.

Kwok Pui-lan gives a concrete example of acknowledging commonalities and differences in expressing her appreciation of Carter Heyward and other lesbians whose work "has helped me to see marginality in yet another way. Women as sensual, erotic, and passionate sexual beings are never accepted anywhere." While recognizing the commonality of this experience of women in general, she recognizes the particularity of lesbian locations: "Women loving women is taboo even to many women. To a certain extent, our lesbian colleagues and friends are outsiders-within their faith communities, the academy, and women's circles." Yet Kwok also holds her "First World" lesbian friends accountable to mutual, interactive solidarity: "As a heterosexual person from the Third World, I benefit from the privileges associated with heterosexism. As I learn to critique my own heterosexual privileges, I also challenge my lesbian friends to extend their passion and love for women to include all women living in inhuman and oppressive situations in the Third World. Our lesbian colleagues will help us immensely if they could show us the way to channel our erotic power in the struggle for justice for all." Eugene et al., "Appropriation and Reciprocity," 105.

72. Eugene et al., "Appropriation and Reciprocity," 117.

73. Ibid., 120–22. For an insightful analysis of the differing impact of North American patriarchy on white and black women, see Delores Williams, "The Color of Feminism: On Speaking the Black Woman's Tongue," *Journal of Religious Thought* 41 (1986): 42–58. Williams explores the impact of the intersection of gender and race in North American society to reveal positive sides of United States patriarchy for white women that are denied black women.

74. Charlene Spretnak, *States of Grace: The Recovery of Meaning in the Postmodern Age* (San Francisco: HarperCollins, 1991), 8.

75. Ibid., 7.

76. Robert Allen Warrior, "Indian Country's Real Struggles," *Christianity and Crisis* 51, no. 14 (October 7, 1991): 303.

77. See Donna Haraway, "The Promise of Monsters: A Regenerative Politics for Inappropriated Others," in *Cultural Studies*, ed. Lawrence Grossberg, Cary Nelson, and Paula A. Treichler (New York and London: Routledge, 1992), 311–15.

78. Wendell Berry, *The Hidden Wound* (San Francisco: North Point Press, 1989). I develop this further in the discussion of ecological location in chapter 10.

79. Rosemary Radford Ruether, *Gaia and God: An Ecofeminist Theology of Earth Healing* (New York: HarperCollins, 1992), 11.

3. Humanity: Rethinking Human Nature and the Natural

1. See, for example, Boswell, *Christianity, Social Tolerance, and Homosexuality*; Brooten, "Paul's Views"; Countryman, *Dirt, Greed, and Sex*; Victor Paul Furnish, *The Moral Teaching of Paul* (Nashville: Abingdon, 1979); McNeill, *The Church and the Homosexual*; Robin Scroggs, *The New Testament and Homosexuality* (Philadelphia: Fortress, 1983); Gerald T. Sheppard, "The Use of Scripture within the Christian Ethical Debate Concerning Same-Sex Oriented Persons," *Union Seminary Quarterly Review* 40 (1985): 13–35. For a brief survey of how the Bible has been used on the issue of homosexuality within some mainline U.S. denominations, see Scroggs, *The New Testament and Homosexuality*, 2–7.

2. Brooten, "Paul's Views," 62.

3. For an ethical analysis of the connections between sexism and heterosexism and their social implications, see Beverly W. Harrison, "Misogyny and Homophobia: The Unexplored Connections," in *Making the Connections*, ed. Robb, 135–51.

4. One important consequence of this framework for ecological ethics is that in identifying women with nature, nature similarly became understood as passive at the hands of active "man." I discuss this further below.

5. In his study of sex and sexuality in ancient Athens, David Halperin makes the same point about the function of essentially asymmetrical relations in maintaining the social order:

> Not only is sex in classical Athens not intrinsically relational or collaborative in character; it is, further, a deeply polarizing experience:

> It serves to divide, to classify, and to distribute its participants into distinct and radically dissimilar categories. . . . 'Active' and 'passive' sexual roles are therefore necessarily isomorphic with superordinate and subordinate social status; hence, an adult, male citizen of Athens can have legitimate sexual relations only with statutory minors (his inferiors not only in age but in social and political status): The proper targets of his sexual desire include, specifically, women, boys, foreigners, and slaves—all of them persons who do not enjoy the same legal and political rights and privileges that he does.

David Halperin, "Sex Before Sexuality: Pederasty, Politics, and Power in Classical Athens," in *Hidden from History: Reclaiming the Gay and Lesbian Past*, ed. Martin Duberman, Martha Vicinus, and George Chauncey, Jr. (New York: New American Library, 1989), 49. See also David Halperin, *One Hundred Years of Homosexuality: And Other Essays On Greek Love* (New York: Routledge, 1990).

6. Brooten notes that while there was widespread consensus in condemning female homoeroticism overall, there was less consensus on whether both participants should be condemned as acting "unnaturally" or only the active partner who usurped the male role. In framing the debate this way, male writers impose an active/passive schema on female homoerotic relations by defining one female as active and thus "contrary to nature."

7. Brooten, "Why Did Early Christians . . . ?," 84.

8. Ibid.

9. For the classic cross-cultural anthropological study of purity and taboo, see Mary Douglas, *Purity and Danger: An Analysis of Concepts of Pollution and Taboo* (London: Routledge and Kegan Paul, 1966). For a helpful overview of the function of purity in religion, see William E. Paden, *Religious Worlds* (Boston: Beacon Press, 1988), 141–59.

10. At this point Countryman argues for a somewhat different interpretation than Brooten of Paul, purity codes, and sexual behavior. Countryman argues that Paul carefully distinguished between physical purity and sin, and was adamant that Christian community entailed the rejection of physical purity requirements such as included in Israel's sexuality codes for entrance into the community. Hence Countryman does not see Paul arguing in Romans 1:24–27 that same sex love is sin, but rather an example of "dirty" behavior that Christians should forgo. See Countryman, *Dirt, Greed, and Sex*, 97–123.

11. Brooten, "Why Did Early Christians . . . ?," 84.

12. See especially Jeffrey Weeks, *Sexuality and Its Discontents: Meanings, Myths, and Modern Sexualities* (New York: Routledge & Kegan Paul, 1985); and Jeffrey Weeks, *Against Nature: Essays on History, Sexuality, and Identity* (London: Rivers Oram Press, 1991).

13. Weeks, *Sexuality and Its Discontents*, x.

14. Ibid., 8.

15. Ibid., 6.

16. Ibid., 69.

17. Ibid., 79.

18. Weeks notes, however, the limits of sexology in that the sexual subjects such as lesbians and gay men have taken the defined categories and subverted them for their own purposes. Here Weeks follows the French gay scholar Michel Foucault in noting that social functions of power often contain within them the seeds of resistance: "The sexologists sought to find the truth of our individuality, and subjectivity, in our sex. In doing so they opened the way to a potential subjection of individuals with the confines of narrow definitions. But these definitions could be challenged and transformed as much as accepted and absorbed. This suggests that the forces of regulation and control are never unified in their operations, nor singular in their impact. We are subjected to a variety of restrictive definitions, but this very variety opens the possibility of resistance and change." Weeks, *Sexuality and Its Discontents*, 95.

19. John A. W. Kirsch and James D. Weinrich, "Homosexuality, Nature, and Biology: Is Homosexuality Natural? Does It Matter?" in *Homosexuality: Research Implications for Public Policy*, ed. John C. Gonsiorek and James D. Weinrich (Newbury Park, Calif.: Sage Publications, 1991), 13–31.

20. Ibid., 17–18.

21. Quoted in ibid., 14. Note Bryant's assumption that natural sexuality is procreative sexuality. Note also her assumption that animals are lower than humans in the hierarchy of nature ("*even* . . . *animals* don't do what homosexuals do"); homosexuals fall outside this hierarchy altogether.

22. From Kirsch and Weinrich, "Homosexuality, Nature, and Biology," 18.

23. Ibid., 18–19.

24. Ibid., 19.

25. Hunt, *Sexual Behavior in the 1970s* (Chicago: Playboy Press, 1974), 299–300 (emphasis added); cited in Kirsch and Weinrich, "Homosexuality, Nature, and Biology," 21. Note the conflation of maleness with masculinity and femaleness with femininity.

26. Kirsch and Weinrich, "Homosexuality, Nature, and Biology," 21.

27. For example, homosexuality might be an expression of "reproductive altruism." This theory poses that homosexual persons in primitive societies, freed from needing to care for their own offspring, may have given a reproductive advantage to their own kin (whose gene pool they share) in helping to care for their offspring and thus increase the likelihood of their survival and transmission of their common genes—including a gene or combination of genes favoring a homosexual orientation. See Kirsch and Weinrich, "Homosexuality, Nature, and Biology," 26–30.

28. Ibid., 30. Kirsh and Weinrich's search for a genetic factor in homosexual behavior could be seen as essentialist, that is, finding the "essence" of homosexuality in nature. They are quick to point out, however, that genetic preconditioning is only one of many factors involved in developing sexual orientation, and sexual orientation is not the same as the sexual identities societies construct to categorize persons and behaviors. These identities are not determined by genes or nature. For our purposes here, the important implication of their work is its demonstration of the ways socially constructed views of sexuality and humanity influence both the interpretation of scientific reports for public policy issues, and the conduct of science itself.

29. Ibid., 30.

4. Earth: Rethinking Nature and the Nature/Culture Split

1. Merchant, *The Death of Nature*, xxi.

2. Ibid., xvi.

3. Carolyn Merchant, *Radical Ecology: The Search for a Livable World* (New York and London: Routledge, 1992), 43.

4. Merchant, *The Death of Nature*, 2.

5. Ibid., 132.

6. Much historical work still needs to be done on this association. For references, especially on the emergence of antisodomy legislation, see *Historical Perspectives on Homosexuality*, ed. Salvatore J. Licata and Robert Petersen (New York: Stein and Day Publishers and The Hayworth Press, 1981).

7. For historical interpretations of this process, see Michel Foucault, *The History of Sexuality, Volume 1: An Introduction* (New York: Pantheon, 1978), and Weeks, *Sexuality and Its Discontents*. Constructs of nature and what is natural have also shaped intersections of sexuality and race in social domination. For an examination of this in North American history, see John D'Emilio and Estelle B. Freedman, *Intimate Matters: A History of Sexuality in America* (New York: Harper and Row, 1988), esp. 85–108; also Joseph Beam, ed., *In the Life: A Black Gay Anthology* (Boston: Alyson Publications, 1986).

8. Patricia Hill Collins, *Black Feminist Thought: Knowledge, Consciousness and the Politics of Empowerment, Perspectives on Gender, Vol. 2* (Boston: Unwin Hyman, 1990), 170.

9. Beverly Guy-Sheftall, *Daughters of Sorrow: Attitudes Toward Black Women, 1880–1920* (Brooklyn, N.Y.: Carlson Publishing, 1990), 13. Bell Hooks describes this shift in attitudes:

> With the shift away from fundamentalist Christian doctrine came a change in male perceptions of women. Nineteenth-century white women were no longer portrayed as sexual temptresses; they were extolled as the "nobler half of humanity" whose duty was to elevate men's sentiments and inspire their higher impulses. The new image of white womanhood was diametrically opposed to the old image. She was depicted as goddess rather than sinner; she was virtuous, pure, innocent, not sexual and worldly. By raising the white female to a goddess-like status, white men effectively removed the stigma Christianity had placed on them. . . . Once the white female was mythologized as pure and virtuous, a symbolic Virgin Mary, white men could see her as exempt from negative sexist stereotypes of the female.

Bell Hooks, *Ain't I A Woman: Black Women and Feminism* (Boston: South End Press, 1982), 31. For further discussion of the differences between white and black attitudes toward nature and wilderness, and the effects of this on how white and black women were viewed, see Delores Williams, *Sisters in the Wilderness: The Challenge of Womanist God-Talk* (Maryknoll, N.Y.: Orbis, 1993), 110–20.

10. Merchant, *The Death of Nature*, 143–44.

11. Janet Biehl, in her in-depth critique of ecofeminism from the perspective of social ecology, distinguishes between "psycho-biologizing"

ecofeminism, which posits an inherent connection or at least a closer affinity between women and nature than that between men and nature, and "social constructionist" ecofeminism, which argues that cultural views of "natural" affinities between women and nature are socially constructed rather than rooted in women's biological identities. See Janet Biehl, *Rethinking Ecofeminist Politics* (Boston: South End Press, 1991). Because of my grounding in social constructionist insights and method, I draw here primarily on social constructionist ecofeminism.

12. Joan L. Griscom, "On Healing the Nature/History Split in Feminist Thought," *Heresies: "Feminism and Ecology"* 13 (1981): 9.

13. Janis Birkeland, "Ecofeminism: Linking Theory and Practice," in *Ecofeminism: Women, Animals, Nature*, ed. Greta Gaard (Philadelphia: Temple University Press, 1993), 22.

14. Greta Gaard, "Ecofeminism and Native American Cultures: Pushing the Limits of Cultural Imperialism?" in *Ecofeminism: Women, Animals, Nature*, ed. Gaard, 304.

15. Stephanie Lahar, "Roots: Rejoining Natural and Social History," in *Ecofeminism: Women, Animals, Nature*, ed. Gaard, 109.

16. Haraway, *Primate Visions* and *Simians, Cyborgs, and Women*.

17. Haraway, *Primate Visions*, 5.

18. Ibid., 108.

19. Ibid., 111.

20. Ibid., 313–14.

21. Ibid., 55.

22. Robert M. Yerkes, *Chimpanzees: A Laboratory Colony* (New Haven: Yale University Press, 1943), 1; cited in Haraway, *Primate Visions*, 61.

23. Human engineering developed in the early twentieth century as the theory and practice of managing society for production and reproduction. It assumed that human beings could be programmed and manipulated to achieve desired social ends. See Haraway, *Primate Visions*, 62f.

24. Ibid., 62.

25. Ibid., 68.

26. Ibid., 146.

27. Ibid., 156. Haraway uses the terms "narrative" and "narrative fields" to refer to scientific claims and the context and material/ideological practices that produce them to emphasize that "there is no *given* reality be-

neath the inscriptions of science, no understandable sacred center to ground and authorize an innocent and progressive order of knowledge" (12). Scientific accounts are persuasive narratives to describe and interpret the world. For Haraway this does not make them relativistic—these narratives can be tested and verified within a scientific framework and through accepted scientific practices—but it does open them to examination as humanly produced and culturally mediated texts and accounts.

28. As Haraway concludes, "It is no wonder Gulf Oil was proud to sponsor that message." Ibid., 185.

29. Ibid., 281. Note that this is the same logic that informs the mechanism of "othering" through association with nature discussed above.

30. Ibid., 375.

31. Haraway, "The Promise of Monsters," 295–337.

32. Ibid., 298.

33. Ibid., 304.

34. Ibid., 327.

35. See Susanna Hecht and Alexander Cockburn, *The Fate of the Forest: Developers, Destroyers, and Defenders of the Amazon* (London and New York: Verso, 1989).

36. Haraway, "The Promise of Monsters," 310.

37. Ibid., 312.

38. Cited in ibid., 312–13.

39. Ibid., 313.

40. For Haraway's arguments for grounding epistemology and claims for objective knowledge in partial, limited, and accountable perspectives over against masculinist claims for objectivity in detached, totalizing, and universalist perspectives, see "Situated Knowledges," 183–201.

41. Haraway, "Situated Knowledges," 193.

42. Donna Haraway, *Simians, Cyborgs, and Women: The Reinvention of Nature* (New York: Routledge, 1991), 155.

43. Haraway, "The Promise of Monsters," 323–24.

44. Ibid., 324.

45. Haraway, "Situated Knowledges," 199.

46. An example of this is Haraway's preference for "regeneration" over "reproduction" as a metaphor to describe the appropriate role of theory. Regeneration acknowledges the ways we have been damaged as well as new possibilities for growth that do not assume the simple replication of

sameness. This is the danger of a nonrelational view of the authority of sacred texts: they will always demand replication of sameness rather than help us to engage in creating new and more appropriate ways of relating.

5. God: Sensing the Divine in Right Relation

1. For an early naming and exploration of compulsory heterosexuality, see Adrienne Rich, "Compulsory Heterosexuality and Lesbian Existence," in *Blood, Bread, and Poetry: Selected Prose, 1979–1985* (New York: W. W. Norton, 1986), 23–75. For an examination of the ethical implications of compulsory heterosexuality, see Harrison, "Misogyny and Homophobia."

2. Carter Heyward explores justice as right relation in *The Redemption of God: A Theology of Mutual Relation* (Washington, D.C.: University Press of America, 1982). She further develops her theo-ethical analysis in *Touching Our Strength*.

3. Hunt, *Fierce Tenderness.*

4. Ibid., 2.

5. Ibid., 3.

6. Ibid., 20. Hunt's statement here has a decidedly anthropocentric ring to it. Elsewhere she is more attentive to including nonhuman nature in the agency of friendship. She argues, for example, that, "Our relationship to the earth needs to be re-evaluated in light of the friendship extended by both humans and other living parts of the cosmos." Ibid., 14.

7. Ibid., 83.

8. Ibid., 14.

9. Ibid., 98f.

10. Hence Hunt pays careful attention to power dynamics along lines of class, race, nationality, and sexual preference present in friendships between women, as well as the different experiences of friendship that emerge from different communities of women constructed along these lines of social difference. See *Fierce Tenderness*, 12f.

11. Ibid., 105.

12. Ibid., 84.

13. This theme is prominent in the religious traditions of other oppressed groups, such as the African American religious experience. In his examination of the spirituals that emerge from the African slave experience, James Cone notes that the slaves saw Jesus as "their friend and com-

panion in slavery," and that this "enabled the people to bear the trouble and endure the pain of loneliness in oppression." See James Cone, *The Spirituals and the Blues* (Maryknoll, N.Y.: Orbis, 1991), 50.

14. Hunt, *Fierce Tenderness*, 132.

15. Ibid., 84.

16. Ibid., 166.

17. Ibid., 166.

18. Ibid., 166.

19. Lorde, "Uses of the Erotic," 53–59. While I focus in this section on reclaiming the erotic as a constructive grounding for moral agency and knowing the divine, feminists also have paid close attention to the ways the erotic has served as a vehicle of violence and abuse against women and children in Western societies shaped by misogyny, sexism, and heterosexism. For an exploration of some of the issues at stake in reclaiming sexuality and the erotic as a grounding for agency, see Carol S. Vance, ed., *Pleasure and Danger: Exploring Female Sexuality* (Boston and London: Routledge and Kegan Paul, 1984). For recent critiques of feminist theological groundings in the erotic see Kathleen M. Sands, "Uses of the Thea(o)logian: Sex and Theodicy in Religious Feminism," *Journal of Feminist Studies in Religion* 8, no. 1 (summer 1992), 7–33; and Carol J. Adams, "Toward a Feminist Theology of Religion and the State," paper delivered to the annual meeting of the American Academy of Religion, San Francisco, November 1992.

20. Lorde, "Uses of the Erotic," 57.

21. Ibid., 57.

22. Ibid., 54.

23. Ibid., 56.

24. Heyward, *Touching Our Strength*, 3.

25. J. Michael Clark, *A Lavender Cosmic Pilgrim: Further Ruminations on Gay Spirituality, Theology, and Sexuality* (Garland, Tex.: Tangelwüld Press, 1990), 53.

26. See Clark, *A Place to Start*, 62, and Heyward, *Touching Our Strength*, 3–4.

27. Heyward, *Touching Our Strength*, 30.

28. For a creative exploration of this dynamic, see John Fortunato, *Embracing the Exile: Healing Journeys of Gay Christians* (New York: Seabury, 1983).

29. Clark, *Lavender Cosmic Pilgrim*, 6. See also Bell Hooks, "Changing the Margin as a Space of Radical Openness," in *Yearnings: Race, Gender,*

and Cultural Politics (Boston: South End Press, 1990), 145–53. Without romanticizing the margins of society, Hooks argues that oppressed people must recognize the margins as sites of both repression *and* resistance, that "understanding marginality as position and place of resistance is critical for oppressed, exploited, and colonized people" (150).

30. Heyward, *Our Passion for Justice,* 84–85.

31. See for example, J. Michael Clark, *Beyond Our Ghettos: Gay Theology in Ecological Perspective* (Cleveland: Pilgrim Press, 1993), 91.

32. Heyward, *Touching Our Strength,* 23.

33. Clark, *Lavender Cosmic Pilgrim,* 8.

34. Clark, *Beyond Our Ghettos,* x. See also Clark, *A Place to Start,* 61–64, for his development of this claim.

35. Heyward, *Touching Our Strength,* 94.

36. Clark, *Beyond Our Ghettos,* 65–69.

37. J. Michael Clark, *Theologizing Gay: Fragments of Liberation Activity* (Oak Cliff, Tex.: Minuteman Press, 1991), 51; *Lavender Cosmic Pilgrim,* 5.

38. Heyward, *The Redemption of God,* 9; *Touching Our Strength,* 104.

39. Clark, *Lavender Cosmic Pilgrim,* 78. In her theo-ethical examination of the voices and agency of marginalized women, especially women of color, Ann Wetherilt notes that divine agency and human agency are often understood more relationally by women than by men: "Many women do not maintain the dichotomy often made by white Western masculinist theologians between God and self." Wetherilt, *That They May Be Many,* 119–20.

40. Clark, *Theologizing Gay,* 49. For an opposing view of the relationship of the sacred to multiple sex partners in gay sex, see the work of gay theologian Ronald E. Long on "ghetto theology," especially "Toward a Phenomenology of Gay Sex: Groundwork for a Contemporary Sexual Ethic," paper delivered to the annual conference of the American Academy of Religion, Washington D.C., November 1993.

41. Clark, *A Place to Start,* 67, 164–75. I note that considering AIDS solely an example of "natural" evil is another example of a too-rigid dichotomizing of the nature/culture boundary. While AIDS as a virus is a "naturally" occurring phenomenon, there can be no adequate understanding of its spread and impact without examining the socially constructed reality of this pandemic. For an excellent assessment of the human factors in the origins and spread of AIDS, see Randy Shilts, *And the Band Played On: Politics, People, and the AIDS Epidemic* (New York: Penguin, 1988).

42. Clark, *A Place to Start*, 68–69.

43. Ibid., 74.

44. Clark, *Lavender Cosmic Pilgrim*, 35. For a development of this perspective of the divine within the tradition of the theology of the cross, see Dietrich Bonhoeffer, *Letters and Papers from Prison* (New York: Macmillan, 1972).

PART 2: CONVERSATIONS WITH COMPANIONS ON THE ROAD

6. Biblical Theology: An Ethic of Prophetic Stewardship

1. This thesis was posed first by Lynn White Jr., in his influential essay "The Historical Roots of Our Ecologic Crisis." For a survey of other perspectives arguing this critique, see Santmire, *The Travail of Nature*, 1–7.

2. See James A. Nash, *Loving Nature: Ecological Integrity and Christian Responsibility* (Nashville: Abingdon, 1991), 68–92.

3. Douglas John Hall is perhaps best known for his extensive efforts to articulate a theology of the cross indigenous to the North American context. Hall believes that only such a theology can help the North American church face honestly the cultural, sociopolitical, and economic crises that confront us. For his theology of the cross, see especially *Lighten Our Darkness: Toward an Indigenous Theology of the Cross* (Philadelphia: Westminster, 1976); and *Thinking the Faith: Christian Theology in a North American Context* (Minneapolis: Augsburg-Fortress, 1989). For a recent example of Hall's work with the churches on this, see his article, "After the American Dream," in *The Lutheran* 6, no. 3 (March 1993): 26–28.

4. Douglas John Hall, *The Steward: A Biblical Symbol Come of Age* (Grand Rapids, Mich.: Wm. B. Eerdmans, and New York: Friendship Press, 1990), 9.

5. Ibid., 19–23.

6. Ibid., 75, 123. For Hall's methodological reflections on the dialogue between biblical theology and contemporary issues, see *Thinking the Faith*, 262–63.

7. Hall, *The Steward*, 49.

8. Ibid., 161.

9. Hall is a frequent speaker in these programs and commonly cites their publications and statements in his own work, such as the extensive use he makes of the 1983 Vancouver Sixth General Assembly of the

World Council of Churches, which adopted the theme, "Justice, Peace, and the Integrity of Creation" (whose three themes coincide with Hall's final three chapters on contemporary global issues). See *The Steward,* 155–57.

10. Ibid., xi.
11. Ibid., 26.
12. Ibid., 34.
13. Ibid., 42.
14. Ibid., 47.
15. Ibid., 81.
16. Ibid., 91.
17. Ibid., 93.
18. Ibid., 95–96.
19. Ibid., 101.
20. Ibid., 117.
21. See Hall, *Thinking the Faith,* 22–39.
22. Hall, *The Steward,* 123.
23. Ibid., 131.
24. Ibid., 153–54.
25. Ibid., 212.
26. Ibid., 74.
27. Ibid., 75.
28. For a more extensive discussion of the relation of the Bible to theology, see Hall, *Thinking the Faith,* 257–63.
29. Hall, *Thinking the Faith,* 258.
30. In chapter 3 of *The Steward,* "What Time Is It?" for example, Hall draws on such writers as Jacques Ellul, W. H. Auden, George Grant, Kurt Vonnegut, Hugh MacLennan, Abraham Heschel, Elie Wiesel, Emil Fackenheim, Walter Miller, Martin Buber, B. F. Skinner, Michael Kammen, Christopher Lasch, Paul Ricouer, and Graham Greene. The only woman cited in the chapter is Eve (cited with Adam)!
31. Given this selection it may not be coincidental that loss of meaning rather than oppression due to sexism, racism, or homophobia emerges for Hall as the central problem facing North American society.
32. Hall, *The Steward,* 36–37; 1 Corinthians 3:21–23: "Let no one boast about human leaders. For all things are yours, whether Paul or Apollos or Cephas or the world or life or death or the present of the future—all belong to you, and you belong to Christ, and Christ belongs to God" (NRSV, 1989).

33. Hall, *The Steward*, 25–26.

34. Ibid., 27–28.

35. Ibid., 26. Here Hall cites the work of John Calvin and Karl Barth who described reflection on the Christian faith as "theoanthropology" rather than theology alone. See, for example, Karl Barth, *Evangelical Theology*, trans. Grover Foley (New York: Holt, Rinehart & Winston, 1963), 12.

36. Hall, *The Steward*, 42 (emphasis added).

37. Ibid., 118.

38. Ibid., 191.

39. Ibid., 200.

40. Ibid., 205.

41. It is telling that Hall does not reflect further at this point on how "common sense" is culturally determined, and in this case may be strongly shaped by the biblical perspective it is being used to buttress. For Native Americans, in contrast, it is common sense that human beings exist wholly within and in continuity with all of nature.

42. Hall, *The Steward*, 209.

43. Ibid., 207–8.

44. Ibid., 211.

45. Ibid., 186–87. See Charles Birch, "Peace, Justice, and the Integrity of Creation: Some Central Issues for the Churches," in *Report and Background Papers of the Meeting of the Working Group, Potsdam, GDR, July 1986* (Geneva: WCC, Church and Society, 1986), 40f.

46. Interestingly, there is little in Hall's work on the God/nature relationship, which seems always to include humanity as mediating steward. One might ask about the nature of this relationship in the roughly fifteen billion years prior to the appearance of human beings on the globe.

47. Anne M. Clifford, in "Feminist Perspectives on Science: Implications for an Ecological Theology of Creation," *Journal of Feminist Studies in Religion* 8, no. 2 (fall 1992): 83–84, critiques Hall's reinterpretation of humanity's relationship with nonhuman nature on three points related to those I have mentioned above. She writes:

> I find Hall's proposal to be problematic on several fronts: (1) The Lordship of Christ has a history of imperial interpretation that has often muted its association with Jesus' sacrificial love. . . . Since human imperialism vis-à-vis nature has resulted in the ecological crisis we are now facing, the imperialism associated with the symbol of Christ's Lordship deserves more serious attention than Hall

gives it. (2) In Hall's interpretation, humanity is placed in a type of redeemer role where nonhuman nature is concerned. Although Hall does argue for humanity's identification (solidarity) with the rest of creation, identification with Christ, the Lord and Savior of the world, lends itself to a triumphalistic otherness. If the human vocation is conformity to Christ's saving Lordship, are humans still really in solidarity with nonhuman nature? It seems to me that nonhuman nature remains unavoidably subordinate and other in its relationship to its human saviors. (3) In my reading of Hall's argument for dominion as stewardship, I do not find that he sufficiently attends to the fact that we humans *are* the ecological crisis. . . . By what right do we envision ourselves as the stewards of creation, when nonhuman nature could take care of itself without us?

48. See Douglas John Hall, *God and Human Suffering: An Exercise in the Theology of the Cross* (Minneapolis: Augsburg, 1986), 157f.

49. See especially Joanne Carlson Brown and Carole R. Bohn, eds., *Christianity, Patriarchy, and Abuse: A Feminist Critique* (Cleveland: Pilgrim Press, 1989). For an insightful critique of the ramifications of atonement theories and emphasis on the cross for African Americans in light of their historical experience of slavery and surrogacy, see Delores Williams, "Black Women's Surrogacy Experience and the Christian Notion of Redemption," in *After Patriarchy: Feminist Transformations of the World Religions,* ed. Paula M. Cooey, William R. Eakin, and Jay B. McDaniel (Maryknoll, N.Y.: Orbis, 1991), 1–14. For Hall's extensive development of the relationship of the theology of the cross to human suffering, see *God and Human Suffering,* esp. 93–121. For an excellent example of how an ecological ethics may be rooted in the tradition of theologies of the cross while paying closer attention to constructionist insights and historical consequences, see Larry Rasmussen, "Returning to Our Senses: The Theology of the Cross as a Theology for Eco-Justice," in *After Nature's Revolt: Eco-Justice and Theology,* ed. Dieter Hessel (Minneapolis: Augsburg-Fortress, 1992), 40–56.

50. See Hall, *Thinking the Faith,* 69–144.

51. Hall, *The Steward,* 43.

52. Hunt, *Fierce Tenderness,* 150.

53. Hall, *The Steward,* 42.

54. A reminder, that following Haraway, the intent of a constructionist analysis is not to discard all such metaphors and themes as "impure."

Rather, in drawing on the tradition we must be critically aware of its origins and functions in order to empower our agency to shape theology and its metaphors for what is needed today. A problem with uncritically appropriating hierarchical attributes for God is the tendency for them to be used to reinforce models of sovereign authority in the human community. See Wetherilt, *Voices of Struggle*, 67–68 and 169–70, for a helpful critique of views of authority predicated on models of sovereignty.

55. Hall, *The Steward*, 180.

7. Christian Liberalism: An Ethic of Realism and Responsibility

1. The classical statement of Christian realism is Reinhold Niebuhr's *Moral Man and Immoral Society* (New York: Scribner's, 1932). For a liberationist critique of Christian realism, see Charles Kammer III, *Ethics and Liberation: An Introduction* (Maryknoll, N.Y.: Orbis, 1988) 167–88.

2. Nash, *Loving Nature*, 19.

3. Ibid., 94.

4. Ibid., 22.

5. Ibid., 93–94.

6. Nash is very aware of the complexity of this process, which will be discussed further below. See *Loving Nature*, 195f., for his discussion of the pitfalls of the process of moving from general theological affirmations to advocating specific policy proposals.

7. Nash, *Loving Nature*, 16.

8. Ibid., 23.

9. Ibid., 21 (emphasis in original).

10. Ibid., 28.

11. Ibid., 39.

12. Ibid., 41.

13. Ethical guidance in response to these questions in Native American communities often took the form of the "Seven Generations Principle": leaders in the community were obligated to consider the impact of their decisions for seven generations. See Mary E. Clark, *Ariadne's Thread: The Search for New Modes of Thinking* (New York: St. Martin's Press, 1989), 354.

14. Nash, *Loving Nature*, 64–66.

15. Lynn White, "The Historical Roots of Our Ecologic Crisis," *Science* 155 (10 March 1967): 1203–7. The substance of the charge is that Christianity's extreme anthropocentrism, which saw nature existing only to serve humankind, paved the way for Western antinature traditions of

technology and science. Nash responds to the charge in two ways. The first is one of confession of sin and ongoing repentance, for "Christianity does bear part of the burden of guilt for our ecological crisis. . . . The ecosphere has generally been perceived as theologically and ethically trivial" (*Loving Nature*, 22). Yet the ecological complaint of Christianity's critics is also a simplistic overgeneralization. As a corrective Nash offers the following five responses (see Nash, *Loving Nature*, 68–92).

1. Drawing on the work of ecofeminist historian Carolyn Merchant, Nash argues that seeing Christianity as the single or most important cause of the ecologic crisis is too simplistic. The major factors in the emergence of antiecological attitudes developed not in response to Christian views, but out of population pressures, the development of expansionist capitalism, and the triumph of Cartesian and Baconian models of science that stressed nature as mechanism and dominion as human mastery over nature.
2. The ecological complaint presumes an exaggeration of religious influences on culture: "On ecological concerns, the Christian traditions probably affected the various cultural forces at work historically, but they were hardly the historical root of our ecological crisis" (78).
3. The Christian faith has coexisted comfortably and coherently with ecological values at many points in its history, including such groups as the Desert Fathers, the Celtic Saints, St. Francis of Assisi, and the whole tradition of Eastern Orthodoxy. These and other traditions provide significant historical evidence against the charge that Christianity is an inherently antiecological faith.
4. It is a fallacy of misplaced comparison to compare the allegedly ecologically sensitive religions of indigenous communities (such as Native American religious values) with those of complex, technologically advanced societies. Rather, the near-universality of ecological problems through time and across the globe suggests that the roots of the crisis lie not in the theological affirmations themselves, but in human character itself.
5. Finally, even if everything in the ecological complaint were true—which Nash clearly does not believe—the Christian tradition of *Semper Reformanda*, "always to be reformed,"

> means the church is not a closed form but always has the possibility of being reformed under the urgings of the Spirit. It can incorporate new elements and reinterpret its teachings to provide a solid grounding for an ecological ethic.

16. Nash, *Loving Nature*, 94. Nash examines nine classical doctrines of mainstream Christian theological tradition: creation, covenant, divine image, incarnation, spiritual presence, sin, judgment, redemption, and the church. Where relevant to the discussion here, specific points from this section are addressed below.

17. Ibid., 98.

18. Ibid., 100.

19. Ibid., 109.

20. Ibid., 110.

21. Ibid., 119. Note that in contrast to his social analysis of the ecological crisis in chapters 1 and 2, where Nash consistently distinguished between the situation of the affluent and the poor, in his understanding of sin the human condition is collapsed and universalized. This is problematic from a liberationist perspective, where often the root sin of marginalized persons is not pride, but refusal to take ourselves seriously—to see our experience as an integral and valid part of nature and human nature. This is discussed further below.

22. Ibid., 132.

23. Ibid., 137–38.

24. Ibid., 139.

25. Nash notes the current and historical dissensus in Christian thought on love, but nevertheless tries to specify some of the basic implications in an ecological context. He thus writes (*Loving Nature*, 145):

> By definition, Christian love, as disposition and/or deed, is always at least caring and careful service, self-giving and other-regarding outreach, in response to the needs of others (human and otherkind), out of respect for their God-endowed intrinsic value and in loyal response to the God who is love and who loves all. It seeks the other's good or well-being and, therefore, is always other-regarding (only the degree is up for debate). This love is expressed through kindness, mercy, generosity, compassion, justice, and a variety of other commendable qualities. Love is a relational concept and initiative; it seeks to establish connections and build caring relation-

> ships. Its ideal forms are expressed in such terms as *reconciliation, communion, community, harmony,* and *shalom.* These features characterize love in every situation, social and ecological.

As in Nash's discussion of sin (see note 21 above), note again the presumption of agency in this definition of love. Absent is any inclusion of conflict connected to love as justice, such as anger. For an alternative, feminist analysis of love, see Harrison, "The Power of Anger in the Work of Love."

26. Nash, *Loving Nature,* 146.

27. Ibid., 146.

28. Ibid., 147.

29. "'Equal regard' for others assumes ontological equality of worth between the lover and the loved. That equality, however, is not evident in a comparison of humans with other species. Morally relevant differences exist that justify disparate and preferential treatment for humans." Nash, *Loving Nature,* 148.

30. Noticeably absent from this list is love as justice. Nash devotes a full chapter to exploring this relationship. In addition, Nash is clear that love as beneficence can never be a substitute for justice. "In my view, beneficence exceeds the expectations of justice; it begins only when the demands of justice have been satisfied." Nash, *Loving Nature,* 153.

31. In contrast to Hall, Nash is ambivalent about using the traditional Christian metaphor of steward to express humanity's relationship to the rest of nature, precisely because of the tendency to see the relationship anthropocentrically. See Nash, *Loving Nature,* 107.

32. Ibid., 152–60.

33. Ibid., 162–63.

34. Ibid., 166.

35. Ibid., 167.

36. Ibid., 181.

37. Ibid., 182.

38. According to Nash, *Loving Nature,* 186–89, the eight "Biotic Rights" include:

1. The right to participate in the natural competition for existence.
2. The right to satisfaction of their basic needs and the opportunity to perform their individual and/or ecosystemic functions.

3. The right to healthy and whole habitats.
4. The right to reproduce their own kind.
5. The right to fulfill their evolutionary potential with freedom from human-induced extinctions.
6. The right to freedom from human cruelty, flagrant abuse, or frivolous use.
7. The right to redress through human interventions, to restore a semblance of the natural conditions disrupted by human activities.
8. The right to a fair share of the goods necessary for the sustainability of one's species.

39. Ibid., 107.

40. In his final chapter, Nash seeks to spell out a series of six principles or middle axioms that can serve to link theo-ethical norms with concrete politics and policy formation. Before doing so, he lists three qualifications to his proposals. First, he explicitly concentrates on political responsibilities for Christians. Emphasis on value and lifestyle changes are important but insufficient in themselves to confront the seriousness of the ecological crisis. Next, the guidelines serve a practical rather than ideological purpose: they help to manage an enormous amount of concrete proposals and suggestions. Finally, in linking Christian theology and ethics to public policy, Nash explicitly is not proposing a new Christendom or Christianization of the social order, but rather a way for Christians to seek solutions together in alliance with others on whatever common moral grounds that may be found. The six principles state that an ecologically sound and morally responsible public policy must:

1. Resolve the economics/ecology dilemma: Nash recognizes the dilemma is real, with ancient roots and unprecedented proportions today. Neither economics nor ecology can any longer be compartmentalized, for they interpenetrate each other and confront us with ecological dilemmas. A critical issue is the moral ambiguity of economic growth: it has made possible numerous positive gains and is critical to attain social justice in the Third World, but it is also culpable of many antiecological outgrowths and is premised on nonsustainable and unecological assumptions. The ideology of growth, nearly unquestioned in governmental centers of power today, is

utopian illusion at best in its assumptions about the indestructibility and inexhaustibility of nature for human use.

While still constrained by whether such change is realistic in today's climate, Nash argues that economic conversion to ecological sustainability is a social, economic, and ecologic necessity. A major moral problem with applying models of economic equilibrium globally is that it does not fit Third World nations that need some economic growth to satisfy basic human needs. Hence models of sustainable development are critical for poorer, less-developed nations.

2. Include public regulations that are sufficient to match social and ecological needs: Nash argues that self-regulation and market competition are inadequate to provide environmental protection and so government must provide the needed regulations. Against widespread classical free-market ideas that examine businesses only in terms of productivity and profitability, Nash argues for a more comprehensive standard for evaluation of performance: "Economic enterprises and systems can be evaluated economically on the basis of their productivity and profitability, but they should also be evaluated socially and ecologically on the basis of their contributions and harms to the well-being of humans and other species in our interdependent relationships" (205). The kind and extent of public regulations that are needed are best determined contextually with other relevant moral considerations.
3. Protect the interests of future generations: Ecological policy based on reasonable anticipations of consequences is critical to respecting the well-being of future human and other life. Against "generational isolationism," Nash argues that our moral community extends not only in space but also in time. Hence future generations can be said to have *anticipatory rights*, and every current generation, therefore, has *anticipatory obligations*. The key moral challenge facing us today is to prevent intolerable choices between present and future generations. To aid in this, Nash presents a list of seven ecological responsibilities.
4. Provide protection for nonhuman species, ensuring the conditions necessary for their perpetuation and ongoing evolution:

For Christians, being guardians of biodiversity is an authentic expression of the vocation of dominion. Nash takes a realist approach in justifying this principle, arguing his case on both explicitly anthropocentric and ecocentric grounds. He concludes that, like responsibility to future generations, guardianship of biodiversity will require a much broader and more radical concept of political representation that includes the nonhuman biosphere.

5. Promote international cooperation as an essential means to confront the global ecological crisis: Nash argues that the current context has made national isolationism impossible, national self-sufficiency obsolete, and national security dependent in part on ecological security. In this context the international community is the lowest social/political unit potentially capable of responding effectively to the global ecological crisis. The dilemma is that while the world is one ecologically, it is fractured politically. Yet global solidarity must soon move beyond future vision to political reality if the crisis is to be confronted successfully.
6. Pursue ecological integrity in intimate alliance with the struggles for social peace and justice: This final principle is an explicit restatement of what is implied throughout Nash's argument: holistic strategies are needed to respond to holistic, interrelated realities. While World Council of Churches documents on "Justice, Peace, and the Integrity of Creation" have been "notable for homiletical exhortations, rather than empirical and ethical analyses," there nevertheless is increasing attention being paid to the need to approach these issues together (218). Nash recognizes that strategically it is impossible for advocates to focus on all aspects of social and ecological crises simultaneously. The key input of this principle is to approach problems with awareness of the other dimensions and to act in ways that at the very least do not cause or aggravate other social and ecological problems.

41. Nash, *Loving Nature*, 221.

42. Ibid., 133.

43. See, for example, contemporary defenses of Western capitalist democracy using a Christian realist analysis in Michael Novak, *The Spirit*

of Democratic Capitalism (New York: Simon and Schuster, 1982); Richard John Neuhaus, *The Naked Public Square: Religion and Democracy in America* (Grand Rapids, Mich.: Wm. B. Eerdmans, 1984); Robert Benne, *The Ethic of Democratic Capitalism* (Philadelphia: Fortress, 1981).

44. Nash, *Loving Nature*, 217.

45. An example of Nash's flexible form of realism is his discussion of the issue of recreational hunting. Seeing such hunting as "morally dubious" at best, he nevertheless counsels on strategic grounds against any present efforts to outlaw sports hunting in the United States except for certain species threatened with extinction. See *Loving Nature*, 211–13.

46. Nash, *Loving Nature*, 100–101, 106.

47. Ibid., 80.

48. For discussions of some of the connections between persecution of earth- and body-centered persons as heretics within Christianity, see for example, Merchant, *The Death of Nature*, 127–48; Mary Daly, *Gynecology: The Metaethics of Radical Feminism* (Boston: Beacon Press, 1978), 177–221; V. L. Bullough, "Heresy, Witchcraft and Sexuality," *Journal of Homosexuality* 1, no. 2 (winter 1974–75): 185–202.

49. Nash, *Loving Nature*, 227n. 14.

50. For example, Nash limits his discussions of justice to Western concepts of justice as distributive justice. See *Loving Nature*, 167–68.

51. Ibid., 146.

52. Ibid., 149.

53. Ibid., 119.

54. Ibid.

55. This is true at least with respect to the human community. It is quite possible, of course, that one can be taking more than one's due in one context—say with respect to other parts of nature—while accepting less than one's due within humanity. To understand multiple locations of exclusion and privilege with respect to both the human community and to the rest of nature, I propose expanding the concept social location to ecological location, which I develop in chapter 10.

56. Nash, *Loving Nature*, 62.

57. Ibid., 141.

58. Ibid., 95.

59. Nash's narrow view of dominion is developed further below.

60. Nash, *Loving Nature*, 106 (emphasis added).

61. Birkeland, "Ecofeminism," 19. For Nash's arguments against an

ethics based primarily on responsibilities and duties, see *Loving Nature*, 170. In his arguments for a rights-based approach to constrain human actions (as opposed to certain feminist approaches that emphasize empowering moral agency), Nash reflects his realist roots.

62. Greta Gaard, "Living Interconnections with Animals and Nature," in *Ecofeminism: Women, Animals, Nature*, ed. Gaard, 2–3.

63. Nash, *Loving Nature*, 130. For Ruether's views on death and ecological ethics, see Rosemary Radford Ruether, *Gaia and God: An Ecofeminist Theology of Earth Healing* (New York: HarperCollins, 1992), 81–84.

64. For an exploration of these themes, see Clark, "Erotic Empowerment, Ecology, and Eschatology."

65. Nash, *Loving Nature*, 96.

66. Ibid.

67. Ibid., 115.

68. Ibid., 111–16, 108–9.

69. Ibid., 105.

70. Ibid., 105–6.

71. Ibid., 108.

72. Ibid., 107.

73. Ibid., 136 (emphasis added).

74. Ibid., 143.

75. Harrison, "The Power of Anger in the Work of Love," 13. Ann Wetherilt notes the importance of recovering an embodied epistemology for theology and ethics: "The identification of knowing with contemplation, evident in Platonic philosophy and its descendants, has had devastating effects on those persons closely identified by dominating discourse as more closely associated with 'nature' and material reality. The association of knowledge with disembodied mind and/or soul . . . must be overcome if multiple voices are to bring their critical wisdom and insight to bear on theo-ethical discourse." Wetherilt, *That They May Be Many*, 89.

76. Harrison, "The Power of Anger in the Work of Love," 14. Carter Heyward makes the same point in *The Redemption of God* (221): "Our passion as lovers is that which fuels both our rage at injustice—including that which is done to us—and our compassion, or empathy with those who violated us, hurt us, and would even destroy us. Rage and compassion, far from being mutually exclusive, belong together."

77. See Sallie McFague, *Models of God: Theology for an Ecological, Nuclear Age* (Philadelphia: Fortress, 1987), esp. 69–77.

78. Nash, *Loving Nature*, 115, 233n. 69.

79. On the need for an ecological Christianity to be open to learning from other systems and traditions see Primavesi, *From Apocalypse to Genesis*, 20–21.

80. Ruether, *Gaia and God*, 253.

81. John B. Cobb Jr., *Sustainability: Economics, Ecology, and Justice* (Maryknoll, N.Y.: Orbis, 1992), 13.

82. While Nash draws on the constructionist work of scholars such as Carolyn Merchant, the few explicit acknowledgments he makes of constructionist insights are dismissive, such as, "The strict constructionist and sectarians who yearn for ideological and verbal purity in the environmental movement do well to pay less attention to words and more to values and commitments." Nash, *Loving Nature*, 107. The reason constructionists "pay attention to words" is precisely to reveal the values and commitments that underlie them. Nash's dismissal of those who do not accept the terms of the dominant discourse as sectarian may also reveal his own allegiance to these terms.

8. Process Theology: An Ethic of Liberating Life and Sustainability

1. For a basic introduction to process theology, see John B. Cobb Jr. and David R. Griffin, *Process Theology: An Introductory Exposition* (Philadelphia: Westminster, 1976).

2. Cobb, *Sustainability*, 4.

3. Ibid.

4. For a detailed examination of the economics/ecology relationship, see Herman E. Daly and John B. Cobb Jr., *For the Common Good: Redirecting the Economy Toward Community, the Environment, and a Sustainable Future* (Boston: Beacon Press, 1989).

5. Birch and Cobb, *The Liberation of Life*, 10.

6. Cobb, *Sustainability*, 5.

7. For examples of Cobb's dialogue with feminist and liberation concerns, see John B. Cobb Jr., "Points of Contact Between Process Theology and Liberation Theology in Matters of Faith and Justice," in *Process Studies* 14, no. 2 (summer 1985): 124–41; and John B. Cobb Jr., "Feminism and Process Thought: A Two-Way Relationship," in *Feminism and Process Thought: The Harvard Divinity School / Claremont Center for Process Studies Symposium Papers*, ed. Sheila G. Davaney (Lewiston, N.Y.: Edwin Mellen Press, 1981), 32–61.

8. Birch and Cobb, *The Liberation of Life*, 20.

9. Ibid., 27.
10. Ibid., 36.
11. Ibid., 51.
12. Ibid., 59.
13. Ibid., 65.
14. Ibid., 70.
15. Ibid., 75.
16. Ibid., 83.
17. Ibid., 86.
18. Birch and Cobb stress that this does not mean there are no distinctions between living and nonliving. They list three criteria: living things locally reverse the increase in entropy; living things have the ability to replicate or reproduce; and living things have the ability to reproduce and transfer information. The point is that any attempt to draw a fixed line is arbitrary, fitting some conditions well, and others poorly. Birch and Cobb, *The Liberation of Life*, 93.
19. Ibid., 98.
20. Ibid., 109.
21. Ibid., 120.
22. Ibid., 123. Note the similarity here to Haraway's understanding of nature as a matrix of actors and actants, human and otherkind, with different forms of agency and subjectivity.
23. Ibid., 131.
24. Following Charles Hartshorne ("Physics and Psychics: The Place of Mind in Nature," in *Mind and Nature*, ed. J. B. Cobb and D. R. Griffin [Washington, D.C.: University Press of America, 1977], 92–93), Birch and Cobb list seven philosophical advantages to this view *(The Liberation of Life*, 133–34):

1. One avoids the problem of how "mere matter" can produce life and minds. Instead the problem becomes, how did higher types of experience develop from lower types?
2. One does justice to the fact that between "life-less" matter and primitive forms of living there is only a relative difference, not an absolute one.
3. Through generalizing the concepts of memory and perception one can construe causal connections between events. What is remembered and perceived is internally related to the remembering and perceiving experience—that is, causes

are *internally* related to effects. This awareness of the connectedness of events is lacking in both materialism and dualism.

4. It is possible now to explore how mind and body are related in animals. The mind/body relationship can be seen as one of sympathy. We share in the feelings of our cells, and they respond to our emotional life.
5. The problem of relating primary and secondary qualities is resolved. Primary qualities are what physics terms causal spatiotemporal relations. Secondary qualities characterize events internally.
6. One can better account for the relationship between perception and behavior and explain why animals act as they do. This entails recognizing both the objective side of animal behavior (such as instinct developed and remembered through evolution) and the subjective side (certain things are experienced subjectively by the animal as pleasant or unpleasant).
7. Thinking of material reality as mere matter, mere mindless and feelingless stuff limits the things for which we can experience sympathy or empathy. But if one recognizes experience in physical nature, then there is a universal community for mutual participation in sympathy.

25. Birch and Cobb, *The Liberation of Life,* 146.

26. Aldo Leopold, "The Conservation Ethic," *The Journal of Forestry* 31 (1933): 643, cited in Birch and Cobb, *The Liberation of Life*, 146.

27. Birch and Cobb, *The Liberation of Life,* 151.

28. Ibid., 152.

29. Ibid., 161.

30. Ibid., 170.

31. Ibid., 201.

32. Ibid., 205.

33. See Cobb, *Sustainability*, 77–78.

34. Birch and Cobb, *The Liberation of Life*, 262.

35. Ibid., 274.

36. Ibid., 277.

37. Ibid., 331.

38. For a detailed explication of Cobb's understanding of postmodernism, see his "Postmodern Christianity in Quest of Eco-Justice," in *After Nature's Revolt,* ed. Hessel, 21–39.

39. Ibid., 25.

40. Ibid., 21–22.

41. Ibid., 186n. 3.

42. Ibid., 27.

43. John B. Cobb Jr., "The Role of Theology of Nature in the Church," in *Liberating Life*, ed. Birch et al., 268.

44. Hence Cobb argues, "When we think self-consciously as Christians we should be free to think as clearly and as vigorously, as openly and as honestly, as it is humanly possible to think. Whether our help comes from those who are called scientists, or those who are called philosophers, or those who are called Hindus, is a quite secondary consideration. It is as important to liberate theology to pursue saving truth wherever it can be found as to liberate particular groups of people from oppression." "The Role of Theology of Nature in the Church," 272.

45. Cobb, "Postmodern Christianity," 38.

46. Cobb, "The Role of Theology of Nature in the Church," 272.

47. Eugene et al., "Appropriation and Reciprocity," 97.

48. The exception is Cobb's attention to nonhuman voices in reorienting ethics from an anthropocentric to an ecocentric grounding.

49. Cobb, *Sustainability*, 94.

50. Ibid., 92.

51. Ibid., 93. Further aspects of Cobb's development of a biblically based theocentric perspective are explored below.

52. For Cobb's reflections on what process theology can learn from liberation theologies see his "Points of Contact Between Process Theology and Liberation Theology."

53. Nash makes passing reference to this in his section on "Resolving the Economics-Ecology Dilemma," but it does not seem to inform his analysis in a primary or integrated way. See Nash, *Loving Nature*, 197–203.

54. It is clear that Cobb is very familiar with Latin American, black, and feminist theologies, and he has acknowledged the contributions and challenges they make to process theology. He is alert to the impact of differing social locations, and that practitioners of process theology are nearly all white, middle-class, and North American. His interest is in developing a process theology that is responsive to the concerns of liberation theologies, but puts its primary energies into developing a compelling vision for the middle class that would allow and facilitate needed social change. This may be appropriate for his social location and audience, but it must be transformed for a liberationist ethics.

55. The phrase, "insurrection of subjugated knowledges" was coined by the Foucault in his focus on resistance to dominant forms of discourse; see Michel Foucault, *Power/Knowledge: Selected Interviews and Other Writings, 1972–1977* (New York: Pantheon Books, 1980), 81. For an example of a feminist theological paradigm that builds on Foucault's insights, see Sharon D. Welch, *Communities of Resistance and Solidarity: A Feminist Theology of Liberation* (Maryknoll, N.Y.: Orbis, 1985), esp. 19f.

56. For an excellent discussion of environmental racism, see Benjamin F. Chavis, "A Place at the Table: A Sierra Roundtable on Race, Justice, and the Environment," *Sierra* 78, no. 3 (May/June 1993): 51–58, 90–91. See also *Toxic Struggles: The Theory and Practice of Environmental Justice*, ed. Richard Hofrichter (Philadelphia: New Society Publishers, 1993).

57. Cobb addresses many of these issues in his article, "Points of Contact," which shows that he is not unaware of their importance; still they are not reflected in any detail in his writings on ecological ethics.

58. Tom Driver makes a similar point to Cobb on agency in nature and the ability of organisms to transcend their environment while remaining fully part of nature: "All organisms transcend their environment precisely by their immersion in it—that is, to the degree they act in it. Their acting is a measure of their engagement. . . . Far from being free of the environment, humans are free in it, to it, and with it, provided only that they recognize how they and their environment coinhere." Tom F. Driver, *Patterns of Grace: Human Experience as Word of God* (San Francisco: Harper and Row, 1977), 164.

59. Birch and Cobb, *The Liberation of Life*, 189.

60. Ibid., 192.

61. Ibid., 196–97.

62. Ibid., 197.

63. Cobb, *Sustainability*, 125.

64. Birch and Cobb, *The Liberation of Life*, 199.

65. Cobb, *Sustainability*, 11.

66. Ibid., 14–15. It is striking that although Birch and Cobb list a commitment to Christianity as one of the common elements that facilitated their collaboration in writing *The Liberation of Life*, Christ is mentioned nowhere in that book.

67. Cobb, *Sustainability*, 93.

68. Ibid., 110.

69. Ibid., 110.

70. Nash, *Loving Nature*, 130.

71. For example, see Pablo Richard et al., *The Idols of Death and the God of Life: A Theology* (Maryknoll, N.Y.: Orbis, 1983).

72. This point was raised by Carol Adams at a conference titled "Theological Education to Meet the Environmental Challenge: Toward Just and Sustainable Communities," held at Stony Point Center, New York, May 13–16, 1993. Cobb develops his own pro-choice position within a process emphasis on God as Life by refusing to grant absolute, equal value to all human life. While both the mother and the fetus have some degree of richness of experience and capacity for richness of experience, that of the mother's is greater than the fetus and deserves ethical consideration in conflicts over whether to carry a pregnancy to term. Reflection about abortion is ethical (and ecological) when it considers the rights of the mother, the father, the fetus, other relatives, and the community. See Birch and Cobb, *The Liberation of Life*, 166–67.

73. How best to include the voices of creatures other than humans remains problematic, as Haraway's discussion on articulation versus representation illustrates. Cobb suggests that animals can best be represented in human discussions by those who know them best and "have devoted themselves to studying how animals suffer at human hands." Cobb, "The Role of Theology and Nature in the Church," 267.

74. See Haraway, "Situated Knowledges."

75. Cobb, *Sustainability*, 14.

76. This is not, however, to minimize the emphasis on justice and sensitivity to the poor and oppressed that permeates Cobb's writings.

9. Ecofeminist Theology: An Ethic of Justice, Interrelatedness, and Earth Healing

1. Ruether, *Gaia and God*, 2.

2. See Irene Diamond and Gloria Feman Orenstein, eds. *Reweaving the World: The Emergence of Ecofeminism* (San Francisco: Sierra Club Books, 1990), ix.

3. Ruether, *Gaia and God*, 2.

4. Ibid., 258.

5. Ibid., 4.

6. Ibid., 1–2.

7. Ibid., 3.

8. Ibid., 3–4.

9. See in particular Rosemary Radford Ruether, *Women-Church: Theology and Practice of Feminist Liturgical Communities* (San Francisco: Harper and Row, 1986), esp. 1–95.

10. Ruether, *Gaia and God,* 206.

11. Ibid., 9, 11.

12. Ibid., 2–3, 11–12.

13. Ibid., 11, 230.

14. Ibid., 15. Note the similarity of Ruether's understanding of creation myths to Donna Haraway's view of origin stories and how they function to sacralize social relations to the benefit of some and to the detriment of others.

15. Ibid., 15–16.

16. Ibid., 17.

17. Ibid., 18.

18. Ibid.

19. In a footnote Ruether acknowledges the efforts of interpreters such as Phyllis Trible who argue that the designation of Eve as a helpmate indicates a relationship of mutuality. Ruether maintains that this ignores the design of the story, which defines the helpmate status in a one-sided auxiliary way. This intent has been made explicit through later rabbinical interpretation. Ibid., 277n. 12.

20. Ibid., 22. Ruether observes that this distinction between making and begetting returns in Christian theology as the primary distinction between the generation of the divine in the Trinity and God's creation of the world.

21. Ibid., 23–24.

22. Ruether notes that neither the Genesis nor Platonic accounts contain the dogma of creatio ex nihilo; both assume that some kind of primordial matter was present in the beginning. Ibid., 26.

23. Ibid., 27.

24. See Santmire, *The Travail of Nature.*

25. Ruether, *Gaia and God,* 28.

26. Ibid., 29.

27. Ibid., 31.

28. Compare to Nash's understanding of human beings as the altruistic predator; Nash, *Loving Nature,* 147f.

29. Ruether, *Gaia and God,* 31.

30. Ibid., 39.

31. Ibid., 45.
32. Ibid., 47.
33. See, for example, Haraway, "The Promise of Monsters."
34. Ruether, *Gaia and God,* 86. Nash makes a similar point in arguing that human moral culpability for the ecological crisis also implies moral capability to respond. See Nash, *Loving Nature,* 32.
35. Ruether, *Gaia and God,* 111.
36. Ibid., 115.
37. Ibid., 116.
38. As discussed earlier, this location of ethics within purity codes that separate holy from unholy based on gender roles has been especially disastrous to homosexual persons as it has been retained and amplified by dominant strands of Christianity and Western morality to justify systematic exclusion and repression of anyone engaging in homoerotic behavior. See chapter 3.
39. Ruether, *Gaia and God,* 123.
40. Ibid., 139.
41. Ibid., 140.
42. Ibid., 143.
43. Ibid., 8.
44. Ibid., 145.
45. Ibid., 151–52.
46. Ibid., 167.
47. Ibid., 183. Ruether here is drawing on the work of Page DuBois, *Centaurs and Amazons: Women and the Pre-History of the Great Chain of Being* (Ann Arbor: University of Michigan Press, 1982), 49–77.
48. Ruether, *Gaia and God,* 197.
49. Ibid., 200.
50. Ibid., 227.
51. Ibid., 253.
52. Ibid., 251.
53. On these themes, see Huey-li Li, "A Cross-Cultural Critique of Ecofeminism," 272–94, and Greta Gaard, "Ecofeminism and Native American Cultures: Pushing the Limits of Cultural Imperialism?" 295–314, in *Ecofeminism: Women, Animals, Nature,* ed. Gaard.
54. Ruether, *Gaia and God,* 86.
55. For example, see Rosemary Radford Ruether, "Homophobia, Heterosexism, and Pastoral Practice," in *Homosexuality in the Priesthood and the Religious Life,* ed. Jeannine Gramick (New York: Crossroad, 1989), 21–35.

56. For studies of the incidence and role of the *berdache* in Native American cultures, see Will Roscoe, *The Zuni Man-Woman* (Albuquerque: University of New Mexico Press, 1991); Walter L. Williams, *The Spirit and the Flesh: Sexual Diversity in American Indian Cultures* (Boston: Beacon Press, 1992); Will Roscoe, ed., *Living the Spirit: A Gay American Indian Anthology* (New York: St. Martin's Press, 1988). Williams in particular discusses the disastrous encounter of European Christianity with its compulsory heterosexual and monogamous two-gender framework with the third-gender berdache tradition that could not fit in. Potential contributions of the berdache tradition to ecological ethics is discussed in chapter 11.

57. Ruether, *Gaia and God*, 5.

58. Ruether cites the principles of interhuman justice and sustainability as two reasons for moving away from a meat-centered diet: "Both interhuman justice and a sustainable relation to the rest of nature demands that humans return to feeding themselves primarily from the first stage of the food chain, from the food produced by plants." Ibid., 52.

59. Interestingly, Cobb and Birch, in positing God as Life, make no reciprocal recovery of what Ruether here terms the masculine dimension of God: the one who mandates justice and intercedes on behalf of the downtrodden. Their gentle, persuasive process God seems closely related to Ruether's Gaia, without the complement of her "God."

60. Ruether, *Gaia and God*, 255.

61. Ibid., 252–54, 273.

62. Ibid., 5.

63. Ibid., 222.

64. Ibid., 207.

65. Ibid., 273.

66. See especially Ruether, *Sexism and God-Talk: Toward a Feminist Theology* (Boston: Beacon Press, 1983), 116–38.

PART 3. GAY AND GAIA: TOWARD AN EROTIC ETHIC OF ECOJUSTICE

10. Shifting Our Grounding: From Social Location to Ecological Location

1. Worster, *The Wealth of Nature*, 3.

2. Esther Boulton Black, *Stories of Old Upland* (Upland, Calif.: Chaffey Communities Cultural Center, 1979), ii.

3. There are other locations, to be sure, such as our planetary or cosmic locations. Such greatly expanded contexts are probably not relevant for social and ecological ethics, aside from the appropriate cosmic humility they generate from recognizing where human beings fit into the cosmic schema. For a discussion of the term "relevant whole" as the necessary context for ethics, see James Gustafson, *Ethics from a Theocentric Perspective, Volume Two: Ethics and Theology* (Chicago: University of Chicago Press, 1984), 219–50.

4. Lahar, "Roots," 96 (emphasis added).

5. Yi-Fu Tuan, *Topophilia: A Study of Environmental Perception, Attitudes, and Values* (New York: Columbia University Press, 1974, 1990), 4.

6. Ibid., 66–69.

7. For a particularly dramatic example of this, see Alfred Crosby's study of the ecological impact of European global migrations, *Ecological Imperialism: The Biological Expansion of Europe, 900–1900* (New York: Cambridge University Press, 1986).

8. Jay B. McDaniel argues that while human beings are unique in having a *moral calling,* other animals may exemplify some degree of moral agency, such as when elephants aid an injured member of their own species, duck or primates adopt and raise members of other species, or porpoises aid humans in distress. See Jay B. McDaniel, *Of God and Pelicans: A Theology of Reverence for Life* (Louisville: Westminster John Knox, 1989), 60–61.

9. Merchant, *Ecological Revolutions,* 8.

10. Ibid., 23.

11. Ibid., 5 (emphasis added).

12. See Lahar, "Roots," 105.

13. For an excellent discussion of the effects of militarization and ecocide in Central America, see Bill Weinberg, *War on the Land: Ecology and Politics in Central America* (London and New Jersey: Zed Books, 1991).

14. Quoted in Patricia S. Cale, "Environmental Lessons from Central America," *Iowa Energy Bulletin* (June/July 1995), 4.

15. Berry, *The Hidden Wound,* 88.

16. Ibid., 105–8 (emphasis in original).

17. Cited in Robert D. Bullard, ed., *Confronting Environmental Racism: Voices from the Grassroots* (Boston: South End Press, 1993), 3.

18. William Cronon, *Changes in the Land: Indians, Colonists, and the Ecology of New England* (New York: Hill and Wang, 1983), 6.

19. Ibid., 15.

20. Worster, *The Wealth of Nature*, 18.

21. Merchant, *Ecological Revolutions*, 2.

22. This is a key insight of the emerging field of bioregionalism, which argues that human culture should adapt to the contours of the bioregions—geographic areas having common characteristics of watersheds, climate, soil, and flora and fauna—rather than forcing nature to adapt to human culture. See Carolyn Merchant, *Radical Ecology: The Search for a Livable World* (New York and London: Routledge, 1992), 217–22, and *Home! A Bioregional Reader*, ed. Van Andruss, Christopher Plant, Judith Plant, and Eleanor Wright (Philadelphia, Santa Cruz, Calif., and Gabriola Island, B.C.: New Society Publishers, 1990).

23. For a creative and helpful exploration of the notion of "ecological identity," see Mitchell Thomashow, *Ecological Identity: Becoming a Reflective Environmentalist* (London and Cambridge, Mass.: MIT Press, 1995).

24. This has come to be symbolized in gay and lesbian culture by the rainbow flag, signifying that lesbians and gay men come from every racial, ethnic, and class group.

25. Berry, *The Hidden Wound*, 88–89.

26. J. Michael Clark explores many of these themes in his gay ecotheology, *Beyond Our Ghettos*. For his discussion of the many types of psychological ghettos we maintain, see esp. 2–4.

27. Clark, *Beyond Our Ghettos*, 114–15 (emphasis in original).

11. Erotic Ecology: Interconnection and Right Relation at All Levels

1. James Nelson, *Body Theology* (Louisville: Westminster/John Knox Press, 1992), 186.

2. Ibid., 23.

3. Carter Heyward, "Embodying the Connections: What Lesbians Can Learn from Gay Men about Sex and What Gay Men Must Learn from Lesbians about Justice," in *Spirituality and Community: Diversity in Lesbian and Gay Experience*, ed. J. Michael Clark and Michael L. Stemmeler (Las Colinas, Tex.: Monument Press, 1994), 138, 140.

4. Walker, *The Color Purple*, 167.

5. As gay theologian Ron Long has argued in critique of J. Michael Clark's close identification of god/dess with nature, "Nature can be quite ruthless with the interests and aspirations of individuals. . . . Not only are

the interests of the whole simply not identical with the interests of (some) individuals, but the interests of some individuals (mice, for example) are at variance with the interests of other individual (in this case, cats); not all individual ideals are existentially possible." Ronald E. Long, "Gay Theology: Almost Home," paper presented to the annual meeting of the American Academy of Religion, Chicago, November 21, 1994.

6. Clark, *Beyond Our Ghettos*, 16.

7. For examples of this in an ecofeminist ethics, see Birkeland, "Ecofeminism." For a gay ecotheological perspective, see Clark, *Beyond Our Ghettos*, 71–72.

8. Clark, *Beyond Our Ghettos*, 71.

9. I limit my observations to capitalist societies where commoditization of sex and the erotic have gone hand-in-hand with increased sexual violence. This is not to suggest that the eroticization of violence does not occur in other contexts, but the correlation between the introduction of international capitalism and its partner, militarism, and an increase in sexual exploitation of women and children and the eroticization of violence has been documented in contexts as diverse as Thailand, the Philippines, the former socialist countries of Eastern Europe, and Brazil. For an ethical analysis of the impact of these changes on women's lives in Asia, see Elizabeth Bounds, "Sexuality and Economic Reality: A First World and Third World Comparison," in *Redefining Sexual Ethics: A Sourcebook of Essays, Stories, and Poems,* ed. Susan E. Davies and Eleanor H. Haney (Cleveland: Pilgrim Press, 1991), 131–44.

10. See for example, Adams, "Toward a Feminist Theology of Religion and the State." For an excellent overview of the debate around the eroticization of violence and grounding feminist theo-ethics in reclaiming eros, see Anne B. Gilson, "Eros Breaking Free: Interpreting Sexual Theo-Ethics" (Cleveland: Pilgrim Press, 1995), 85–106.

11. To emphasize this point, J. Michael Clark uses the terms *hetero*patriarchy and *hetero*male privilege to extend ecofeminist analysis of sexism to include heterosexism: "Gay theology argues that not male privilege generally, but male privilege specifically defined as *both* not feminine and *not homosexual,* is the real determinant of western culture and theology. To reflect that difference, 'heteropatriarchy' evokes the idea of straight male privilege." *Beyond Our Ghettos,* x.

12. Here the experience of lesbians is distinct from that of gay men, for lesbians are both devalued as women and disvalued as homosexuals,

while gay men as men—and particularly white gay men who remain in the closet—still have access to male privilege unavailable to lesbians as women. While insisting on attention to heterosexism as key to understanding how Western white capitalist heteropatriarchy exploits nature and whole groups of people, an ecojustice ethics of Gay and Gaia also must pay close attention to the particularity of experience within lesbian and gay male communities along lines of social and cultural difference such as gender, race, class, and physical ability.

13. Clark, *Beyond Our Ghettos*, 16–17 (emphasis in original).

14. Ibid., 17 (emphasis in original).

15. Even this claim must be nuanced in light of biotechnology's increasing ability to create living creatures that may have a solely destructive purpose, such as pathogens for biochemical weapons. And Donna Haraway's attention to cyborgs as creature/machine hybrids that blur the line between living creatures and created artifacts further complicates the claim for intrinsic value in all ("naturally" occurring) living beings.

16. Clark, *Beyond Our Ghettos*, 70.

17. Ibid., 72.

18. Ibid., 79.

19. I state this even though it pains this "geologian" not to recognize absolute, equal intrinsic value in all things geological.

20. Long, "Gay Theology: Almost Home," 6.

21. This is illustrated by the example Clark used earlier in the debate over logging in the Pacific Northwest. While Clark focused on the short-term profits that logging companies may make, the loggers and families involved who stand to lose their livelihoods, homes, and communities can make a compelling case for further logging on the basis of human necessity. This indicates the need for additional criteria that can guide resolution of such ecological/economic dilemmas.

22. McDaniel, *Of God and Pelicans*, 79.

23. Note the similarity of this to Cobb's position, reflected in his interpretation of Genesis 1 where God *sees* that creation is good.

24. McDaniel, *Of God and Pelicans*, 80.

25. Ibid., 80–81.

26. Ibid., 84.

27. See Jay B. McDaniel, *Earth, Sky, Gods, and Mortals: Developing an Ecological Spirituality* (Mystic, Conn.: Twenty-Third Publications, 1990), 67–68.

28. Recognizing difference in intrinsic value also can provide guidance in the difficult area of animal testing. Here Clark's criterion of degree of human necessity is relevant. Inflicting pain and suffering on rabbits is morally different when the goal is testing for an AIDS vaccine rather than for trying out different formulas for mascara.

29. For a review of the history of gay and lesbian attention to the berdache, see Williams, *The Spirit and the Flesh*, 201–7. For a number of reasons, including the near-total absence of written records and ethnographic field work, far less is known about *amazons* in Indian cultures—females who eschewed traditional women's roles in favor of masculine or androgynous roles and identity. For discussion of amazons, see Williams, *Amazons of America: Female Gender Variance in the Flesh and the Spirit*, 233–51; Paula Gunn Allen, *The Sacred Hoop: Recovering the Feminine in American Indian Traditions* (Boston: Beacon Press, 1986); Evelyn Blackwood, "Sexuality and Gender in Certain Native American Tribes: The Case of Cross-Gender Females," *Signs: Journal of Women in Culture and Society* 10 (1984): 27–42; Midnight Sun, "Sex/Gender Systems in Native North America," in *Living the Spirit*, ed. Roscoe; and Will Roscoe, "Strange Country This: Images of Berdaches and Warrior Women" in *Living the Spirit*, ed. Roscoe, 32–76. I limit my discussion here to the berdache, about which more is known and whose impact on the gay and lesbian liberation movement has been substantial. For a study of the life of We'wha, a famous Zuni berdache, see Roscoe, *The Zuni Man-Woman*. For a critique of the attempt to find gay role models in the berdache see Ramón Gutiérrez, "Must We Deracinate Indians to Find Gay Roots," *Outlook* 1, no. 4 (winter 1989), 61–67.

30. Williams, *The Spirit and the Flesh*, 127.

31. Hence Williams writes, "We have to understand berdache gender status, especially their economic role in the family, as a reflection of the fact that women also were persons of consequence. . . . Since women had high status, there was no shame in a male taking on women's characteristics. He was not giving up male privilege, or 'debasing' himself to become like a woman, simply because the position of women was not inferior. It may be accurate to suggest that the status of berdaches in a society is directly related to the status of women." *The Spirit and the Flesh*, 65–66.

32. See Williams, *The Spirit and the Flesh*, 252–75.

33. J. Michael Clark, "An Ecological Berdache: 'Third Gender' Reflections on Native American Wisdom" (unpublished paper, 1993), 4.

34. Ibid., 3.

35. Ibid., 11.

36. Williams notes that at the time of the Spanish conquest, "Sodomy was a serious crime in Spain, being considered second only to crimes against the person of the king and to heresy. It was treated as a much more serious offense than murder." *The Spirit and the Flesh*, 132. Williams also sees militarized cultures that pursue an ideal of absolute machismo that condemns effeminacy as contributing to hostility toward berdachism. In this he includes the Spanish conquistadores and the Aztec, who, unlike their conquered neighbors, seemed to have had taboos against homosexuality. See *The Spirit and the Flesh*, 148. One can see a similar dynamic at play in recent debates over whether to lift the ban on homosexuals in the U.S. military.

37. Williams, *The Spirit and the Flesh*, 187–92. Williams notes a generational difference here. Often the oldest members of a tribe who remember the berdache tradition are most accepting of lesbian and gay youth. The generations in between, affected by the massive physical and cultural dislocation of Native American communities since the 1930s, are the most ignorant of the berdache tradition and most homophobic in their response to gay and lesbian youth. Even here, however, Williams argues that because of Native American respect for one's character as connected to nature, there tends to be more acceptance of lesbian and gay persons in Native American families than in Anglo families and society.

38. Williams, *The Spirit and the Flesh*, 188–89. Here Williams cites the work of Peggy Reeves Sanday whose cross-cultural anthropological work has demonstrated the importance of religious mythology in justifying the equal status of women to men. See also Peggy Reeves Sanday, *Female Power and Male Dominance: On the Origins of Sexual Inequality* (Cambridge: Cambridge University Press, 1981).

12. Gay and Gaia: Features of an Erotic Ethic of Ecojustice

1. Karen Herseth Wee, "Coming Out," in *The Book of Hearts* (Goodhue, Minn.: The Black Hat Press, 1993), 23.

2. Respecting the bodily and ecological integrity of other creatures inevitably brings one up against the reality of predation. How do humans participate in and respect the integrity of the ecological web of relations when all creatures depend on the consumption of other creatures and resources for survival? I discuss this further in the section below on Gay and Gaia and relation to animals.

3. For example, see John C. Ryan, "Conserving Biological Diversity," in *State of the World 1992*, ed. Lester R. Brown et al. (New York: W. W. Norton, 1992), 10–26.

4. Ruether, *Gaia and God*, 172.

5. Warren J. Blumenfeld, ed., *Homophobia: How We All Pay the Price* (Boston: Beacon Press, 1992); see esp. 8–14.

6. As I write this, the Republican "Contract with America" steamroller is carrying this free-market worldview to its logical conclusion in trying to roll back the past twenty-five years' gains in environmental legislation and protection. Similar rollbacks are occurring throughout the globe in the name of economic growth.

7. For an excellent discussion of the shift from a market economic mentality to a market morality, see Larry Rasmussen, *Moral Fragments and Moral Community: A Proposal for Church in Society* (Minneapolis: Fortress Press, 1993).

8. Vandana Shiva, "Development as a New Project of Western Patriarchy," in *Reweaving the World*, ed. Diamond and Orenstein, 192.

9. Clark, *Beyond Our Ghettos*, 68 (emphasis in original).

10. For an excellent collection of essays addressing these issues, see *Toxic Struggles*, ed. Hofrichter, and *Global Ecology*, ed. Wolfgang Sachs (London and New Jersey: Zed Books, 1993). For a Marxist analysis of global capitalism and its effects on both people and nature, see David Pepper, *Eco-Socialism: From Deep Ecology to Social Justice* (London and New York: Routledge, 1993).

11. Chris Glaser, "Indispensable Passion," *Christianity and Crisis* 48, no. 8 (16 May 1988): 173–74.

12. Ibid.

13. Senator Jesse Helms, quoted in the *New York Times*, July 5, 1995.

14. J. Michael Clark, "From Gay Men's Lives: Toward a More Inclusive, Ecological Vision," *Journal of Men's Studies* 1, no. 4 (May 1993): 354.

15. Clark, "Erotic Empowerment, Ecology, and Eschatology," 6.

16. Here the work of gay theologian Ron Long in calling attention to the centrality of male beauty in the gay male community is particularly important for countering heterosexist constructions of masculinity and grounding ethical relationship in positive experiences of the erotic. See Ronald E. Long, "The Sacrality of Male Beauty and Homosex: The Neglected Factor in the Understanding of Contemporary Gay Reality," *Journal of Men's Studies* 4, no. 3 (February 1996): 225–42.

17. For examples of the ecological problems generated by, and alternatives to, consumeristic practices, see John E. Young, "Reducing Waste, Saving Materials," in *State of the World, 1991*, ed. Lester R. Brown et al. (New York: W. W. Norton, 1991), 39–55.

18. Also, because white gay men by and large have set the standards of gay male culture, there often is a spillover effect of consumerist practices into other segments of the gay and lesbian communities as well.

19. Michael Callen, "Come One, Come All! Michael Callen Hails the Second Gay Sexual Revolution," *QW* 28 (May 10, 1992): 59 (emphasis added).

20. For a brief overview of these practices, see McDaniel, *Earth, Sky, Gods, and Mortals*, 63–65. For more thorough discussions see Tom Regan, *The Case for Animal Rights* (Berkeley, Calif: University of California Press, 1983); and Jim Mason and Peter Singer, *Animal Factories* (New York: Crown Publishers, 1980).

21. See Carol J. Adams, "The Feminist Traffic in Animals," in *Ecofeminism: Women, Animals, Nature*, ed. Gaard, 195–218.

22. McDaniel, *Earth, Sky, Gods, and Mortals*, 74–75.

23. See Weinberg, *War on the Land*; Ingemar Hedström, ed., *La Situación Ambiental en Centroamérica y el Caribe* (San Jose, Costa Rica: DEI, 1989). For an examination of the church's response to this phenomenon in Costa Rica, see Daniel T. Spencer, "La Iglesia Costarricense frente la crisis ecológica," *Vida y Pensamiento* 2, no. 2 (Julio-Diciembre 1982): 71–88. Imported animal products are not the only social justice concern related to agricultural imports from Third World countries, of course. Many of our winter vegetables are grown on Third World lands that once fed their local populations. And chemical poisoning of farmworkers, their families, and lands to produce export crops continues to be a serious problem.

24. Even in these cases ecological damage to local ecosystems often was severe, revealing the bucolic picture of the stable farm to be more mythology than reality. See, for example, the comments of Wendell Berry in *The Unsettling of America: Culture and Agriculture* (New York: Avon, 1977) and *A Continuous Harmony: Essays Cultural and Agricultural* (New York: Harcourt Brace Jovanovich, 1972), and Alfred Crosby in *Ecological Imperialism*.

25. For the overall effects of livestock production, see Alan Thein Durning and Holly B. Brough, "Reforming the Livestock Economy," in

State of the World, 1992, ed. Lester Brown et al. (New York and London: W. W. Norton, 1992), 66–82; on soil damage, see Hans Jenny, "The Making and Unmaking of a Fertile Soil," in *Meeting the Expectations of the Land: Essays in Sustainable Agriculture and Stewardship*, ed. Wes Jackson, Wendell Berry, and Bruce Coleman (San Francisco: North Point Press, 1984), 42–55; on energy in agriculture and animal production, see Amory B. Lovins, L. Hunter Lovins, and Marty Bender, "Energy and Agriculture," in *Meeting the Expectations of the Land*, 68–86; on pollution from livestock production, see Carol J. Adams, "Feeding on Grace: Institutional Violence, Christianity, and Vegetarianism," in *Good News for Animals? Christian Approaches to Animal Well-Being*, ed. Charles Pinches and Jay B. McDaniel (Maryknoll, N.Y.: Orbis, 1993), 150–51.

26. Living in Iowa I have been struck by the sheer volume of animals raised and slaughtered for human consumption, and of the near total orientation of the state's ecosystems and economy toward this end. Daily radio reports give the numbers of animals "slaughtered" (ironically, the animal industry's own term), which regularly surpass 1.5 million hogs and between three and four hundred thousand cattle *each week*. That comes to one hog slaughtered for every two to three human Iowans each week. Yet aside from the radio reports, the volume and practices of this organized animal slaughter go virtually unnoticed by most of the human population, which makes little to no connection between this process and the meat products purchased at the grocery store.

27. In the gay male community, in particular, attention to nutrition has played a critical part of the community's response to HIV and AIDS.

28. See, for example, Harvey and Marilyn Diamond, *Fit for Life* (New York: Warner Books, 1985); Francis Moore Lappé, *Diet for a Small Planet* (New York: Ballantine, 1971, 1982); Jane Brody, *Jane Brody's Nutrition Book* (New York: W. W. Norton & Co., 1981).

29. Adams, "Feeding on Grace," 144.

30. For an analysis of the social construction of animals as consumable objects, see Adams, "Feminist Traffic in Animals." Adams's essay is an excellent example of how animal rights activists "politicize the natural" in order to expose social constructions of human/animal relations that justify exploitation of animals by "naturalizing the political." On the use of "meat" for animal flesh, for example, she argues, "Because of the reign of 'meat' as a mass term, it is not often while eating 'meat' that one thinks: 'I am now interacting with an animal.' We do not see our own personal

'meat'-eating as contact with animals because it has been renamed as contact with food. But what is on the plate in front of us is *not* devoid of specificity. It is the dead flesh of what was once a living, feeling being. The crucial point here is that we make some*one* who is a unique being and therefore not the appropriate referent of a mass term into some*thing* that is the appropriate referent of a mass term" (202).

31. McDaniel, *Earth, Sky, Gods, and Mortals*, 76. Attention to the material practice of meat production also exposes the fallacy of those who point to indigenous hunting cultures or peasant cultures in the Third World that involve occasional consumption of animal protein to justify our meat-eating habits. By decontextualizing meat eating in the United States, these arguments serve to further obscure rather than highlight the morally objectionable practices involved in utilizing animal suffering to meet unnecessary human wants. Contextualizing the issue by paying attention to actual material practices also exposes the problematic nature of James Nash's claims about human beings as altruistic predators. Arguing that "humans are predators by necessity" (*Loving Nature*, 147), Nash makes no distinction between predation that is necessary for human survival, which would include consuming plants, and that which is not necessary and thus optional, such as predation of other animals (including humans). This distinction is critical for good ethical evaluation. In addition, a focus on predation needs to expose rather than collapse the different material practices at stake between a "natural ecology of predation" with human beings in an ecologically sustainable and therefore numerically reduced role as omnivores (the food chain cannot support an ever-increasing number of animal-eating creatures without extensive ecological damage and resultant checks and balances on population), and the artificially constructed animal industry (which is not a consequence of natural necessity and human need for animal protein, but a response to constructed human desires).

32. McDaniel, *Earth, Sky, Gods, and Mortals*, 76. This is not to conflate a preferential option for the poor with a preferential option for animals, but rather to emphasize that they can be complementary rather than competing or antithetical commitments. Further discussion of a vegetarian diet as a part of an ecojustice ethics would require analysis of class components, as well. Because of the current higher costs of organic produce and the heavy government subsidies of the dairy, beef, and poultry industries, a vegetarian diet often is more expensive and hence more out of reach for the poor than an animal protein-based diet.

33. Clark, "Leathersexuality," in *Theologizing Gay*, 17–18.

34. Heyward, "Embodying the Connections," 142.

35. Note how "leather" here operates in similar ways to "meat" in food production, as a mass category that obscures both the reality that the hides of an individual animals are involved and the material practices that produce them.

36. Clark, *Beyond Our Ghettos*, 92–93.

37. See Clark, *A Place to Start*.

Conclusion

1. Nash, *Loving Nature*, 221.

2. The phrase "earth and its distress" is Larry Rasmussen's, and was the title of his Stringfellow Lecture delivered at Drake University, Des Moines, Iowa, February 12, 1996.

3. Ibid.

4. Helder Camara, "Earth, Sister Earth," in *Sister Earth: Ecology and the Spirit* (London: New City, 1990), 83.

5. Walker, *The Color Purple*, 242.

Appendix: A Postscript on Method

1. Quoted in Patricia S. Cale, "Environmental Lessons from Central America," *Iowa Energy Bulletin* 20, no. 3 (June/July 1995): 4.

2. Here I am indebted to the methodological reflections of Donna Haraway, who articulates four positions similar to these to ground her investigations of primatology. See Haraway, *Primate Visions*, 6–8. These methodological claims are developed further in chapter 2.

3. For a helpful recasting of these perspectives as "minoritizing" and "universalizing" views of sexuality that both have important contributions to make to sexual theorizing and politics, see the introduction in Sedgwick, *Epistemology of the Closet*. Further discussion of constructionist methodology is taken up in chapter 2. For a detailed critique of the constructionist perspective, see Richard D. Mohr, *Gay Ideas: Outing and Other Controversies* (Boston: Beacon Press, 1992), 221–41.

4. See Beverly W. Harrison, "Sexism and the Language of Christian Ethics," in *Making the Connections*, ed. Robb, 30.

5. See Heyward, *Touching Our Strength*, 193–94.

6. Sedgwick, *Epistemology of the Closet*, 29.

7. Harrison, "Misogyny and Homophobia," 136. See also Heyward, *Our Passion For Justice*, esp. chaps. 5, 6, 11, 22.

8. Clark, *A Place to Start*, 29.

9. G. Goodman et al., *No Turning Back: Lesbian and Gay Liberation in the '80s* (Philadelphia: New Society, 1983); cited in Clark, *A Place to Start*, 29.

10. Harrison, "Misogyny and Homophobia," 136; Clark, *A Place to Start*, 31–32. For an excellent exposition of the source and influence of dualisms on Christian theology and tradition, see James Nelson, *Embodiment: An Approach to Sexuality and Christian Theology* (Minneapolis: Augsburg, 1979).

11. Donald Worster, *Nature's Economy: The Roots of Ecology* (San Francisco: Sierra Club Books, 1977), 204, 378.

12. For a discussion of different social constructions of nature, see Ruether, *Gaia and God*, 5.

Bibliography

Adams, Carol J. "Feeding on Grace: Institutional Violence, Christianity, and Vegetarianism." In *Good News for Animals? Christian Approaches to Animal Well-Being*, edited by Charles Pinches and Jay B. McDaniel, 142–59. Maryknoll, N.Y.: Orbis, 1993.

———. "The Feminist Traffic in Animals." In *Ecofeminism: Women, Animals, Nature*, edited by Greta Gaard, 195–218. Philadelphia: Temple University Press, 1993.

———. *The Sexual Politics of Meat: A Feminist-Vegetarian Critical Theory*. New York: Continuum, 1990.

———. "Toward a Feminist Theology of Religion and the State." Paper presented to the annual meeting of the American Academy of Religion, San Francisco, November 1992.

Allen, Paula Gunn. *The Sacred Hoop: Recovering the Feminine in American Indian Traditions*. Boston: Beacon Press, 1986.

Amon, Aline. *The Earth Is Sore: Native Americans on Nature*. New York: Athencum, 1981.

Beam, Joseph, ed. *In the Life: A Black Gay Anthology*. Boston: Alyson Publications, 1986.

Berry, Wendell. *A Continuous Harmony: Essays Cultural and Agricultural*. New York: Harcourt Brace Jovanovich, 1972.

———. *The Hidden Wound*. San Francisco: North Point Press, 1989.

———. *The Unsettling of America: Culture and Agriculture*. New York: Avon, 1977.

Birch, Charles. "Peace, Justice, and the Integrity of Creation: Some Central Issues for the Churches." In *Report and Background Papers of the Meeting of the Working Group, Potsdam, GDR, July 1986*. Geneva: WCC, Church and Society, 1986.

Birch, Charles, and John B. Cobb Jr. *The Liberation of Life: From the Cell to the Community*. Cambridge: Cambridge University Press, 1981.

Birch, Charles, William Eakin, and Jay B. McDaniel, eds. *Liberating Life: Contemporary Approaches to Ecological Theology*. Maryknoll, N.Y.: Orbis, 1990.

Birkeland, Janis. "Ecofeminism: Linking Theory and Practice." In *Ecofeminism: Women, Animals, Nature*, edited by Greta Gaard, 13–59. Philadelphia: Temple University Press, 1993.

Blackwood, Evelyn. "Sexuality and Gender in Certain Native American Tribes: The Case of Cross-Gender Females." *Signs: Journal of Women in Culture and Society* 10 (1984): 27–42.

Blumenfeld, Warren, ed. *Homophobia: How We All Pay the Price*. Boston: Beacon Press, 1992.

Boff, Leonardo. *Ecology and Liberation*. Maryknoll, N.Y.: Orbis, 1995.

Bonino, José Míguez. *Toward a Christian Political Ethics*. Philadelphia: Fortress Press, 1983.

Boswell, John. *Christianity, Social Tolerance, and Homosexuality: Gay People in Western Europe from the Beginning of the Christian Era to the Fourteenth Century*. Chicago: University of Chicago Press, 1980.

Bounds, Elizabeth. "Sexuality and Economic Reality: A First World and Third World Comparison." In *Redefining Sexual Ethics: A Sourcebook of Essays, Stories, and Poems*, edited by Susan E. Davies and Eleanor H. Haney, 131–44. Cleveland: Pilgrim Press, 1991.

Brooten, Bernadette J. "Paul's Views on the Nature of Women and Female Homoeroticism." In *Immaculate and Powerful*, edited by Clarissa Atkinson, Constance Buchanan, and Margaret Miles, 61–86. Boston: Beacon Press, 1985.

———. "Why Did Early Christians Condemn Sexual Relations Between Women?" Paper presented to the annual meeting of the Society for Biblical Literature, San Francisco, November 22, 1992.

Brown, Robert McAfee. *Gustavo Gutiérrez: An Introduction to Liberation Theology*. Maryknoll, N.Y.: Orbis, 1990.

Chavis, Benjamin F. "A Place at the Table: A Sierra Roundtable on Race, Justice, and the Environment." *Sierra* 78, no. 3 (May/June 1993): 51–58, 90–91.

Clark, J. Michael. *Beyond Our Ghettos: Gay Theology in Ecological Perspective*. Cleveland: Pilgrim Press, 1993.

———. "An Ecological Berdache: 'Third Gender' Reflections on Native American Wisdom." Unpublished paper, 1993.

———. "Erotic Empowerment, Ecology, and Eschatology, or Sex, Earth, and Death in the Age of AIDS." Paper presented to the annual meeting of the American Academy of Religion, Washington, D.C., November 1993.

———. "From Gay Men's Lives: Toward a More Inclusive, Ecological Vision." *The Journal of Men's Studies* 1, no. 4 (May 1993): 347–58.

———. *A Lavender Cosmic Pilgrim: Further Ruminations on Gay Spirituality, Theology, and Sexuality*. Garland, Tex.: Tangelwüld Press, 1990.

———. *A Place to Start: Toward an Unapologetic Gay Liberation Theology*. Dallas: Monument Press, 1989.

———. *Theologizing Gay: Fragments of Liberation Activity*. Oak Cliff, Tex.: Minuteman Press, 1991.

Clark, Mary E. *Ariadne's Thread: The Search for New Modes of Thinking*. New York: St. Martin's Press, 1989.

Clifford, Anne M. "Feminist Perspectives on Science: Implications for an Ecological Theology of Creation." *Journal of Feminist Studies in Religion* 8, no. 2 (fall 1992): 65–90.

Cobb, John B., Jr. "Feminism and Process Thought: A Two-Way Relationship." In *Feminism and Process Thought: The Harvard Divinity School / Claremont Center for Process Studies Symposium Papers*, edited by Sheila G. Davaney, 32–61. Lewiston, N.Y.: Edwin Mellen Press, 1981.

———. "Points of Contact Between Process Theology and Liberation Theology in Matters of Faith and Justice." *Process Studies* 14, no. 2 (summer 1985): 124–41.

———. "Postmodern Christianity in Quest of Eco-Justice." In *After Nature's Revolt: Eco-Justice and Theology*, edited by Dieter T. Hessel, 21–39. Minneapolis: Fortress, 1992.

———. "The Role of Theology of Nature in the Church." In *Liberating Life: Contemporary Approaches to Ecological Theology*, edited by Charles Birch, William Eakin, and Jay B. McDaniel, 261–72. Maryknoll, N.Y.: Orbis, 1990.

———. *Sustainability: Economics, Ecology, and Justice.* Maryknoll, N.Y.: Orbis, 1992.

Cobb, John B., Jr., and David R. Griffin, *Process Theology: An Introductory Exposition*. Philadelphia: Westminster, 1976.

Collins, Patricia Hill. *Black Feminist Thought: Knowledge, Consciousness and the Politics of Empowerment. Perspectives on Gender, Vol. 2*. Boston: Unwin Hyman, 1990.

Comstock, Gary D. *Gay Theology without Apology*. Cleveland: Pilgrim Press, 1993.

Countryman, L. William. *Dirt, Greed, and Sex: Sexual Ethics in the New Testament and Their Implications for Today*. Philadelphia: Fortress, 1988.

Cronon, William. *Changes in the Land: Indians, Colonists, and the Ecology of New England.* New York: Hill and Wang, 1983.

Crosby, Alfred. *Ecological Imperialism: The Biological Expansion of Europe, 900–1900.* New York: Cambridge University Press, 1986.

Daly, Herman E., and John B. Cobb Jr. *For the Common Good: Redirecting the Economy Toward Community, the Environment, and a Sustainable Future.* Boston: Beacon Press, 1989.

Diamond, Irene, and Gloria Feman Orenstein, eds. *Reweaving the World: The Emergence of Ecofeminism.* San Francisco: Sierra Club Books, 1990.

Dolce, Joe. "And How Big Is Yours? An Interview with Simon LeVay." *The Advocate* 630, 1 June 1993, 38–44.

Driver, Tom F. *Patterns of Grace: Human Experience as Word of God.* San Francisco: Harper and Row, 1977.

DuBois, Page. *Centaurs and Amazons: Women and the Pre-History of the Great Chain of Being.* Ann Arbor: University of Michigan Press, 1982.

D'Emilio, John, and Estelle B. Freedman. *Intimate Matters: A History of Sexuality in America.* New York: Harper and Row, 1988.

Eugene, Toinette M., et al., "Appropriation and Reciprocity in Womanist/Mujerista/Feminist Work." *Journal of Feminist Studies in Religion* 8, no. 2 (fall 1992): 91–122.

Fortunato, John. *Embracing the Exile: Healing Journeys of Gay Christians.* New York: Seabury, 1983.

Gaard, Greta, ed. *Ecofeminism: Women, Animals, Nature.* Philadelphia: Temple University Press, 1993.

Gilson, Anne B. *Eros Breaking Free: Interpreting Sexual Theo-Ethics.* Cleveland: Pilgrim Press, 1995.

Glaser, Chris. "Indispensable Passion." *Christianity and Crisis* 48, no. 8 (16 May 1988): 173–74.

Gutiérrez, Ramón. "Must We Deracinate Indians to Find Gay Roots." *Outlook* 1, no. 4 (winter 1989): 61–67.

Guy-Sheftall, Beverly. *Daughters of Sorrow: Attitudes Toward Black Women, 1880–1920.* Brooklyn, N.Y.: Carlson Publishing, 1990.

Hall, Douglas John. "After the American Dream." *The Lutheran* 6, no. 3 (March 1993): 26–28.

———. *God and Human Suffering: An Exercise in the Theology of the Cross.* Minneapolis: Augsburg, 1986.

———. *Lighten Our Darkness: Toward an Indigenous Theology of the Cross.* Philadelphia: Westminster, 1976.

———. *The Steward: A Biblical Symbol Come of Age*. Grand Rapids, Mich.: Eerdmans, 1990.

———. *Thinking the Faith: Christian Theology in a North American Context*. Minneapolis: Augsburg Fortress, 1989.

Halperin, David. *One Hundred Years of Homosexuality: And Other Essays On Greek Love*. New York: Routledge, 1990.

———. "Sex Before Sexuality: Pederasty, Politics, and Power in Classical Athens." In *Hidden from History: Reclaiming the Gay and Lesbian Past*, edited by Martin Duberman, Martha Vicinus, and George Chauncey Jr., 37–53. New York: New American Library, 1989.

Haraway, Donna. *Primate Visions: Gender, Race, and Nature in the World of Modern Science.* New York and London: Routledge, 1989.

———. "The Promise of Monsters: A Regenerative Politics for Inappropriated Others." In *Cultural Studies*, edited by Lawrence Grossberg, Cary Nelson, and Paula A. Treichler, 295–337. New York: Routledge, 1992.

———. *Simians, Cyborgs, and Women: The Reinvention of Nature*. New York: Routledge, 1991.

Hartsock, Nancy. "The Feminist Standpoint: Developing the Ground for a Specifically Feminist Historical Materialism." In *Discovering Reality: Feminist Perspectives on Epistemology, Metaphysics, Methodology, and Philosophy of Science*, edited by Sandra Harding and Merrill Hintikka, 283–310. Dordrecht, Holland: D. Reidel Publishing Company, 1983.

Hedström, Ingemar. *Volverán las Golondrinas? La Integración de la Creación desde una Perspectiva Latinoamericana*. San José, Costa Rica: DEI, 1990.

———, ed. *La Situación Ambiental en Centroamérica y el Caribe*. San José, Costa Rica: DEI, 1989.

Henderson, William Haywood. *Native*. New York: Dutton, 1993.

Heyward, Carter. "Embodying the Connections: What Lesbians Can Learn from Gay Men about Sex and What Gay Men Must Learn from Lesbians about Justice." In *Spirituality and Community: Diversity in Lesbian and Gay Experience*, edited by J. Michael Clark and Michael L. Stemmeler, 133–45. Las Colinas, Tex.: Monument Press, 1994.

———. *Our Passion for Justice: Images of Power, Sexuality, and Liberation.* New York: Pilgrim Press, 1984.

———. *The Redemption of God: A Theology of Mutual Relation*. Washington, D.C.: University Press of America, 1982.

———. *Touching Our Strength: The Erotic as Power and the Love of God.* San Francisco: Harper and Row, 1989.

Hofrichter, Richard, ed. *Toxic Struggles: The Theory and Practice of Environmental Justice.* Philadelphia: New Society Publishers, 1993.

Hooks, Bell. *Ain't I a Woman: Black Women and Feminism.* Boston: South End Press, 1982.

———. *Yearnings: Race, Gender, and Cultural Politics.* Boston: South End Press, 1990.

Hunt, Mary E. *Fierce Tenderness: A Feminist Theology of Friendship.* New York: Crossroad, 1991.

Kammer, Charles, III. *Ethics and Liberation: An Introduction.* Maryknoll, N.Y.: Orbis, 1988.

Katz, Jonathan Ned. *Gay/Lesbian Almanac: A New Documentary History.* New York: Harper and Row, 1983.

Kirsch, John A. W., and James D. Weinrich. "Homosexuality, Nature, and Biology: Is Homosexuality Natural? Does It Matter?" In *Homosexuality: Research Implications for Public Policy,* edited by John C. Gonsiorek and James D. Weinrich, 13–31. Newbury Park, Calif: Sage Publications, 1991.

Lahar, Stephanie. "Roots: Rejoining Natural and Social History." In *Ecofeminism: Women, Animals, Nature,* edited by Greta Gaard, 91–117. Philadelphia: Temple University Press, 1993.

LeVay, Simon. *The Sexual Brain.* Cambridge, Mass.: MIT Press, 1993.

Li, Huey-li. "A Cross-Cultural Critique of Ecofeminism." In *Ecofeminism: Women, Animals, Nature,* edited by Greta Gaard, 272–94. Philadelphia: Temple University Press, 1993.

Licata, Salvatore J., and Robert Petersen, eds. *Historical Perspectives on Homosexuality.* New York: Stein and Day Publishers and Hayworth Press, 1981.

Long, Ronald E. "Gay Theology: Almost Home." Paper presented to the annual meeting of the American Academy of Religion, Chicago, November 21, 1994.

———. "The Sacrality of Male Beauty and Homosex: The Neglected Factor in the Understanding of Contemporary Gay Reality." *Journal of Men's Studies* 8, no. 3 (February 1996): 225–42.

———. "Toward a Phenomenology of Gay Sex: Groundwork for a Contemporary Sexual Ethic." Paper presented to the annual meeting of the American Academy of Religion, Washington, D.C., November 1993.

Lorde, Audre. "Uses of the Erotic: The Erotic as Power." In *Sister Outsider: Essays and Speeches by Audre Lorde*, 53–59. Freedom, Calif: Crossing Press, 1984.

Lovelock, James E. *The Ages of Gaia: A Biography of Our Living Earth*. New York: Norton, 1988.

———. *Gaia: A New Look at Life on Earth*. Oxford: Oxford University Press, 1979.

Lugones, María. "On the Logic of Pluralist Feminism." In *Feminist Ethics*, edited by Claudia Card, 35–44. Lawrence: University of Kansas Press, 1991.

Macy, Joanna, Arne Næss, Pat Fleming, and John Seed. *Thinking Like a Mountain: Towards a Council of All Beings*. Philadelphia: New Society, 1988.

Margulis, Lynn, and Dorian Sagan. *Microcosmos: Four Billion Years of Evolution from Our Microbian Ancestors*. New York: Summit Books, 1987.

McDaniel, Jay B. *Earth, Sky, Gods, and Mortals: Developing an Ecological Spirituality*. Mystic, Conn.: Twenty-Third Publications, 1990.

———. *Of God and Pelicans: A Theology of Reverence for Life*. Louisville: Westminster John Knox, 1989.

McFague, Sallie. "Imaging a Theology of God's Nature: The World as God's Body." In *Liberation Theology: An Introductory Reader*, edited by Curt Cadorette, Marie Giblin, Marilyn Legge, and Mary H. Snyder, 269–89. Maryknoll, N.Y.: Orbis, 1992.

———. *Models of God: Theology for an Ecological, Nuclear Age*. Philadelphia: Fortress, 1987.

McNeill, John. *The Church and the Homosexual*. Boston: Beacon Press, 1988.

Merchant, Carolyn. *The Death of Nature: Women, Ecology, and the Scientific Revolution*. San Francisco: HarperCollins, 1980.

———. *Ecological Revolutions: Nature, Gender, and Science in New England*. Chapel Hill: University of North Carolina Press, 1989.

———. *Radical Ecology: The Search for a Livable World*. New York and London: Routledge, 1992.

Mosala, Itumeleng J. *Biblical Hermeneutics and Black Theology in South Africa*. Grand Rapids, Mich.: Eerdmans, 1989.

Nash, James A. *Loving Nature: Ecological Integrity and Christian Responsibility*. Nashville: Abingdon, 1991.

Nelson, James. *Body Theology*. Louisville: Westminster John Knox Press, 1992.

———. *Embodiment: An Approach to Sexuality and Christian Theology*. Minneapolis: Augsburg, 1979.

Niebuhr, Reinhold. *Moral Man and Immoral Society*. New York: Scribner's, 1932.

———. *The Nature and Destiny of Man*. New York: Charles Scribner's Sons, 1941.

Primavesi, Anne. *From Apocalypse to Genesis: Ecology, Feminism, and Christianity*. Minneapolis: Fortress, 1991.

Rasmussen, Larry. *Moral Fragments and Moral Community: A Proposal for Church in Society*. Minneapolis: Fortress, 1993.

———. "Returning to Our Senses: The Theology of the Cross as a Theology for Eco-Justice." In *After Nature's Revolt: Eco-Justice and Theology*, edited by Dieter Hessel, 40–56. Minneapolis: Augsburg Fortress, 1992.

Rich, Adrienne. "Compulsory Heterosexuality and Lesbian Existence." In *Blood, Bread, and Poetry: Selected Prose, 1979–1985*, 23–75. New York: Norton, 1986.

Richard, Pablo, et al. *The Idols of Death and the God of Life: A Theology*. Maryknoll, N.Y.: Orbis, 1983.

Robb, Carol, ed. *Making the Connections: Essays in Feminist Social Ethics*. Boston: Beacon Press, 1985.

Roscoe, Will. *The Zuni Man-Woman*. Albuquerque: University of New Mexico Press, 1991.

———, ed. *Living the Spirit: A Gay American Indian Anthology*. New York: St. Martin's Press, 1988.

Ruether, Rosemary Radford. *Gaia and God: An Ecofeminist Theology of Earth Healing*. New York: HarperCollins, 1992.

———. "Homophobia, Heterosexism, and Pastoral Practice." In *Homosexuality in the Priesthood and the Religious Life*, edited by Jeannine Gramick, 21–35. New York: Crossroad, 1989.

———. *New Woman, New Earth: Sexist Ideologies and Human Liberation*. New York: Seabury, 1975.

———. *Sexism and God-Talk: Toward a Feminist Theology*. Boston: Beacon Press, 1983.

———. *Women-Church: Theology and Practice of Feminist Liturgical Communities*. San Francisco: Harper and Row, 1986.

Sachs, Wolfgang, ed. *Global Ecology*. London and New Jersey: Zed Books, 1993.

Sands, Kathleen M. "Uses of the Thea(o)logian: Sex and Theodicy in Religious Feminism." *Journal of Feminist Studies in Religion* 8, no. 1 (summer 1992): 7–33.

Santmire, H. Paul. *The Travail of Nature: The Ambiguous Ecological Promise of Christian Theology*. Philadelphia: Fortress, 1985.

Scroggs, Robin. *The New Testament and Homosexuality*. Philadelphia: Fortress, 1983.

Sedgwick, Eve Kosofsky. *Epistemology of the Closet*. Berkeley and Los Angeles: University of California Press, 1990.

Segundo, Juan Luís. *The Liberation of Theology*. Maryknoll, N.Y.: Orbis, 1976.

Shilts, Randy. *And the Band Played On: Politics, People, and the AIDS Epidemic*. New York: Penguin, 1988.

Spencer, Daniel. "How Shall We Name Ourselves? The Ethical and Political Implications of [De]Constructing Gay and Lesbian Identities in the Current Ecclesial-Political Context." Unpublished field examination, New York: Union Theological Seminary, 1991.

Spretnak, Charlene. *States of Grace: The Recovery of Meaning in the Postmodern Age*. San Francisco: HarperCollins, 1991.

Thomashow, Mitchell. *Ecological Identity: Becoming a Reflective Environmentalist*. London and Cambridge, Mass.: MIT Press, 1995.

Tuan, Yi-Fu. *Topophilia: A Study of Environmental Perception, Attitudes, and Values*. New York: Columbia University Press, 1974, 1990.

Vance, Carole S., ed. *Pleasure and Danger: Exploring Female Sexuality*. Boston and London: Routledge and Kegan Paul, 1984.

Walker, Alice. *The Color Purple*. New York: Harcourt Brace Jovanovich, 1982.

Warrior, Robert Allen. "Indian Country's Real Struggles." *Christianity and Crisis* 51, no. 14 (7 October 1991): 302–4.

Weeks, Jeffrey. *Against Nature: Essays on History, Sexuality, and Identity*. London: Rivers Oram Press, 1991.

———. *Sexuality and Its Discontents: Meanings, Myths, and Modern Sexualities*. New York: Routledge and Kegan Paul, 1985.

Weinberg, Bill. *War on the Land: Ecology and Politics in Central America*. London and New Jersey: Zed Books, 1991.

Welch, Sharon D. *Communities of Resistance and Solidarity: A Feminist Theology of Liberation*. Maryknoll, N.Y.: Orbis, 1985.

Wetherilt, Ann K. *That They May Be Many: Voices of Women, Echoes of God*. New York: Continuum, 1994.

White, Lynn, Jr. "The Historical Roots of Our Ecologic Crisis." *Science*, 10 March 1967, 1203–7.

Williams, Delores. "Black Women's Surrogacy Experience and the Christian Notion of Redemption." In *After Patriarchy: Feminist Transformations of the World Religions*, edited by Paula M. Cooey, William R. Eakin, and Jay B. McDaniel, 1–14. Maryknoll, N.Y.: Orbis, 1991.

———. "The Color of Feminism: On Speaking the Black Woman's Tongue." *Journal of Religious Thought* 41 (1986): 42–58.

———. *Sisters in the Wilderness: The Challenge of Womanist God-Talk*. Maryknoll, N.Y.: Orbis, 1993.

Williams, Walter L. *The Spirit and the Flesh: Sexual Diversity in American Indian Cultures*. Boston: Beacon Press, 1992.

Worster, Donald. *Nature's Economy: The Roots of Ecology*. San Francisco: Sierra Club Books, 1977.

———. *The Wealth of Nature: Environmental History and the Ecological Imagination*. New York: Oxford University Press, 1993.

Index